THIS SPECIAL EDITION OF

THE ANATOMY OF

THE HUMAN ORBIT

AND ACCESSORY ORGANS OF VISION

by S. Ernest Whitnall

HAS BEEN PRIVATELY PRINTED

FOR THE MEMBERS OF

THE CLASSICS OF OPHTHALMOLOGY LIBRARY

THE ANATOMY OF
THE HUMAN ORBIT

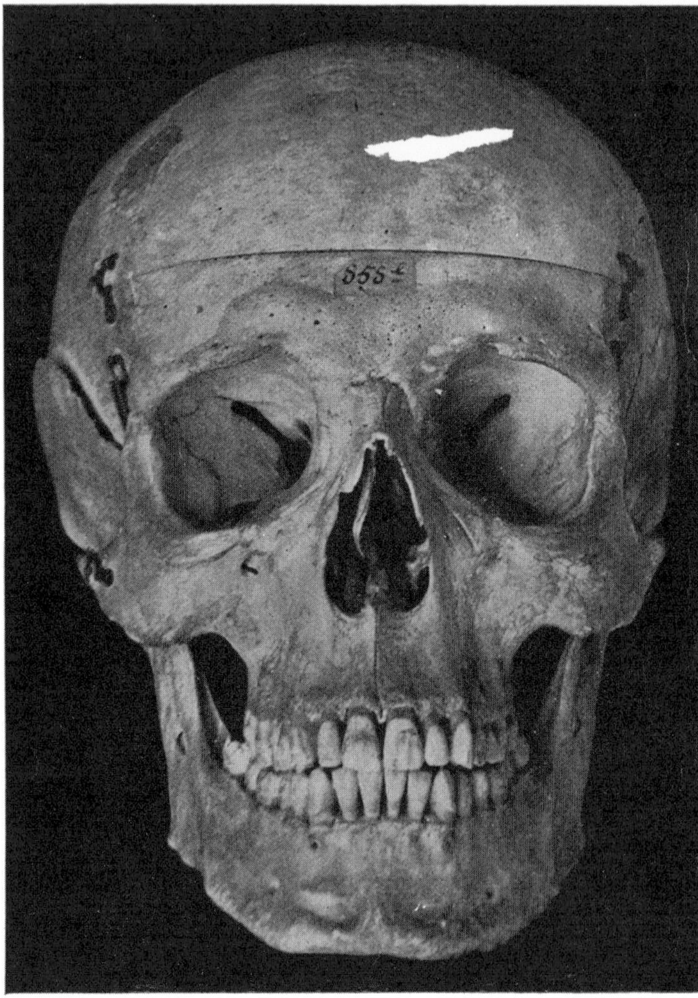

FIG. 1.—Adult Male Skull, English. Cephalic index, 83; orbital index, 86·3; for other orbital measurements see p. 96. $\times \frac{2}{3}$.

OXFORD MEDICAL PUBLICATIONS

THE ANATOMY OF
THE HUMAN ORBIT
AND ACCESSORY ORGANS OF VISION

BY

S. ERNEST WHITNALL

M.A., M.D., B.CH. (OXON.), M.R.C.S., L.R.C.P. (LOND.)
PROFESSOR OF ANATOMY, MCGILL UNIVERSITY, MONTREAL;
LATE UNIVERSITY DEMONSTRATOR OF HUMAN ANATOMY
OXFORD

Illustrated largely by Photographs of Actual Dissections

SECOND EDITION

HUMPHREY MILFORD
OXFORD UNIVERSITY PRESS
London Edinburgh Glasgow Leipzig New York
Toronto Melbourne Capetown Bombay
Calcutta Madras Shanghai
1932

First Edition . . . 1921
Second Edition . . . 1932

PRINTED IN GREAT BRITAIN

TO

ARTHUR THOMSON

IN MEMORY OF MUCH HAPPINESS

IN WORK

PREFACE TO SECOND EDITION

As is customary in preparing a new edition, the text has been revised throughout, though little alteration in the subject matter has been found necessary. Additions, however, have been numerous, especially in the later sections, and over 190 papers published since 1921 have been read and incorporated.

Again, a grant from the Cooper Fund of McGill University has been of assistance in enabling the use of colour to enhance many of the old illustrations and in the preparation of new ones. The source of several borrowed figures is acknowledged in the legends. My thanks for these matters are also extended to my secretary, Miss V. H. Roberts, for her great help in bibliographical research, correction of proofs, and checking of index.

McGILL UNIVERSITY,
MONTREAL,
January 1932.

PREFACE TO FIRST EDITION

THE subject-matter of this work originally formed the substance of a series of lectures given to candidates for the Oxford Diploma of Ophthalmology, and here is presented in an amplified and completed form. The work has suffered many interruptions since its initiation in 1913, which must be held as an excuse for any fluctuations in the manner of its treatment. Primarily intended for the use of ophthalmologists, it is hoped that the necessary incorporation of matter relative to the neighbouring nasal region will also prove of interest to the rhinologist, and that as a regional study the whole will commend itself to the general anatomist.

It was intended that the chief value of the lectures should lie in the demonstration of the large series of dissections and preparations (now in the Oxford anatomical museum) which by the kindness of Professor Arthur Thomson the author was able to make while working under him. Similarly, it is hoped that the chief feature of value in this book will be the photographic illustrations of such of these preparations as lend themselves to the purpose. For any success this object may attain I am greatly indebted to the technical skill and tireless interest of the departmental assistant, Mr. Chesterman, who took all the photographs, with the exception of Nos. 3, 25, 27, 55, 61, 136, which were prepared at McGill. The source of the borrowed illustrations is acknowledged in the legends. The dates following the authors' names in the text refer to the year of publication of articles specially bearing upon the subject which will be found in the bibliography.

To my Associate Professor of Histology and Embryology, Professor J. C. Simpson, I am indebted for the redrawing of the majority of my diagrams, and their presentation in a more artistic form. It is regretted that the cost of colour illustration prohibits the more general use of this means of simplifying the diagrams. To my wife I wish to express my indebtedness for the careful transcribing of my pencil MSS.

The final completion of the work has been greatly facilitated by a grant from the Cooper Fund of McGill University, to the Trustees of which I tender my thanks.

McGILL UNIVERSITY,
MONTREAL,
April 1921.

CONTENTS

PART I

OSTEOLOGY

THE Orbits (orbital cavities, *fossae orbitales*) are two large and deep cavities situated one on each side of the root of the nose. They lie between and are formed by the bones of the face and those of the calvaria (Fig. 2). The wide anterior opening of the orbit onto the face is called the **base** or entrance (*aditus orbitae*), and is circumscribed by the **margin** (*margo orbitalis*); the cavity is enclosed by four **walls,** the roof, floor, lateral and medial walls, connected by rounded **angles,** and it is terminated posteriorly by the **apex.** The continuity of the walls is interrupted in the skull by one large orifice, the optic foramen (*foramen opticum*), and by several smaller foramina, by two wide fissures, the superior and inferior orbital fissures (or sphenoidal and spheno-maxillary), and by the opening of the naso-lacrimal canal (*canalis naso-lacrimalis*); all these openings are closed in life by structures which pass through them. The walls are crossed by sutures between the bones which form them, and lined with periosteum, here termed the **periorbita.**

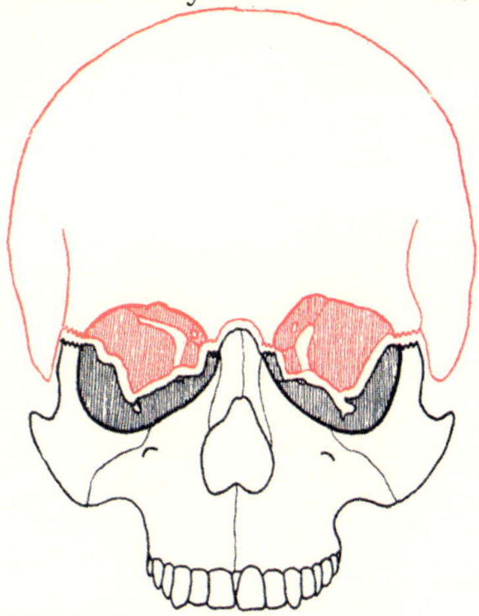

FIG. 2.—Diagram to illustrate the **position of the Orbits** between the bones of the Calvaria (red) and those of the Face (black). (After Poirier, modified.)

Above the orbit is situated the anterior cranial fossa of the skull; below lies the maxillary air-sinus; medially the lateral mass of the ethmoidal bone and the lacrimal bone intervene between it and the nasal fossa in front, whilst behind the two orbits are separated by the sphenoidal bone; the temporal region and middle cranial fossa are in relation to its lateral aspect; it is bounded in front by the eyelids.

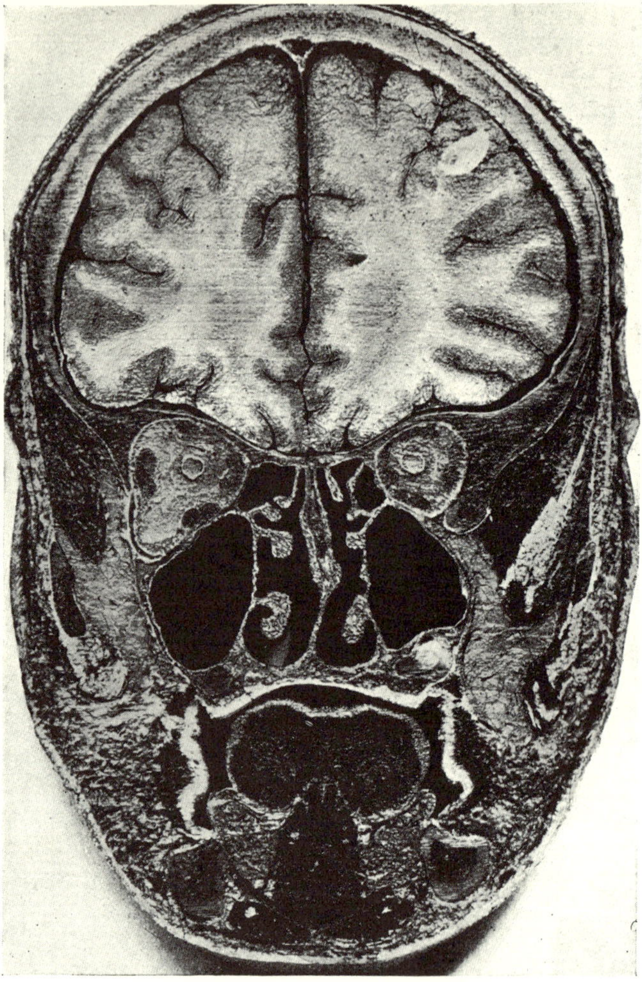

FIG. 3.—Coronal section of an Adult Male Head to show the general **relations of the Orbits;** the section cuts through them at about the junction of their posterior third and anterior two-thirds. Compare with the horizontal section through the head seen in Fig. 136, p. 254. From the McGill Anatomical Museum. $\times \frac{2}{3}$.

It is convenient to study the osteology of the orbit on the following plan:

1. The **form** of the orbit.
2. The orbital **margin.**
3. The **walls** of the orbit.
4. The **accessory air-sinuses** of the nose in their relation to the orbit.
5. The osseous **lacrimal passages.**
6. Table of **apertures** of the orbit.
7. The **periorbita.**
8. **Mensuration.**
9. **Development** and growth.
10. **Variations,** Asymmetry, and Deformities.

1. THE FORM OF THE ORBIT

The form of the orbit is usually likened to either a cone or a four-sided pyramid with rounded angles, of which figure the base corresponds to the wide opening directed forwards onto the face and circumscribed by the orbital margin, and the apex to the narrowed portion leading backwards within the skull.

An examination of plaster casts of the orbit (Fig. 4) shows that neither of these descriptions is accurate as regards the shape. No common geometrical figure will serve to illustrate it without missing several important features; moreover, it has been found that the mathematical volume of a corresponding pyramid is less by one-third than the volume of the orbit calculated empirically by means of filling it with lead pellets (Gayat, 1873).

The first point to be noticed is that owing to the overhang of the orbital walls in front the largest part of the cavity does not lie at the facial entrance, but is just behind it, and in consequence of this narrowing of the margin a cast cannot be withdrawn without partial removal of the bones. Thus the roomiest part of the orbit is found about one centimetre behind the margin, and since here is lodged the eyeball the walls are roughly moulded upon it. The primitive spheroidal shape is, however, distorted by outward bulgings of the walls in several regions, to accommodate the lacrimal gland, the

a

b

c

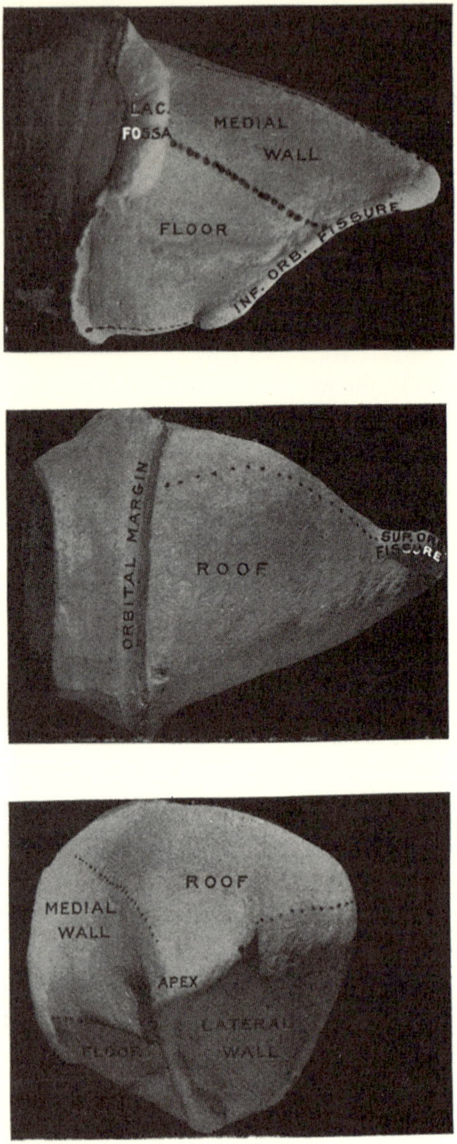

FIG. 4.—**Casts** of Right Orbit. Adult male, natural size.
(*a*) Medial aspect. (*b*) Superior aspect. (*c*) Posterior aspect.

lacrimal sac, and the pulley of the superior oblique muscle. The growth of these structures within the orbit, together with the development of the surrounding parts, causes the bulk of the cast to present a quadrilateral shape with rounded angles, as is seen in a section through the skull itself taken just behind the orbital margin (Fig. 20, p. 47). The shape of this forepart of the cavity allows the orbit to be conveniently described as contained by four walls, but the cast emphasizes important differences between them as they are traced from the base to the apex. The medial wall is oblong in outline; the roof and lateral wall are the shape of an isosceles triangle; the floor is also triangular (equilateral), but is shorter than the other boundaries, extending only as far back as the junction between the anterior two-thirds and posterior third of the orbit. Behind this point is the region of the apex, and here the absence of the floor is compensated by the inclination of the lateral towards the medial wall. There are, therefore, only three walls forming the apex, as seen in the cast and also shown by the more triangular outline of a section of the skull through this region (Fig. 3). This three-sided pyramidal form of the apex is accentuated by the presence of the superior and inferior orbital fissures, between which lies the posterior extremity of the lateral wall. The apex lies wholly behind the eyeball, and becomes narrow in conformity with its less bulky contents, which are chiefly the optic nerve and the origin of the ocular muscles.

The comparison of the form of the orbital cavity to that of a cone or four-sided pyramid thus misses the facts shown so well by a cast, namely, that the figure has a constricted base, a roughly four-sided anterior portion in which the eyeball is lodged, and a well-defined three-sided pyramidal apex wherein lie the optic nerve and muscles. If the form must be likened to some common object, it may be said to resemble that of a pear with the thin end (apex) and stalk (optic nerve) curving slightly towards the mesial plane.

Although not sharply marked, the division of the cavity into an anterior two-thirds or ocular portion and a posterior third or retro-ocular part is to be emphasized, since these two regions differ not only in regard to their chief contents but also in their external relations, and moreover, practically in their clinical interests.

2. THE ORBITAL MARGIN

From a practical standpoint the superficial position and prominence of the margin of the orbit are sufficient justifications for its detailed description apart from the walls of the cavity. The surface anatomy is better considered after such description, when the features which can be made out by digital examination will more readily be recognized.

The **shape** of the margin may be regarded first. Since the orbital walls are moulded upon the previously formed eyeball, the primitive outline of the margin is circular and comparatively flat, as is seen in the earlier foetal skulls up to the age of six months, where also the margin is more closely applied to the globe and only extends as far as its equator (Fig. 47, p. 100). After this period the outline usually assumes an oval contour, due probably to the more rapid expansion of the brain-containing part of the skull, and later on the striking changes in the infantile face consequent to the development of the nose and jaws react upon the margin, altering it in both frontal and sagittal planes. In the former plane the circular outline may be altered to oval, square, or oblong (as seen in the photographs on pp. 94, 105, 107), the quadrangular contour being found the generalized type in modern male skulls by Cameron (1920); in the latter plane the flattened margin becomes more undulating. (See also under fronto-sphenoidal process, p. 13, and refer p. 102 on the development.)

It should be noticed that even in foetal skulls the orbital margin does not form a completed circle, owing to the presence of the fossa for the lacrimal sac on the nasal side. This fossa is bounded by two crests, of which the anterior is clearly the unbroken continuation of the infra-orbital margin, whilst the posterior can be traced, though less definitely, as a downward prolongation of the supra-orbital margin. The whole outline therefore assumes the form of one coil of an undulating spiral, the two ends of which overlap on the nasal side and enclose the fossa. This overlap is not only obvious from the lateral aspect but in most cases can be seen from the front, owing to the pushing laterally of the upper underlap of the spiral, that is, the posterior lacrimal crest, by the development of the ethmoidal air-cells, as is seen in Figs. 6, 10, 24.

It is convenient for descriptive purposes to subdivide the margin into four segments corresponding to the four walls of the orbit, namely, the supra-orbital, lateral, infra-orbital, and medial margins, though the limits of these regions can be defined less strictly than the walls, owing to the fact that they may be united by either curves or well-marked angles, according to type.

Four bones form the orbital margin: the frontal, zygomatic (malar), maxilla, and lacrimal:

(i) The supra-orbital margin is formed entirely by the frontal.
(ii) The lateral margin by frontal and zygomatic.
(iii) The infra-orbital margin by zygomatic and maxilla.
(iv) The medial margin by maxilla and frontal, with the addition of the lacrimal posterior to the former.

Three sutures interrupt its continuity:

(*a*) The zygomatico-frontal (fronto-malar) laterally.
(*b*) The fronto-maxillary on the medial side.
(*c*) The zygomatico-maxillary on the infra-orbital margin.

In the following description attention will also be drawn to features of interest which lie beyond the margin itself.

(i) **The supra-orbital margin** (*margo supraorbitalis*) is curved to a degree which varies in individuals and races. It is sharp in the lateral two-thirds, that is, as far as the supra-orbital notch, but is rounded in the medial third, where it is crossed by vessels and nerves. It is formed wholly by the frontal bone, and corresponds to the junction of its vertical and horizontal portions (*pars frontalis* and *pars orbitalis*), as seen in Fig. 13, *a*. It forms an arcade extending between the medial and lateral angular processes of that bone (*pars nasalis* and *processus zygomaticus*), which processes form also the upper parts of the medial and lateral orbital margins respectively.

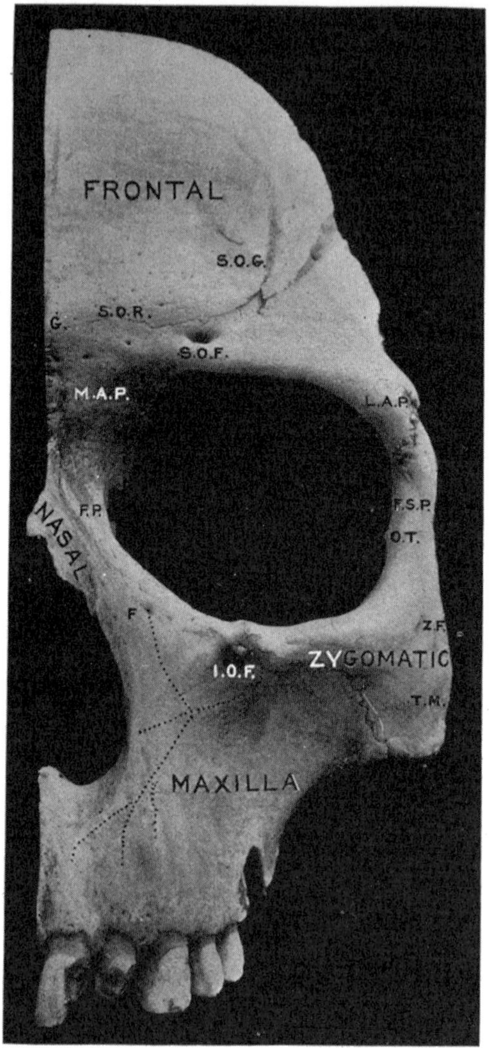

FIG. 5.—The **Orbital Margin.** Frontal section of left half of skull passing just behind orbital margin. Natural size. Orbital height = 32 mm., orbital width = 41 mm., orbital index, 78·05.

S.O.G. = Supra-orbital grooves.
S.O.R. = Supra-orbital ridge.
S.O.F. = Supra-orbital foramen.
G. = the Glabella.
M.A.P. = Medial angular process of frontal bone.
L.A.P. = Lateral angular process of frontal bone.
F.S.P. = Fronto-sphenoidal process of zygomatic bone.
F.P. = Frontal process of maxilla.
O.T. = Orbital tubercle of zygomatic bone.
Z.F. = Zygomatico-facial foramen.
T.M. = *Tuberositas malaris*.
I.O.F. = Infra-orbital foramen, surmounted by an infra-orbital tubercle.

The dotted lines show the position of the minute osseous canals through

At the junction of the medial third with the lateral two-thirds the supra-orbital margin is cut by the **supra-orbital notch** (*incisura supraorbitalis*), which in about a quarter of cases is converted into a foramen (*foramen supraorbitale*). The notch is bridged in life by the thickened edge of the septum orbitale, and transmits from within the orbit to the forehead the supra-orbital nerve and vessels, as shown in Fig. 59, p. 127; it is situated about 25 mm. or an inch (varying from 13 to 40 mm.; Schwegel, 1859) from the mid-line of the skull, is from 3 to 5 mm. broad, and its lateral boundary is sharper than the medial; the size, form, and depth are very variable. In rare cases the supra-orbital margin is unmarked by grooves, a smooth condition ('pithekoid') typical of the higher apes. On the other hand, the nerve sometimes cuts the margin so deeply that it lies in a supra-orbital canal (*canalis supraorbitalis*), which may run from 5 to 30 mm. deep to the rim, and be from 5 to 15 mm. long, as seen in Figs. 5, 6, 11. Leading from the supra-orbital notch, there are occasionally to be seen long grooves, which pass upwards and lateralwards over the face of the frontal bone, as in Fig. 5. These 'supra-orbital grooves' contain the nerve, artery and vein, or may lodge only one of them; they are not dependent upon the presence or absence of the supra-orbital notches, and they may be found at all ages (even in the foetal skull, Fig. 48, p. 101), and their relative frequency is marked in some races, but not in others. Zoja (1884) found them in 16·4 per cent. of 908 skulls, and they have been seen roofed over to form long canals. Dixon (1904), who has described them at length, considers them caused by the growth of the nerves not having been proportionate to the expansion of the cranium. Opening into or near the supra-orbital notch there is nearly always a minute foramen, termed the 'supra-ciliary canal' by Ward (1858), who found it constant in the foetal

---

which run the anterior superior alveolar (dental) nerve and accompanying artery. The ascending canal contains an arterial twig only and it emerges from a foramen in the *sutura notha* to anastomose with the facial artery at F. The horizontal canal contains both nerve and arterial twigs which supply the nasal mucous membrane, and the foramen through which they emerge on the nasal aspect of the frontal process of the maxilla can be seen in Figs. 37 and 38. Compare with Fig. 20.

skull and in 72 out of 105 adult skulls. Zweibach (1900) found it present in 50 per cent. of 1,062 skulls. It may be single or double, and it transmits a nutrient artery into the bone and affords exit to the anterior or frontal diploic vein, which drains blood from the diploë of the frontal bone and mucous membrane of the frontal air-sinus. A nerve filament ('nerve of Kobelt') also enters it; it may open into the frontal sinus and afford a channel for the transmission of disease; an example is seen in the second bone in Fig. 31, p. 64.

A short distance to the medial side of the supra-orbital notch there occurs in about 50 per cent. of cases a second one, the **frontal notch** (*incisura frontalis*), which likewise, though much more rarely, may either be converted into a foramen by ossification of the periosteal ligament which bridges it (2½ per cent.), or form a short canal, as in Fig. 6, or be continued into a groove over the frontal bone. It transmits the medial frontal division of the supra-orbital nerve when the latter has divided within the orbit, together with corresponding blood-vessels. Both these notches have been studied specially by Zweibach (1900), from whose account, as cited by both Merkel (1901) and Ledouble (1906), the following table is taken.

On examination of 1,062 skulls there were found—

	Orbits.			Orbits.	
	Right.	Left.		Right.	Left.
	%	%		%	%
Supra-orbital notch	70·5	67·7	Frontal notch .	47·5	47
,, foramen	22·2	22·3	,, foramen	2·5	2·4
,, canal .	2·8	2·8			
Margin smooth .	1	5	Neither . .	48·3	48·5

with numerous variations and combinations on either side. From Ledouble's figures, quoting various other authors, there was found a frontal notch in 25·7 per cent. and a frontal foramen in 1·7 per cent. of 1,738 orbits.

There may also be found instances of supernumerary notches, foramina and canals, which are usually smaller than the normal ones and lie to their lateral side. There appear to be no sexual or racial peculiarities in their disposition. As in the case of the supernumerary infra-orbital canals, the variation is no doubt due to the nerve dividing abnormally into separate branches within the orbit instead of after leav-

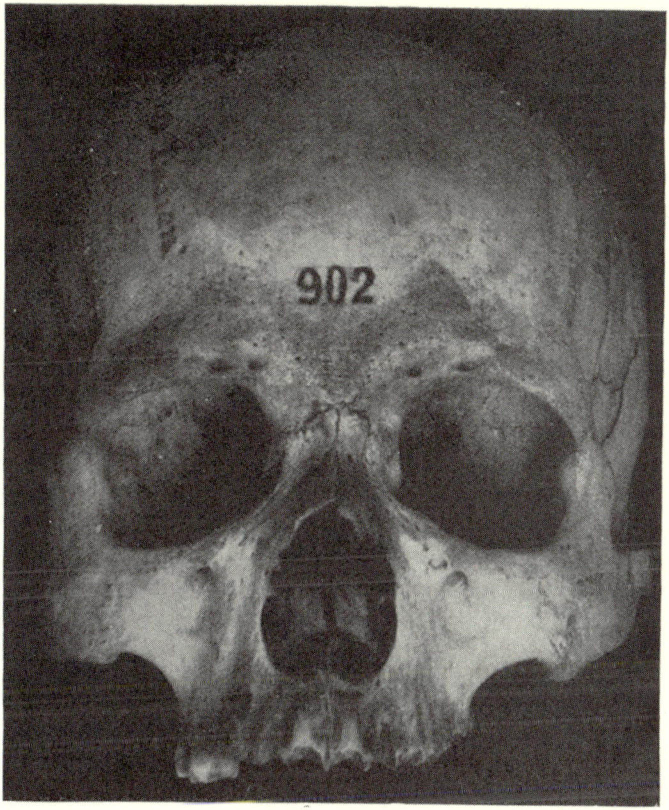

FIG. 6.—Skull showing both Frontal (medial) and Supra-orbital (lateral) **Foramina**, well-developed Supra-orbital Ridges, two Infra-orbital Foramina on left side, and a well-marked **Orbital Tubercle** of the Zygomatic bone. Orbital width = 42 mm., orbital height = 33·5 mm., orbital index = 79·76. Kansas Indian, adult male. × about $\frac{2}{3}$.

ing that cavity. Practically, it is important to remember that the nerve can thus cross the orbital margin in separate divisions, and it should be noted that there is yet another branch of the frontal nerve (from which the supra-orbital is derived), namely the supra-trochlear, which leaves the orbit and crosses the margin still nearer the mid-line of the skull (Fig. 9).

Above the inner half of each supra-orbital margin in the male skull there is usually a prominence of the bone called the **supra-orbital ridge** (*arcus superciliaris*) or superciliary arch. These ridges, which are variable in size, form a characteristic difference between the sexes (see Fig. 49, p. 105); at the mid-line they meet to form a low eminence, the 'glabella' (from the Latin *glaber*, smooth, hairless), and between them and the outstanding lateral angular processes lie shallow and obliquely placed depressions, along which there sometimes pass the supra-orbital nerves and vessels, and, when present, the supra-orbital grooves, as seen in Fig. 5, p. 8. Whilst the size of the supra-orbital ridges is due to the degree of development of the underlying frontal air-sinuses, it is important to note that it does not follow that the sinuses are small or absent when the ridges are but slightly marked.

(ii) **The lateral orbital margin** (outer, external, or temporal border, *margo lateralis*) is the strongest part of the orbital rim; it is curved with a backwardly directed concavity, more especially marked in the lower part, thereby permitting a greater extension of the lateral range of vision on the ground where it is most required, whereas in the upper region the prominence of the lateral angular process of the frontal bone serves to protect the eye from injury on the side where blows are most likely to fall (Humphry). It is formed by parts of two bones, the lateral angular or zygomatic process of the frontal above, and the fronto-sphenoidal process and half of the orbital border of the zygomatic bone below. The zygomatico-frontal suture uniting these processes has been found duplicated in 17 out of 400 skulls by Nicola (1903), a condition explained by non-union of the lateral angular process, for which a secondary centre of ossification is described (Fig. 46, p. 99). This margin may also be cut by a second suture in those very rare cases where the zygomatic bone consists of two pieces (*mala bipartita* or *os japonicum*, found present by Ledouble in 0·16 per cent. of European and in 21·1 per cent. of Japanese skulls).

The lateral angular or zygomatic process of the frontal

bone is strong and prominent in the male, narrow and more delicate in the female, forming a characteristic sexual difference; just beneath it, and between it and the globe over which it arches, is lodged the lacrimal gland. The sharp posterior margin of this process marks the beginning of the temporal ridge (*linea temporalis*), to which is attached the fascia covering the temporal muscle. Just below the zygomatico-frontal suture the margin is sometimes marked by fine hair-like grooves corresponding in position to the terminal branches of the lacrimal nerve which supply the skin in this region.

The breadth of the fronto-sphenoidal process of the zygomatic bone, as seen from the side, varies greatly. Its degree of recession, which appears to be a racial trait in man, is combined with a proportionately pronounced erection in the horizontal sense, and so has an effect upon the frontal inclination of the orbital margin (Oetteking, 1919). On its free or temporal border there may be developed to a variable degree the *processus marginalis* of Sömmering, also called the *spina zygomatica* by Broca (Fig. 23, p. 51). According to Ledouble, this process is more pronounced on the right side and in the male; it increases in old age, and is a muscular process, but is not well developed in Australian skulls (A. Thomson).

On the orbital surface of the fronto-sphenoidal process of the zygomatic bone, just within the lateral orbital margin at its centre and about 11 mm. below the suture, is a small tubercle, which may be termed the lateral palpebral or 'orbital tubercle' (*tuberculum orbitale*) of the zygomatic bone; it is illustrated in Fig. 7, and is well seen in the orbits shown in Figs. 5, 6, 44. Often inconspicuous, it can generally be appreciated more readily by the touch, by inserting the finger-tip deeply between the orbital margin and the eyeball opposite to the lateral commissure of the eyelids and rubbing the finger up and down. It is of importance as marking the point of attachment of (*a*) the 'check ligament' of the lateral rectus muscle; (*b*) the suspensory ligament of the eyeball; (*c*) the lateral palpebral ligament; and (*d*) the aponeurosis of the levator palpebrae superioris muscle (these confluent attachments forming the *retinaculum oculi laterale* of Hesser,

1913, as shown in Fig. 60, p. 128). It was found present in more than 95 per cent. of 2,000 skulls of all races (Whitnall, 1911).

The facial aspect of the zygomatic bone may present at its centre a slight tuberosity (*tuberositas malaris*). This underlies the most prominent part of the cheek and marks the attachment of the zygomaticus major muscle, as indicated in Fig. 65, p. 136; it can sometimes be felt in life, and is in line with the infero-lateral angle of the orbital margin and the long axis of the inferior orbital fissure. On this surface of the bone there is also to be seen the single or double opening of the zygomatic canal (*foramen zygomatico-faciale*, Figs, 7, 11, 23, 24), through which there passes from within the orbit to the skin of the cheek the zygomatico-facial branch of the zygomatic nerve from the maxillary division of the fifth cranial nerve, seen in Fig. 58, p. 126, together with an arterial twig.

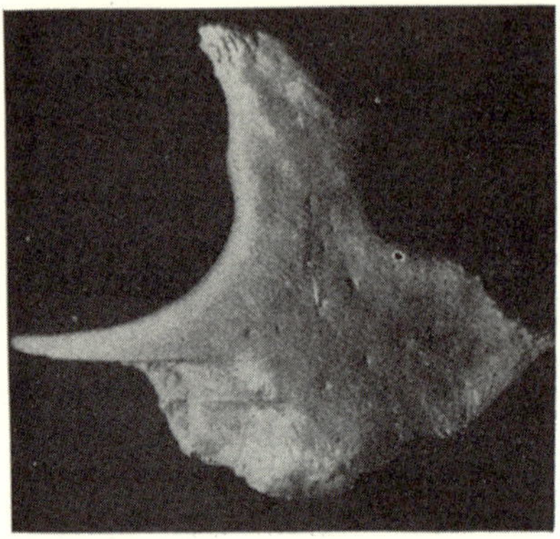

FIG. 7.—Left Zygomatic Bone seen from in front in natural position, showing the **Orbital Tubercle** (*tuberculum orbitale*). Natural size.

(iii) **The infra-orbital margin** (*margo infraorbitalis*) is flattened, its edge is blunt and projects but little over the orbital cavity. Laterally it curves widely round into the lateral margin; medially it slopes upwards, to become continuous with the anterior lacrimal crest of the frontal process of the maxilla. It is formed by the confluent 'orbital borders' of two bones, the zygomatic on the lateral side, the maxilla on the medial side, each contributing a half, so that the zygomatico-maxillary suture uniting them lies at about the middle of the margin (p. 44). At this point there is frequently an elevation which can be felt in life, the 'infra-orbital tubercle' (*tuberculum infraorbitale marginale*, Fig. 5); Zuckerkandl found it more frequently present in Malays (19 out of

30 skulls) than in Europeans. The infra-orbital part of the orbital border of the zygomatic bone usually forms a long thin spur, the 'marginal process', which rests upon the maxilla lateral to the suture (Figs. 7, 14); it may be broken off by injury, or, very rarely, exist as a separate *ossiculum infra-orbitale marginale* (Grüber, 1877).

Directly below the suture, and about 10 mm., but varying from 4 to 12 mm. ($\frac{1}{4}$ to $\frac{1}{2}$ inch) from the margin and in rare cases even less, is an opening on the face of the maxilla, the **infra-orbital foramen** (*foramen infraorbitale*), which leads into the infra-orbital sulcus on the floor of the orbit; this foramen varies in shape from a circular opening to a vertically placed slit; usually it is oval in outline, from 4 to 5 mm. deep by 3 to 4 mm. wide; its lateral edge is sharp, and its opening is directed downwards and medially, as is seen in Fig. 24, p. 53. There is to be found rarely a second or 'supernumerary infra-orbital foramen' (*foramen infraorbitale supernumerarium* of Grüber), as in the case of the supra-orbital notch, and similarly explained by an early division of the nerve (Fig. 6); the two foramina may lie as far as 1 cm. apart; also 'supplementary infra-orbital foramina' (*foramina infraorbitalia anomala*) may occur lateral or medial to and smaller than the main foramen. Finally, there may be seen minute 'false infra-orbital foramina' on the anterior face of the maxilla which do not lead into the orbit but into either the maxillary antrum, the nasal fossa, or a nutrient dental canal. Ledouble found a double infra-orbital foramen in about 11 per cent. of nearly 3,000 skulls, and more often on one side than on both; three, four, or even five extra openings have been noted; Poirier in 217 skulls found 35 double, 2 triple, and 1 case of quadruple foramina.

Through the foramen pass the infra-orbital nerve (the main continuation of the maxillary division of the fifth nerve) and artery (a branch of the internal maxillary), and possibly lymph vessels. At this point of emergence upon the face the nerve lies in a hollow much deeper from the surface than might be supposed, an average depth being 10 mm. or nearly $\frac{1}{2}$ inch; it is embedded in fat and covered by the *levator labii superioris* muscle (*caput infraorbitale* of the *quadratus labii superioris*, the origin of which is sometimes marked by a slight ridge just above the foramen) as well as by the orbicularis oculi muscle and the skin. Below the foramen arises the *levator anguli oris* (*caninus*) muscle (Fig. 65, p. 136). The bridge of bone between the infra-orbital margin and the foramen may be marked by a fine vertically directed suture (*sutura infraorbitalis verticalis vel sagittalis*), indicating the line of fusion of the maxilla as it closed up round the nerve in development. It usually cuts the margin at a point medial to the zygomatico-maxillary suture, and in most cases is obliterated early in life; an example is seen in the female skull, Fig. 49, p. 105. The face of the maxilla below the foramen forms the anterior wall of the maxillary air-sinus (antrum of Highmore), and embedded in a minute canal in its substance is the anterior superior alveolar (dental) nerve. The nerve arises from the *lateral* aspect of the forepart of the infra-orbital nerve, and in company with a small artery winds round below the foramen to gain its *medial* side before bending downwards at an angle to reach the two incisors and canine teeth which it supplies. From the artery and nerve twigs pass through minute bony channels to the mucous membrane lining the nasal cavity, and a communication is effected with the facial artery by a vessel which lies in a groove (the *sutura notha*) on the face of the frontal process of the maxilla at the infero-medial angle of the orbital margin, as indicated in Fig. 5, p. 8. By these intra-osseous paths it is possible for disease of dental origin, especially from the canine tooth, to spread to ocular or lacrimal regions (see p. 84). Occasionally the infra-orbital margin is faintly grooved by the passage over it of a branch of the infra-orbital artery (Fig. 25). Other variations in the bones forming this border concern the medial half, and may result in exclusion of the maxilla from it, either by extension of the marginal process of the zygomatic as a long spur up to the lacrimal bone, as is constant in lower mammals with enclosed orbits (the extremely rare condition in man of a 'lacrimo-jugal suture' so formed was found present by Reid (1910) in two out of 4,500 skulls at Cambridge), or by the intervention of a hamular process of the lacrimal bone (Fig. 22, p. 50). These abnormalities have been studied by Grüber, and are tabulated by Merkel (1901). The varieties of the infra-orbital suture will be found in Ledouble's treatise (1906), and are also the subject of papers by Turner (1885) and Davida (1913).

(iv) **The medial orbital margin** (internal or nasal border, *margo medialis*) is the most ill-defined segment of the margin, since it is not only rounded and indistinct in its upper part, but is complicated below by the presence of the fossa for the lacrimal sac (*fossa sacci lacrimalis*). It is formed by means of three bones: in its upper part by the medial angular process (*pars nasalis*) of the frontal bone; in its lower part by the lacrimal bone behind, together with the frontal (nasal or ascending) process of the maxilla in front. The supra-orbital margin medial to the notch loses its sharpness, but though interrupted by slight vascular grooves is still traceable as a rough line passing downwards and slightly backwards into the sharp posterior lacrimal crest (*crista lacrimalis posterior*) of the lacrimal bone: on the other hand, the infra-orbital margin traced medially passes upwards, to become continuous with the anterior lacrimal crest (*crista lacrimalis anterior*) of the frontal process of the maxilla. Hence, as explained on p. 6, the ring formed by the whole orbital margin is broken on the medial side, and the two ends which form the lacrimal crests separate to enclose the fossa of the lacrimal sac. Opinions differ as to whether the anterior or posterior lacrimal crest should be described as forming the exact orbital margin in this region, the fixation being chiefly important for purposes of mensuration (p. 91). Merkel took the posterior crest, since to it is attached the *septum orbitale*, a membrane which arises from the lip of the whole margin, and forms the true limit to the orbital contents (Fig. 65, p. 136): the fossa then is extra-orbital in position, with which opinion Evatt concurs; Topinard, on the other hand, considered the whole lacrimal apparatus to lie within the orbit. Morphologically, the position of the fossa is interesting, since in most mammals it is undoubtedly placed outside the orbit, but of the primates in lemurs alone is it so placed, being considered to lie within it in all others (Flower). From the attachments of the septum orbitale, however, the posterior lacrimal crest must be taken as the limit of the orbital cavity on the medial side, and the fossa for the lacrimal sac in man is situated *on* the margin, which is broadened out or divides to accommodate it; so also it will be found that parts of the chief constituents of the eyelids (the medial palpebral ligament of the tarsal plate and the orbicularis oculi muscle) divide, and are attached to the margin both in front and behind the fossa. The fossa for the lacrimal sac is the chief feature of interest in this margin, but in view of the fact that it is partly formed by the lacrimal bone, and moreover, presents important deep relations to the ethmoidal air-cells, which structures come under the heading of the medial wall, its description, together with that of its continuation the naso-lacrimal canal, is best deferred until later and will be found on p. 67.

There remain a few minor points to be noticed. At the junction of the medial and inferior orbital margins, that is, at the base of the anterior lacrimal crest, there is in many cases a small 'lacrimal tubercle', also considered later in its relation to the lacrimal fossa on p. 68. On the lateral aspect of the frontal process of the maxilla, in front of the lower part of the lacrimal crest and running parallel with it, there is constantly to be seen a fine groove, the *sutura notha* or *sutura longitudinalis imperfecta* of Weber (Fig. 11). It is a vascular groove lodging a branch of the infra-orbital artery, from which minute twigs pass through a row of small foramina into the bone and to the nasal mucous membrane. In rare instances it divides off a strip of bone from the margin to form the *ossiculum maxillo-frontale* of Macalister (who found it in 1 per cent. of English skulls), or, since the fragment helps to form the lacrimal fossa, the *os de la gouttière lacrymale* of Ledouble (1·6 per cent. of 2,723 skulls), who suggests a

---

Description of Fig. 8*a, b.*

1. Frontal bone. 1*a*. Frontal sinus. 2. Frontal tuberosities (eminences). 3. Crista galli. 4. Nasion (frontonasal suture). 5. Supra-orbital ridge. 6. Infra-orbital ridge. 7. Ethmoid cell. 8. Inner canthus of eye. 9. Optic foramen, approximate location in the sagittal plane. 10. Outer canthus of eye. 11. Superior orbital fissure. 12. Orbit. 13. Maxillary sinus. 14. Zygomatic bone. 15. Infra-orbital foramen. 16. Canine fossa. 17. Maxilla. 18. Middle concha. 19. Inferior concha. 20. Nasal septum. 21. Nasal cavity. 22. Anterior nasal spine. 23. Temporal fossa. 24. Coronal suture. 25. Glenoid cavity for mandible. 26. Condyloid process of mandible. 27. Coronoid process of mandible. 28. Ramus of mandible. 29. Angle of mandible. 30. Body of mandible. 31. Mental canal. 32. Mental foramen. 33. Mental protuberance. 34. External acoustic meatus. 35. External ear (auricle). 36. Lateral angular process of frontal bone. 37. Radiographic base line. 38. Mastoid cells. (*Eastman & Co.*)

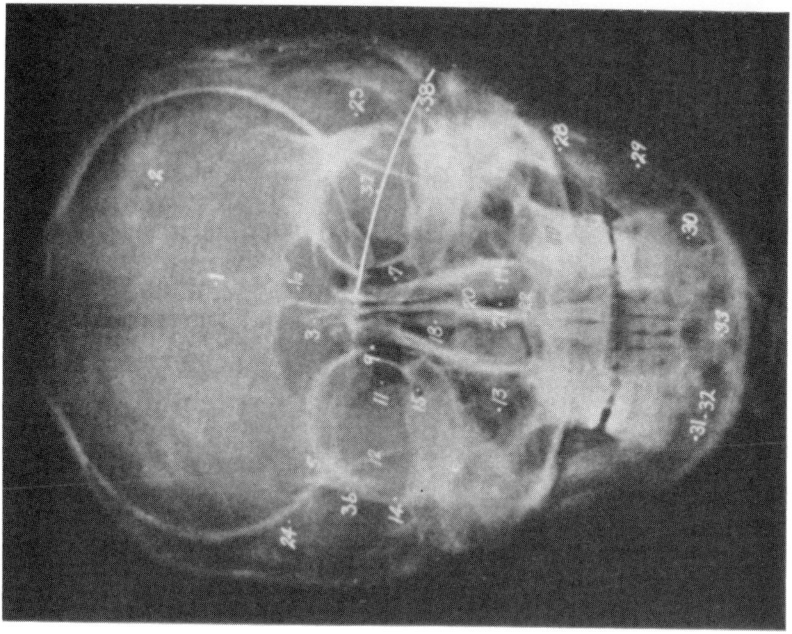

8b

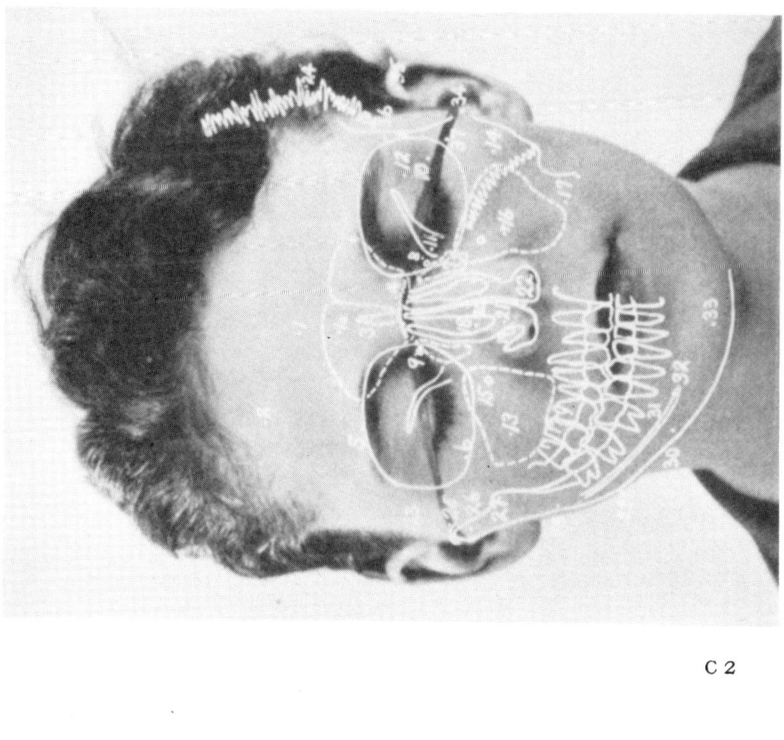

8a

FIGS. 8*a*, *b*. **Surface Markings** and **X-ray of Face.**

For description see opposite

Eastman & Co.)

possible secondary centre of ossification for this piece. Ledouble also records and figures three cases of separation of the medial angular process of the frontal bone to form an *os maxillo-naso-lacrimo-frontal*; this he regards as a Wormian bone, but it has been held to represent the pre-frontal of lower vertebrates; it may be explained by non-union of a part of the bone developed from a secondary centre of ossification (Fig. 46, p. 99).

The relations of the whole orbital margin to the eyeball are considered under the section devoted to the latter and will be found on p. 258, and its dimensions are given under Mensuration on p. 92.

The Surface Anatomy of the Orbital Margin.—The whole margin can be defined in life and its main features identified by the finger. In the mid-line above the hollow at the root of the nose the **glabella** can be felt, and lateral to this on either side the prominence of the **supra-orbital ridge;** overlying the bone between the ridge and the actual orbital margin is the 'head' of the eyebrow. The medial third of the supra-orbital margin is rounded and indistinct as it passes downwards into the medial angular process of the frontal bone, but may present a frontal notch about a finger's breadth from the mesial plane; lateral to this, at the junction of the medial third and lateral two-thirds of the margin and about an inch from the mid-line of the skull, the **supra-orbital notch** may be felt, and the presence of the corresponding nerve realized by pressure just above it. A line drawn downwards from this notch to the interval between the upper two bicuspid teeth crosses the infra-orbital foramen. From the supra-orbital notch the sharp lateral two-thirds of the margin can be traced round the supero-lateral angle to the **zygomatico-frontal suture** and the upper part of the lateral orbital margin. This suture can always be felt owing to a notch on its hinder edge. It lies at a slightly higher leve than the corresponding fronto-maxillary suture on the opposite margin of the orbit. Above the suture the prominent **lateral angular process** of the frontal bone can be felt, and the temporal ridge traced upwards from its posterior edge;

the 'tail' of the eyebrow lies over this process and the lacrimal gland lies beneath it, though nothing can be felt of this organ. Below the suture the lateral margin is formed by the fronto-sphenoidal process of the zygomatic bone; on its outer side there may be felt the *processus marginalis* projecting

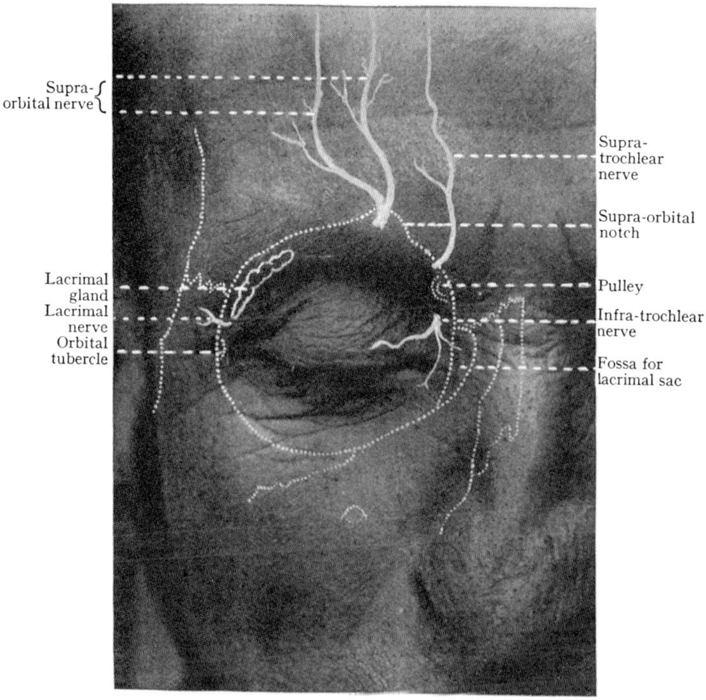

FIG. 9.—To illustrate the **surface anatomy** of the Orbital Margin.

backwards towards the temporal fossa; just below it the zygomatico-temporal branch of the zygomatic nerve reaches the skin. On the orbital aspect of the process, and just within the centre of the lateral margin, lies the **orbital tubercle** of the zygomatic bone: as previously noted, it can be felt by inserting the finger-tip deeply between the orbital margin and the eyeball opposite to the lateral commissure (external canthus) of the eyelids and rubbing the finger up and down. This margin is backwardly curved to such an extent that it lies on a plane 22 mm. posterior to the root of the nose, 12 mm. behind the anterior lacrimal crest, and 7 mm. behind the fovea trochlearis (Maddox). Below the suture the lateral margin can be traced as a well-defined edge downwards round the infero-lateral angle of the base and along the infra-orbital margin until at about its centre the **zygomatico-maxillary suture** is felt. The site of the suture may be more obvious owing to the presence of an infra-orbital tubercle, and about 1 cm. below it and the same distance from the skin there is situated the infra-orbital foramen. Medial to this suture, the margin slopes upwards into the **anterior lacrimal crest,** at the commencement of which there may be felt a 'lacrimal tubercle'. It is generally easy to feel with the finger-tip the depression of the **fossa for the lacrimal sac** behind this crest and behind and above the tubercle if present, and also to identify the posterior lacrimal crest which bounds the fossa behind. In the upper part of the fossa the outlines are rendered indistinct by the medial palpebral ligament (*tendo oculi internus*) which crosses it. The fronto-maxillary suture cannot be felt; above it the margin curves up into the supra-orbital margin, so forming the supero-medial angle of the base. This angle is full and rounded, and is crossed by the supra-trochlear nerve and by the important anastomoses between the ophthalmic blood-vessels and the angular vessels of the face. By pressing the end of the thumb within this angle the **trochlea** or pulley of the superior oblique muscle can readily be felt on the orbital roof.

As regards the relation of the orbital margin to the brain, it may be noted that the anterior pole of the temporal lobe lies about 2 to 2½ cm. or 1 inch behind the zygomatico-frontal suture of the lateral margin, and about the same distance above the zygoma; the inferior surface of the frontal lobe lies about half an inch above the suture, and its medial lower edge is on a level with the fronto-nasal suture (Fig. 205, p. 395).

The position of the vessels and nerves which cross the orbital margin in various regions is shown in Fig. 9 and Fig. 89, p. 167. Overlying the whole margin is the orbital part of the orbicularis oculi muscle, covered by superficial fascia and skin, which are thickened to form the eyebrow in the supra-orbital region.

3. THE WALLS OF THE ORBIT

For convenience of description the orbital cavity is regarded as being bounded by four walls, but the four angles formed between them are so rounded off, especially in front, that certain features may be described as belonging to either one or other of two adjacent walls. The walls are:

(i) The roof or superior wall (*paries superior*).
(ii) The lateral or external wall (*paries lateralis*).
(iii) The floor or inferior wall (*paries inferior*).
(iv) The medial or internal wall (*paries medialis*).

Of these, the medial wall alone is quadrilateral in shape; the others are triangular in outline, with the apices behind and the bases in front at the margin; the floor is the shortest and extends only to two-thirds of the depth of the cavity. The lateral wall is best demarked owing to the presence of the superior and inferior orbital fissures above and below its posterior part; the medial wall may be limited above and below by prolonging forwards and backwards the lines of the upper and lower sutures of the ethmoidal bone.

The bones are chiefly those which have already been seen to form the margin, namely, the frontal, zygomatic, maxilla, and lacrimal; in addition, portions of the sphenoidal, ethmoidal, and palatine help posteriorly; of these bones the frontal, sphenoidal, and ethmoidal are common to both orbits.

(i) The roof is formed by the frontal, and the lesser wing (*ala parva*) of the sphenoidal bone.

(ii) The lateral wall is formed by the zygomatic, and greater wing (*ala magna*) of the sphenoidal, together with a small part of the frontal bone.

(iii) The floor is formed by the maxilla, together with the palatine and part of the zygomatic bone.

(iv) The medial wall is formed by the frontal, lacrimal, ethmoidal, and body of the sphenoidal bone.

(i) **The Roof** (superior or cerebral wall, vault, *paries superior*) is separated from the medial wall in greater part by the fronto-ethmoidal suture; a backward prolongation of this line passes just below the optic foramen (which lies, therefore, in the apex of the roof) and cuts into the widest part of the superior orbital fissure, whilst its forward prolongation runs across the medial angular process of the frontal bone and approximately hits off that point at the root of the nose where the nasal bones articulate with the frontal (known in craniometry as the point 'nasion'). The

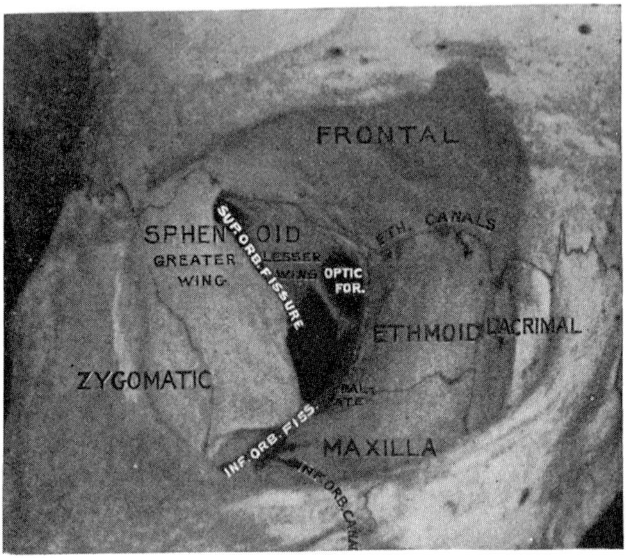

FIG. 10.—Right Orbit, showing the **bones** which form it, and the **apertures.** ×2.

line of separation from the lateral wall is indicated by continuing the direction of the upper limb of the superior orbital fissure until it cuts the supero-lateral angle of the margin at the most prominent point of the lateral angular process of the frontal, a short distance above its articulation with the zygomatic bone. The roof presents in outline the shape of an isosceles triangle. It is nearly flat at the apex behind, and of increasing concavity in front; it forms a smooth dome, well arched from side to side, but less so from behind forwards. It is formed in greater part by the orbital plate (*pars orbitalis*) of the frontal bone, and is completed behind at the apex by a small triangular piece formed by the lesser or orbital wing (*ala parva*) of the sphenoidal bone. This piece is about a half to three-quarters of an inch in length and is pierced by the optic foramen (Fig. 12). The features to be noticed in the roof are—the fossa for the lacrimal gland, the *fovea trochlearis*, the fronto-sphenoidal suture, the optic foramen, the architecture, and the relations to adjacent parts.

The **fossa for the lacrimal gland** (*fossa glandulae lacrimalis*) is a depression in the frontal bone at the antero-lateral angle of the roof of the orbit; it is overhung by the lateral angular process. In some skulls the fossa is nearly conical in shape, as in the foetal condition, but generally a uniform hollow is found, the fossa being merely a deeper part of the general concavity of the roof and best appreciated by feeling the bone with the finger; its limits are defined below only by the marked ridge running along the zygomatico frontal suture on the lateral orbital wall, as seen in the right orbit of Fig. 11. Whether the marked difference in depth found in various skulls (not necessarily in the two sexes) is accounted for by a corresponding development of the gland is questionable, since the fossa not only lodges the gland but is filled up by a mass of orbital fat behind it; this posterior part of the fossa is called by Rochon-Duvigneaud (1903) the *loge accessoire*, and in some cases a faint undulation of the surface may be felt to demark the two regions.

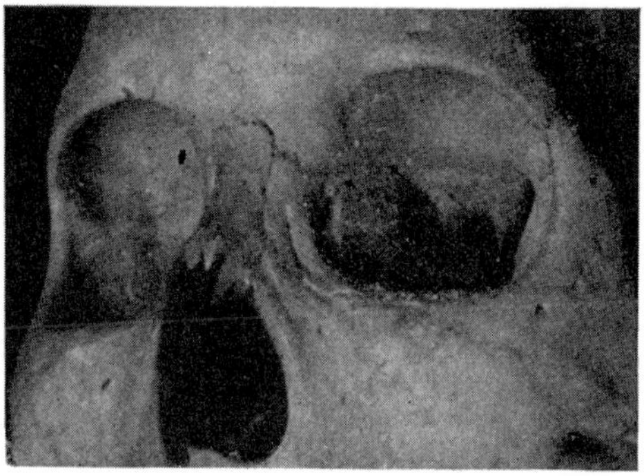

FIG. 11.—Features in the orbital walls. The Left Orbit shows the *fovea trochlearis* as a dimple in the antero-medial corner of the roof; there is a supra-orbital canal, a well-marked *sutura notha*, and triple zygomatico-facial foramina. The Right Orbit shows the fossa for the lacrimal gland beneath the lateral angular process of the frontal bone, limited below by a ridge along the line of the zygomatico-frontal suture. Calcutta Indian skull. × ½.

The **fovea trochlearis** is a small circular pitting or dimple of the bone in the antero-medial angle of the roof, 4 to 5 mm. from the orbital margin; its position is said to vary according to the shape of this angle of the orbit, being found higher when the angle is narrow, lower when the angle is broader (Stanculeanu, 1902). It is present in 80 per cent. of cases on one or both sides, and marks the point of attachment of the cartilaginous pulley (*trochlea*) of the superior oblique muscle of the globe. Sometimes it is surmounted by a small spine, the *spina trochlearis*, as seen in Fig. 24, p. 53, for which a secondary centre of ossification has been described (Keibel and Mall, 1912). Sperino (1904) pointed out that it may occupy a variable position with regard to the fovea, though most often developed above and behind it; in the observations quoted by Ledouble the spine was found present in nearly 16 per cent. of 1,516 skulls, more often on one side only and then on the right, and in all races, at any age, and in either sex. In 1 to 2 per cent. of skulls there is a second spine situated below the fovea, and even a complete bony ring has been observed. These spines are formed by ossification of the fibrous bands which attach the ends of the semilunar cartilage of the pulley to the edges of the fovea, and the more frequent occurrence of the upper one is explained by the greater traction exerted upon the corresponding ligament, giving rise to an activity of the osteogenetic periosteum. The bone may be grooved above the fovea trochlearis by the supra-trochlear nerve, and below by the infra-trochlear nerve and terminal branches of the ophthalmic blood-vessels. A description of the pulley itself will be found on p. 277 in connexion with the tendon which passes through it, and its position is considered on p. 106. The bony wall on which it is situated is undermined in one-third of cases by the frontal air-sinus, as is well shown in Fig. 69, p. 141.

The **fronto-sphenoidal suture** (posterior limb) crosses the apex of the roof obliquely from the anterior end of the superior orbital fissure on the lateral side to a point just behind the posterior ethmoidal foramen on the medial wall, as is well seen in Fig. 10; it cuts off the lesser wing of the sphenoidal bone with the enclosed optic foramen, and is the only suture which normally crosses the roof, but is usually obliterated in the adult.

The **optic foramen** (*foramen opticum*).—This name is

commonly applied to the anterior or orbital opening of a short canal (*canalis nervi optici, canalis opticus*), which has also a posterior or cerebral opening into the middle cranial fossa. It lies in the apex of the roof of the orbit at the hinder end of the angle between the roof and the medial wall; it is nearly in line with the two ethmoidal foramina, which lie along this angle, and the nasion, and a horizontal plane drawn through this last point at the root of the nose would pass just above it. The distance of this opening from the supra-orbital margin is about 50 mm. or 2 inches, but the precise measurements from this and other points on the margin will be found under mensuration on p. 95, since the optic foramen is the point from which the lengths of the orbital walls are usually measured. Below it there may be present a minute 'infra-optic tubercle' (p. 266). The distance between the two foramina is affected by the type of skull.

The **optic canal** is formed by the union of the two roots of the lesser wing of the sphenoidal bone which spring from the side of its body. The floor projects posteriorly, the roof anteriorly: that is to say, its two ends are obliquely cut from above downwards and backwards, as shown in Fig. 33, p. 69, but the bony roof is prolonged behind for a few millimetres, or in rare cases even entirely formed, by a fibrous arcade or 'falciform fold' of the dura mater. The length of the canal varies from 4 to 9 mm., and its long axis is inclined posteriorly and medially (continuing the line of the lateral orbital wall) so that the axes of the two canals, which are separated at their orbital openings by a distance of about 25 mm. or 1 inch, would, if projected backwards, meet at the centre of the dorsum sellae of the body of the sphenoid; this central meeting-point lies 27 to 29 mm. behind the orbital end of the canal along its projected long axis. The canal lies opposite the middle and widest part of the zygoma and a finger's breadth above it; its horizontal distance from the centre and most prominent part of this arch was 40 mm. in a skull of five years, 43 mm. in one of twelve years, 49·5 mm. in one of sixteen years, 52·5 mm. in one of nineteen years, 52 mm. in an adult female, and 60·5 in an adult male skull, as found by Weiss (1890, 1897); the position relative to the zygoma is seen in Fig. 197, p. 379. The anterior opening is oval in outline and measures 5 to 6 mm. in its longest (the vertical) diameter; the central part of the canal is circular with an average diameter of 5·5 mm., and the posterior opening is generally considerably flattened from above downwards. Recently, the canal has been studied by radiography; by White (1924) in 164 skulls, Van der Hoeve (1925), Goalwin (1927) in 806 living cases and 194 skulls, and Del Duca (1929), who gives an extensive table of measurements of these dimensions in 90 skulls.

The contents of the canal are the optic nerve and the ophthalmic artery, the latter being accompanied by minute branches from the carotid plexus of the sympathetic nervous system. The optic nerve is enclosed in sheaths prolonged from the pia mater, arachnoidea, and dura mater of the brain. The artery is embedded in the dural sheath on the lower and lateral side of the nerve, and ossification of the fibrous septum between them accounts for the rare abnormality of a 'doubled optic foramen' (Fig. 24, p. 53), the vessel or one of its branches passing through the lower and smaller opening; Ledouble notes 12 cases in the literature, single, bilateral, and in either sex. White saw 3 instances; more commonly the vessel grooves the hinder part of the floor of the canal, and Del Duca finds it moulding the shape of the optic canal both internally and externally.

The relations of the optic canal are as follows: superiorly a fairly strong plate of bone separates it from the base of the olfactory tract on the under surface of the frontal lobe of the brain. In some instances the roof of the canal is hollowed out by an extension of the sphenoidal air-sinus; indeed, according to Howell Evans (1908), this 'optic projection' of the sinus into the lesser wing or the clinoid process of the sphenoidal bone occurs as often as 1 in 3. On the infero-lateral aspect the optic canal is separated from the superior orbital fissure by a bridge of bone formed by the lateral root of the lesser wing of the sphenoid; this bridge, which also may be hollowed out by an extension of the sphenoidal sinus or of a posterior ethmoidal air-cell, sometimes presents a minute tubercle in its lower part (Fig. 16), to which is attached the tendon of Zinn (p. 266). Ledouble has seen two cases in which the bridge was absent, so that the optic canal communicated with the superior orbital fissure, a condition found in birds, marsupials, and whales. On the medial side the canal is in relation with the sphenoidal air-sinus or occasionally with a posterior ethmoidal air-cell. According to Loeb (1909) the optic chiasma is nearly always

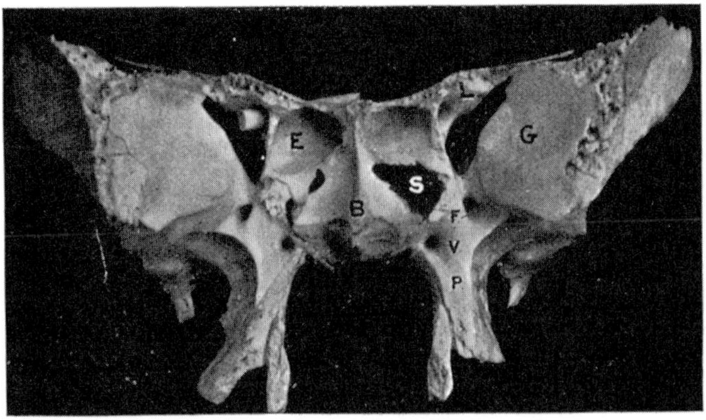

FIG. 12.—**Sphenoidal Bone** from in front (orbital aspect). Adult. × ⅔.

B. = Body of sphenoidal bone.
G. = Orbital surface of greater wing separated from the lesser wing by the superior orbital fissure.
L. = Lesser wing pierced by the optic canal, through which a rod has been passed on the right side.
P. = Pterygoid process.
V. = Vidian canal.
E. = Posterior ethmoidal cell into which the optic canal projects.
S. = Sphenoidal sinus showing relation to F., the *foramen rotundum*.

related in front to both the sphenoidal sinuses and never to the posterior ethmoidal cells, whilst the optic nerve is usually related to the sphenoidal sinus but in a few cases to an ethmoidal cell; Young (1922) and Westlake (1923) have especially considered this last relation. Onodi (1903) even classifies thirty-eight different forms and twelve groups of relations of these air-spaces to the optic canal, which indicates the great variation that may occur in their disposition. Fig. 12 shows a sphenoidal bone in which the optic canal is related to the hindermost ethmoidal cell. In one remarkable specimen of the Oxford collection the sphenoidal sinus surrounds the optic canal on all sides, so that the latter passes freely through the middle of the cavity as a bony tube; and in the McGill Museum is a specimen in which a large posterior ethmoidal cell presents a similar relation, the sphenoidal sinus being much reduced in size, and another specimen where a large right sphenoidal sinus is related to both optic canals. The frontal sinus has been seen by the writer to undermine the roof of the orbit so far back as to come close to the optic canal, but the maxillary sinus was never found to extend nearer than 7 mm. to it.

The most important fact in regard to these relations is the delicacy of the bony wall which separates the optic canal from the neighbouring air-space, especially when this is formed by an ethmoidal cell; it is most often as thin as 0·5 mm., but has been found from 5 to 7 mm. thick (Stanculeanu, 1902). Francis and Gibson (1911) found a very thin wall in 38 per cent. of cases; this wall, moreover, may be the site of senile atrophy, dehiscences having been recorded by Barraguer (1896), Berger (1894), Holmes, C. R. (1896, two cases in 50 skulls), and Gallemaerts (two in 200), the optic nerve in its sheath being then in contact with the mucous membrane lining the sinus; Del Duca (1929) found many such perforations in his series of 90 skulls; one such case has been seen by the writer. The clinical importance of minute diploic venous channels passing from the air-sinuses through this wall to reach the orbital system of veins is noted by Bordley (1921), Young (1922), Syme (1924), and Hays (1926), and the ocular involvement in sinus disease is discussed by Thomson (1928, 9). The embryology of the optic canal is the subject of a paper by Fileti (1928); it has definite osseous formation as early as the fifth foetal month and full size is reached at three years. Del Duca found it to be 7·5 mm. long at 10 to 11 years.

The relations of the ocular muscles at their origin to the anterior optic foramen are illustrated in Fig. 142, p. 262, and described on p. 263; here it may be noted that the superior and medial recti muscles are closely related to the superomedial side of the foramen, with the superior oblique and levator palpebrae superioris muscles lying just above them.

The **architecture of the roof of the orbit.**—The roof presents more variations in thickness than any other wall;

it is usually about 1 mm. thick, but is often of such paper-like delicacy that a tap with the finger-nail will break through it, and dehiscences due to atrophy of the bone in old age may be present, the periorbita then lying in contact with the dura mater; on the other hand, in one case it was found to be 4 mm. thick (Fig. 13, *b*). It may be doubled in lesser or greater degree by extensions from the medial side of the frontal and ethmoidal air-cells between the two tables of the bone, as indicated in Fig. 31, p. 64, and well shown on p. 379. Witt (1908) finds such extensions in one-third of cases and describes three types—(*a*) formed by the frontal sinus alone, (*b*) by the frontal sinus and one ethmoidal cell, (*c*) by the frontal sinus and two ethmoidal cells, each cell having its own opening into the nose; he terms them 'orbito-ethmoidal cells'. Cryer (1907) calls them 'supra-orbital sinuses' and notes that the posterior ethmoidal

a

b

c

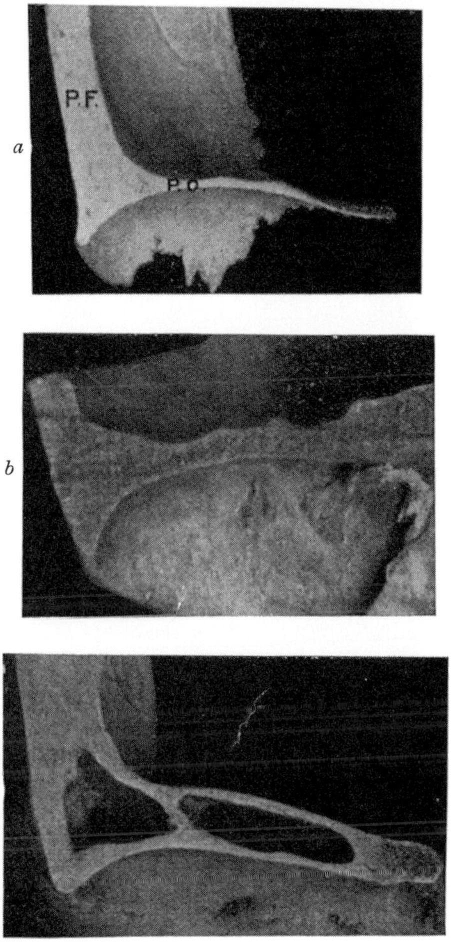

FIG. 13.—**Architecture** of Orbital **Roof,** three types. × $\frac{2}{3}$.

a = Thin, normal. P.F. = *pars frontalis;* P.O. = *pars orbitalis* of frontal bone.
b = Thick, solid.
c = Doubled by extensions of air-cells with thick walls (compare Fig. 33, where the walls are thin, making a fourth type).

cells also may occasionally form them. The two walls thus formed may be very thin or fairly thick, and any variation between these extremes, which are illustrated in Fig. 13 and Fig. 33, p. 69, will be found on examination of a series of bones.

The posterior two-thirds of the roof is usually the thinnest part, and on holding the bone up to the light there may be seen the lines of certain ridges formed on the cerebral aspect. These ridges (*juga cerebralia*) and the hollows between them (*impressiones digitatae*) are moulded by the sulci and convolutions of the super-imposed inferior surface of the frontal lobe of the brain (Fig. 176, p. 330). The anterior third of the roof is always thicker and is not transparent.

Abnormal sutures may divide the roof into many pieces, and under the designation of 'accessory frontal bones' Augier (1912) classifies six varieties. Tenchini (1905) observed an abnormal 'orbito-frontal emissary vein', which passed from the superior longitudinal sinus of the brain through the roof of the orbit to join the superior ophthalmic vein.

The bone in the antero-lateral region of the roof is sometimes marked by an area of numerous fine holes, termed *cribra orbitalia* by Welcker (1887), which may even extend medially across the greater part of the roof. This porous condition affects only the orbital table of the bone, and is caused not by a pathological process but by a fault in development of the bone whereby the venous blood of the diploë, instead of being collected by the normal diploic channels, passes into the veins of the periorbita by numerous venules, hence the bony tissue of the orbital wall cannot develop as a compact layer and becomes cribriform. In rare cases a spongy swelling may be formed which projects into the orbital cavity and resembles an osteophyte. Ahrens (1904) found cribra orbitalia chiefly in young adults and in about 25 per cent. of 861 skulls of all races, in 17 per cent. of Europeans, but more commonly in lower races. Koganei (1913), in a more detailed account, differentiates three types according to the degree of development. He found them present more often in children (24·4 per cent. of 84 skulls) than in adults (11 per cent. of 372 skulls), and always on both sides. A well-marked type is illustrated in Fig. 14, where in the right orbit it forms a well-marked osteophyte; less developed examples are seen in two of the orbital roofs in Fig. 31, p. 64. It may be added that the condition is also found, though much more rarely, in the parietal and occipital bones of the skull.

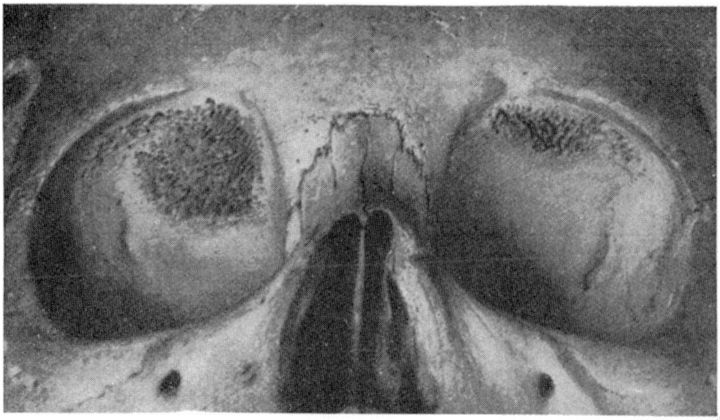

FIG. 14.—Roofs of Orbits showing condition of **cribra orbitalia.** The right orbit shows an advanced condition forming an exostosis. Compare with the slight condition shown in Fig. 31, p. 64 (second and third bones). New Zealand skull. × $\frac{2}{3}$.

The **relations of the roof.**—Internally, the periorbita is thin and easily detachable, except along the orbital margin and at the fovea trochlearis, where it binds the trochlea down to the bone. It is also firmly attached along the fronto-sphenoidal suture, so that a fluid introduced between it and the bone will not pass backwards beyond this line to reach the optic nerve or enter the cranial cavity. Closely applied to the roof and traversing its whole length is the frontal nerve, but the supra-orbital artery comes into relation with its anterior half only. Beneath the nerve the levator palpebrae superioris and underlying superior rectus muscles lie close against the roof, with the possible intervention of a little fat. Along the medial margin lies the superior oblique muscle, separated posteriorly by the fourth cranial nerve, and against the anterior and lateral part is lodged the lacrimal gland.

Externally, the roof separates the orbital cavity from the anterior cranial fossa of the skull in which rests the frontal lobe of the brain, separated from it only by the meninges; so close is the relation between them that the bone is moulded by the cerebral convolutions and fissures, causing the 'digital impressions' noted on p. 32; on the other hand, this region of the brain is slightly hollowed by the domed roof of the orbit and presents on its surface the composite fissure termed the *sulcus orbitalis* (Fig. 175, p. 329). The olfactory tract crosses obliquely the extreme posterior end of the roof on its medial side. The air-spaces which may enter into relation with the roof have been described under the architecture on p. 31.

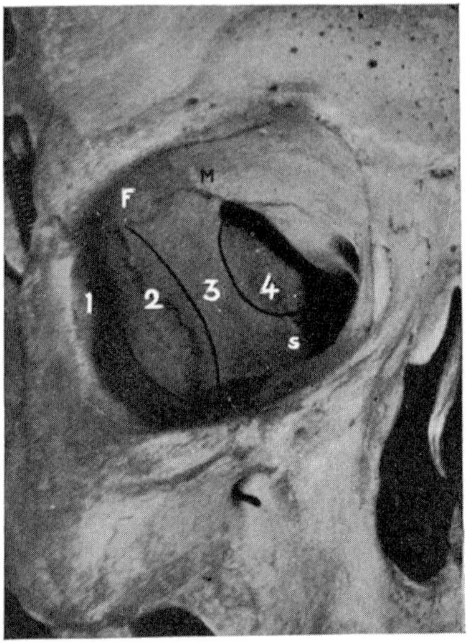

FIG. 15.—**Architecture of Lateral Wall** of Right Orbit. The four numbers show areas which differ in strength and external relations (compare with the horizontal section of this wall shown in Fig. 17). Natural size.

F. = Fossa for lacrimal gland. M. = Meningeal foramen. S. = *Spina recti lateralis.*

(ii) **The Lateral Wall** (outer, external or temporal wall; *paries lateralis*).—The limits of this wall are well defined, since the posterior part is bounded by the two orbital fissures, the superior dividing it from the roof, the inferior from the floor, and by prolonging their long axes forwards to the orbital margin the outline is completed; the upper line so taken ends on the most prominent part of the lateral angular process of the frontal bone, the lower one at the infero-lateral angle of the margin. The wall is flat, save for a slight concavity just behind the margin, and triangular in shape, with a free apex directed superiorly and medially and pointing towards the base of the mastoid process of the opposite side of the skull (see p. 42). It is formed chiefly by two bones, the greater wing (*ala magna*) of the sphenoidal in its posterior two-thirds and the upper half of the orbital surface of the zygomatic in its anterior third, but a small part of the frontal completes this wall above the zygomatic area. The features to be noticed are—three sutures, the orbital tubercle, the *spina recti lateralis*, the openings of the zygomatic canals and the meningeal foramen, the architecture and relations.

The **sutures** are three in number, two placed horizontally and one vertically, and they meet at a point towards the upper limit of the wall not far from its centre. (i) The zygomatico-frontal is a finely denticulated suture running backwards from the orbital margin and usually raised to form a ridge which marks the lower limit of the fossa for the lacrimal gland, as seen in Fig. 11. Nicola (1903) has studied the variations of these sutures and states that an obliquity from above downwards and from behind forwards of this one in particular is most often found in lower races. Sollas finds that in many human skulls the line descends steeply, and in many anthropoids it is horizontal. (ii) The fronto-sphenoidal suture (anterior limb) continues the line of the zygomatico-frontal suture backwards to the anterior extremity of the superior orbital fissure. It divides the frontal bone from the greater wing of the sphenoidal, but is usually obliterated in the adult. (iii) The spheno-zygomatic suture runs up the lateral wall from the anterior end of the inferior orbital fissure to the point of junction of the above two sutures, meeting their line nearly at right angles; it lies between the orbital plate of the zygomatic bone and greater wing of the sphenoidal, and is often serrated and marked by small isolated ossicles analogous to the wormian bones of the cranium.

The **orbital tubercle** of the zygomatic bone has been described above on p. 13 with reference to the lateral orbital margin, close to which it is situated.

The *spina recti lateralis* (Merkel) is a small spur frequently found at or near the apex of the lateral wall, and projecting backwards towards the widest part of the superior orbital fissure opposite the optic foramen; it is slightly marked in the above figure, but is exceptionally well developed in Fig. 16. It may be rounded or spinous, and affords an attachment to the lateral rectus muscle. Below and in front of this spine there may occasionally be found a second small

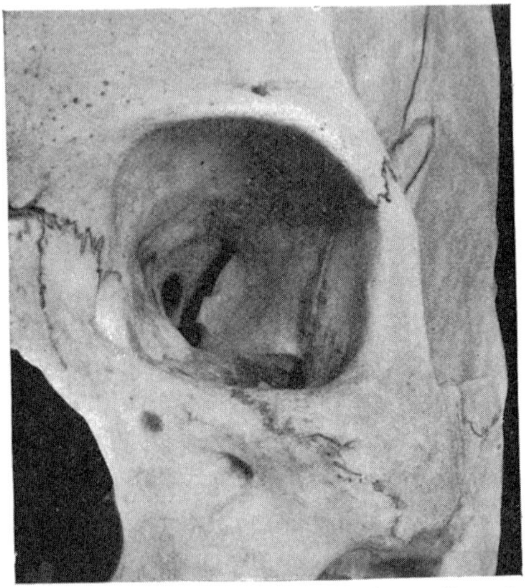

FIG. 16.—Left Orbit of an African Negro, showing well-marked crescentic **spina recti lateralis.** The minute 'infra-optic tubercle' can just be discerned below the optic foramen.

tubercle or rugosity possibly affording additional origin to the same muscle. W. Krause (1879) refers to these spines as *spinae orbitales,* superior and inferior, and regards the former as an extra origin of the lateral rectus and the latter as the origin of the orbital muscle of Müller.

The **zygomatic canals.**—On the orbital surface of the zygomatic bone there is usually to be seen the opening of a single canal (*canalis zygomatico-orbitalis*), which divides in the substance of the bone into two divergent branches, one of which opens on the facial aspect at the zygomatico-facial foramen, whilst the other ascends to open into the temporal fossa behind the frontal process at the zygomatico-temporal foramen. These canals contain the zygomatic or orbital nerve with its two terminal divisions, accompanied by branches of the infra-orbital artery. Though seldom absent, they are very irregular in size, number, and course. In one-third of 800 skulls Ledouble found two complete canals, and in rare cases three or four have been seen; one or more may end blindly within the bone and contain only nutrient vessels; there may be a single canal containing the facial branch of the nerve, the temporal branch of which, accompanied by a twig from the lacrimal artery, then runs up the surface of the wall, sometimes in a groove along the spheno-zygomatic suture, to reach the temporal fossa through a foramen in its upper part (the spheno-zygomatic foramen, which in the adult may be wholly confined to the zygomatic bone). The lower edge of the greater wing of the sphenoidal bone may also be grooved by the main stem of the orbital nerve before it divides; the canals are formed in early development by an upgrowth of the zygomatic bone round the nerves, and indicate the line of fusion between the two chief ossific centres.

The **meningeal foramen** appears fairly constantly near the anterior end of the superior orbital fissure in or near the anterior fronto-sphenoidal suture. It transmits the orbital branch of the middle meningeal artery, accompanied, according to Testut, by a vein. Examples of the foramen are seen in Figs. 10, 14, 15.

The **architecture.**—The lateral is the strongest of the orbital walls, though it is by no means of uniform thickness. It is strong anteriorly, where the two processes of the frontal and zygomatic bones form the orbital margin; behind the margin the anterior half of the wall up to the spheno-zygomatic suture is usually thin, and separates the orbit from the temporal fossa. Posterior to the suture the wall is strengthened on its outer side by its confluence with the lateral wall of the cranium, and still further back the apex of the wall consists of a thin plate of bone which separates the orbit from the middle cranial fossa of the skull. Thus there are two strong regions alternating with two weak ones, and the lateral wall may be divided into four areas (Figs. 15, 17). The weaker temporal and cranial areas are often the site of senile atrophy, the wall being then exceedingly thin in these regions. Ono (1928) illustrates the different degrees of strength here, typically strong, medium, and weak.

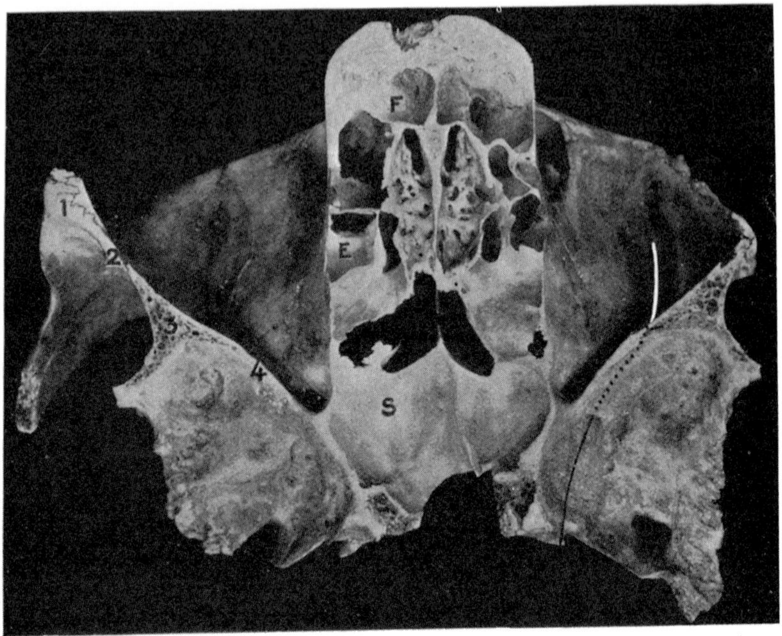

FIG. 17.—Horizontal section through Orbits to illustrate **architecture and relations of walls;** lower half seen from above. The preparation shows the parallelism of the medial walls and their relation to the frontal sinus (F.) in front (the infundibulum is seen on the right side), the ethmoidal cells (E.) in the middle, and the sphenoidal sinus (S.) behind. Between the ethmoidal cells the cribriform plate is well displayed.

The lateral walls show the difference of thickness in the four regions: (1) thick at the orbital margin, (2) thin in relation to the temporal fossa, (3) thick at the confluence with the lateral wall of the cranium, and (4) thin in relation to the middle cranial fossa. The lateral wall forms an angle with the middle wall of approximately 45°, and with its fellow of the opposite side of 90° (compare with Figs. 45 and 136).

The dotted line on the right side follows the curved course of the maxillary nerve through the *foramen rotundum* and its continuation as the infra-orbital nerve along the infra-orbital sulcus of the orbital floor. Natural size.

The **relations** of the Lateral Wall.—Internally the lateral rectus muscle is in close contact with this wall from the spina recti lateralis posteriorly to the orbital tubercle anteriorly, where its 'check ligament' is attached. The lacrimal artery and nerve run along the upper border of the muscle close to the wall. The relations of the zygomatic or orbital nerve and its two branches are indicated in the description of the zygomatic canals: there is an anastomotic loop between the lacrimal nerve and the zygomatico-temporal branch which is closely applied to the bone (Fig. 77, p. 154). The lacrimal gland extends as low down as the fronto-zygomatic suture on the forepart of this wall, and is in close contact with it. Externally the lateral wall presents different relations in its anterior and posterior regions. The forepart of the wall or temporal area separates the orbit from the temporal fossa, in which lies the temporal muscle and the middle temporal artery with its zygomatico-orbital branch. It may be of interest to note that the orbital plate of the zygomatic bone which forms this partition is developed in higher primates only, and its double relation to the frontal and sphenoidal bones is a distinctive character of the primates as opposed to all other mammals, since in the latter the orbit communicates openly with the temporal fossa (as indeed it does in the early human foetus). A more important relation is that of the posterior part of the wall behind the strong junction which it forms with the cranium. Here the orbit is separated from the fore end of the middle cranial fossa in which is lodged the temporal lobe of the brain, the apex of the wall lying against its middle temporal gyrus. These relations of the lateral wall to the soft parts are well displayed in Fig. 136, p. 254. The distance of this wall from the skin surface is about 15 mm. just behind the margin, and from 45 to 50 mm., or nearly 2 inches, at its posterior free end, as measured in the preparation just referred to.

An illustration of the saw-cuts made through this wall in Kronlein's operation may be of interest (Fig. 18). The globe and nearly half the optic nerve can thus be reached.

The **superior orbital fissure** (sphenoidal fissure; *fissura orbitalis superior*; *foramen lacerum anterius*) may most fittingly be described here since it separates the roof and lateral walls in their posterior thirds. It is the gap between the greater and lesser wings of the sphenoidal bone and is closed at its anterior extremity by the frontal bone. It consists of two segments: a narrow lateral limb lying between the lateral wall and roof of the orbit, and a medial broader part lying between the apex of the lateral wall and the bridge of bone which bounds the optic foramen. Below the lower extremity of the fissure is the foramen rotundum, a relation clearly seen on looking from inside a sectioned skull but not

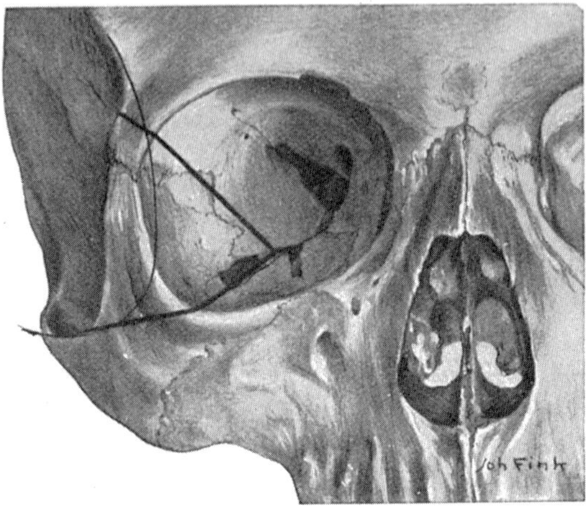

FIG. 18.—To illustrate bone cuts in lateral orbital wall used in **Kronlein's Operation.** (After Haab, from de Schweinitz, *Diseases of the Eye*, 1924.)

so obvious on looking into the orbit from in front, since at this point the superior orbital fissure becomes confluent with the inferior and the foramen lies deep to their junction. The shape of the fissure varies greatly, being dependent upon the formation of its anterior boundary. It is said to be longer in dolichocephalic than in brachycephalic skulls. It may be asymmetrical in the two orbits and is larger in the aged through absorption of the edge of the greater wing of the sphenoid. Gallemaerts (1897) summarizes three types in 160 skulls of various races: (*a*) The commonest type, with two segments, one narrow, the other broad, as described above and as seen in Fig. 15; (*b*) a type with both segments broad; and (*c*) a triangular type, as in Fig. 12. The lateral segment is said to be smaller in negroes, and in two skulls of this type the fissure was found reduced to a round hole, as in the gorilla and gibbon. He finds the length of the fissure to be most often 15 or 16 mm. (Panas and also Krause give a mean of 22 mm.), and that of its lateral segment 9 or 10 mm.; the distance of its fore end from the orbital margin was found by the writer to vary from 12 to 18 mm., with an average of 15·5 mm. in 10 male skulls, and from 12 to 19 mm. with an average of 14·3 in 10 female skulls. It puts the orbit in communication with the middle cranial fossa, and on looking at the orbit from in front there can be seen through the widest part of the fissure the depression on the apex of the petrous bone which lodges the semilunar or Gasserian ganglion of the fifth nerve.

The following structures pass through the superior orbital fissure:

(i) The third, fourth, and sixth cranial nerves.
(ii) The ophthalmic division of the fifth nerve in three branches: nasal, frontal, and lacrimal.
(iii) The ophthalmic vein.
(iv) The orbital branch of the middle meningeal artery (when it does not pass through the meningeal foramen).
(v) The sympathetic root of the ciliary ganglion, and some filaments from the cavernous plexus of the sympathetic system.

The central part of the fissure is crossed by the 'annulus of Zinn', from which arise the recti muscles of the eyeball, and the position of the above nerves relative to it is shown diagrammatically in Fig. 142, p. 262. The importance of the superior orbital fissure is to be emphasized since, with the exception of the optic and maxillary, there pass through it all the nerves in the orbit—nerves which control the movements of the eyeball, the pupil, the accommodation of the lens, and the levator palpebrae superioris muscle, and nerves conveying sensation from the cornea, iris, chorioid, conjunctiva, lacrimal apparatus, frontal region, upper eyelid and forepart of nose.

The **inferior orbital fissure** (spheno-maxillary fissure, *fissura orbitalis inferior*) lies along the infero-lateral angle of the orbit, and separates the lateral wall from the floor in their posterior two-thirds. It is bounded above by the inferior margin of the greater wing of the sphenoid, and below by the lateral margins of the orbital surface of the palatine bone behind and that of the maxilla in front. Posteriorly, the foramen rotundum opens into it, and just above this point it is continuous with the lower end of the superior orbital fissure. It is narrower in the centre than at the extremities, the constriction being sometimes emphasized by a tongue-like process of the floor of the orbit hollowed out by an extension of the maxillary sinus or a posterior ethmoidal cell (Fig. 24, p. 53); its anterior end is the wider, and is completed by the zygomatic bone in over 60 per cent. of cases (Ledouble), this bone being otherwise excluded by the junction of the sphenoidal bone with the maxilla.

The fissure is about 20 mm. long, extending further forwards than does the superior fissure, and reaching to within 15 to 20 mm. or $\frac{3}{4}$ inch of the orbital margin. It lies on a level with the upper edge of the zygoma, is directed forwards and outwards from the apex of the orbit, and its long axis is in line with the zygomatic tuberosity on the cheek in one direction, the optic foramen in the other. This line corresponds with the long axis of the optic canal, and beyond this can be followed in a sectioned skull along the superior petrosal groove of the opposite side to a surface point at the centre of the base of the mastoid process; the line also indicates the inclination of the lateral wall of the orbit.

By the inferior orbital fissure communication is established with the pterygo-palatine (spheno-maxillary) fossa behind and the infra-temporal (zygomatic) fossa in front. In life it is closed by the periorbita, here containing the involuntary orbitalis muscle (Fig. 30). Along its posterior part pass the maxillary nerve and the infra-orbital artery on their way to the infra-orbital sulcus on the floor of the orbit; it also transmits the zygomatic or orbital nerve, a few twigs from the spheno-palatine ganglion, and vessels passing from the inferior ophthalmic vein to the pterygoid plexus. According to Treves, after violent blows upon the temple, blood may enter the orbit through this fissure and cause a sub-conjunctival ecchymosis; on the other hand, the writer has found a small hernia of the orbital fat protruding through it into the infra-temporal fossa.

In man and the higher mammals this fissure represents the wide communication between the orbit and the temporal fossa found in lower vertebrates, as noted above on p. 39; it is relatively larger in the skull at birth, and also in the aged from absorption of its bony margins; either from this cause or from mal-development, its anterior end may be abnormally large and encroach upon the lateral wall of the orbit, as in an Australian skull illustrated by Duckworth (1904); in rare instances a 'spheno-zygomatic fissure' may thus be formed along the line of the corresponding suture.

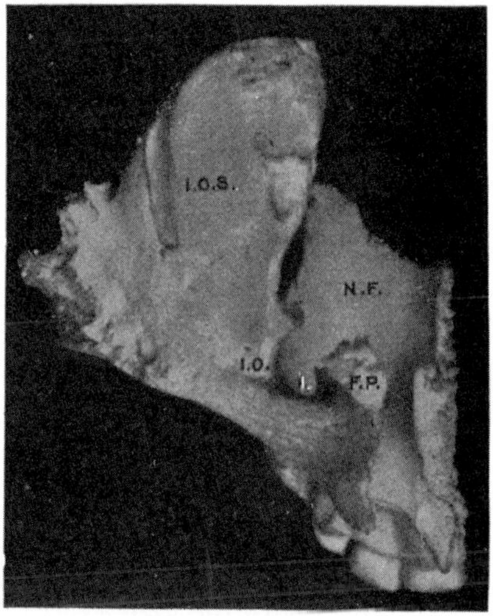

FIG. 19.—Right Maxilla seen from above to illustrate features on **Floor of Orbit.** Natural size.

I.O.S. = Infra-orbital sulcus.
I.O. = Fossa of origin of inferior oblique muscle.
I. = *Incisura lacrimalis* of maxilla, forming part of superior orifice of naso-lacrimal canal.
F.P. = Frontal process of maxilla, pointing up towards the reader.
N.F. = Floor of nasal fossa.

(iii) **The Floor of the Orbit** (maxillary or inferior wall, *paries inferior*).—The floor is limited on the lateral side by the inferior orbital fissure with a line prolonged from its anterior extremity to the orbital margin; medially it curves smoothly into the medial wall, but may be demarked from it by extending the lower border of the ethmoidal bone, a line marked from before backwards by the edge of the naso-lacrimal canal, and the sutures between the lacrimal and maxilla, ethmoidal and maxilla, and sphenoidal and palatine bones. It is the shortest wall, extending only two-thirds of the depth of the orbit; the outline is that of an equilateral triangle, and the surface is slightly concave anteriorly and convex posteriorly, owing to the upward expansion of the maxillary sinus. It is formed in greater part by the orbital plate (*facies orbitalis*) of the maxilla, completed antero-laterally by part of the orbital surface of the zygomatic and posteriorly by a small triangular area (the *processus orbitalis*) of the palatine bone (Fig. 10). Of the **sutures** which unite these bones, that between the maxilla and the palatine bone is usually obliterated in the adult skull, and in consequence the palatinal area of the floor is hard to define. The suture between the maxilla and zygomatic bone appears both on the orbital floor and on the cheek, and is **V**-shaped, with its apex at the middle of the infra-orbital margin; the orbital limb is inclined forwards and medially from the fore end of the inferior orbital fissure, running wide of the infra-orbital sulcus, to cut the margin at its mid-point, where it has already been noticed (p. 14); the facial limb slopes from this point in the opposite direction, backwards and then abruptly downwards to cut the lower border of the zygomatic arch at the fore end of the infra-temporal fossa (Fig. 5, p. 8). The features to be noted on the floor are the infra-orbital sulcus and canal, the fossa of origin of the inferior oblique muscle, the architecture and relations.

The **infra-orbital sulcus** or groove (*sulcus infra-orbitalis*) is a shallow gutter running nearly the whole length of the floor from the lateral side of its apex towards the centre of the margin. Its anterior half or two-thirds is converted into a canal (*canalis infraorbitalis*), the roofing over being caused in greater part by an extension of its lateral wall in the form of a very thin plate (the *lamina infraorbitalis* of Henle), as seen in Fig. 20; this union of the walls is marked along the supero-medial angle of the roof of the canal by a fine line, the *sutura canalis infraorbitalis longitudinalis*, which is continuous with the *sutura infraorbitalis verticalis* of the orbital margin. Anteriorly the canal sinks down beneath the margin, to emerge on the face at the infra-orbital foramen. The sulcus and canal lie entirely in the maxilla, and indicate the line of confluence between the maxillary and malar centres of the bone (see Fig. 46). The sulcus runs parallel to the mesial plane of the skull and not to the long axis of the orbital floor, which latter has an antero-lateral inclination, owing to the development of the underlying maxillary sinus (Fig. 17). The total length of both sulcus and canal is about 20 mm., with an average width of 4 mm. and a depth of 1·5 to 2 mm. From the lateral aspect arise the superior alveolar (dental) canals, the middle leaving the sulcus far back and running to the bicuspid teeth, whilst the anterior (usually double) leads out of the canal not far from its facial opening, and is continued on to the canine and incisor teeth (p. 16).

The contents are the infra-orbital nerve and artery, and possibly lymph channels. Stanculeanu (1902) described a vein here, but Sesemann could not find it, and Festal considered it exceptional. Czermak (1895), in an atlas of sections through the frozen orbit, figures the artery lateral to the nerve and a vein medial to it, but in a few cases examined by the writer the artery lay above and medial to the nerve, and no vein was found. The sulcus and canal are lined, and the former with its contents is also covered, by the periorbita; the infra-orbital nerve therefore does not lie freely within the peri-orbital boundary, as does the supra-orbital nerve.

The variations are explained by reference to the mode of development. In the embryo the infra-orbital nerve lies at first above the orbital surface of the maxilla, and passes freely over the orbital margin; it comes in contact with them towards the end of the second month, forming a groove on the bone, the walls of which finally close over it anteriorly to form the canal (Fawcett, 1911). The *sutura infraorbitalis*, with its longitudinal (orbital) and vertical (marginal) portions, marks the line of junction of this overlap. Later on the nerve cannot accommodate itself quickly enough to the more rapidly growing underlying maxillary sinus, and sinks into its roof; the groove is curved laterally by this expansion, instead of lying in a straight line between the foramen rotundum and infra-orbital foramen as in the skull at birth. In rare cases the anterior part of the sulcus is not ultimately transformed into a canal, but persists as an open groove passing over the infra-orbital margin on to the face, as seen in three adult skulls by von Langer (Merkel); on the other hand, the degree of ossification of the peri-orbital roof of the sulcus determines the length of the canal, which may occasionally be found to extend backwards the whole length of the floor of the orbit. An early or intra-orbital division of the nerve into its terminal branches accounts for the presence of branching or supernumerary infra-orbital canals; a medial canal containing the nasal and palpebral branches was found in eighteen and a lateral one containing the labial division of the nerve in twelve out of 2,700 skulls by Ledouble; other variations in number will be found in conformity with those of the infra-orbital foramen (p. 15); finally, Ledouble has seen the canal duplicated in its whole extent like a double-barrelled gun, an anomaly caused by ossification of the fibrous tissue which separates the nerve from the artery.

The **fossa of origin of the inferior oblique muscle** may occasionally be seen, but is usually better felt. It is a very shallow depression, situated in the extreme antero-medial angle of the floor just behind the margin and lateral to the orifice of the naso-lacrimal canal (see also p. 278). This orifice is sometimes described as a feature of the floor of the orbit, but will be dealt with under a later section in conjunction with the osseous lacrimal passages (pp. 78, 79). Its lateral margin, formed by the incisura lacrimalis of the maxilla, notches the antero-medial angle of the floor, and is seen in Fig. 19 above.

The **architecture.**—The floor is always composed of a thin ($\frac{1}{2}$ to 1 mm.) plate of bone, though it is stronger just behind the margin. Its thinnest part is along the floor of the infra-orbital sulcus and canal, and here dehiscences may occur, whereby the contents are only separated from the mucous membrane lining of the maxillary sinus by the periorbita; gaps are also to be seen in rare cases along the medial margin of the orbital floor between the maxilla and the ethmoidal bone. The zygomatic portion of the floor sometimes contains small air-cells or a cavity (*sinus zygomaticus*), which may or may not communicate with the maxillary sinus (Ledouble). A small vein has been noted by Gaillard to pass through the floor from the maxillary sinus to the orbital venous system, but it was not found by Stanculeanu.

The **relations** of the floor: internally, the periorbita both

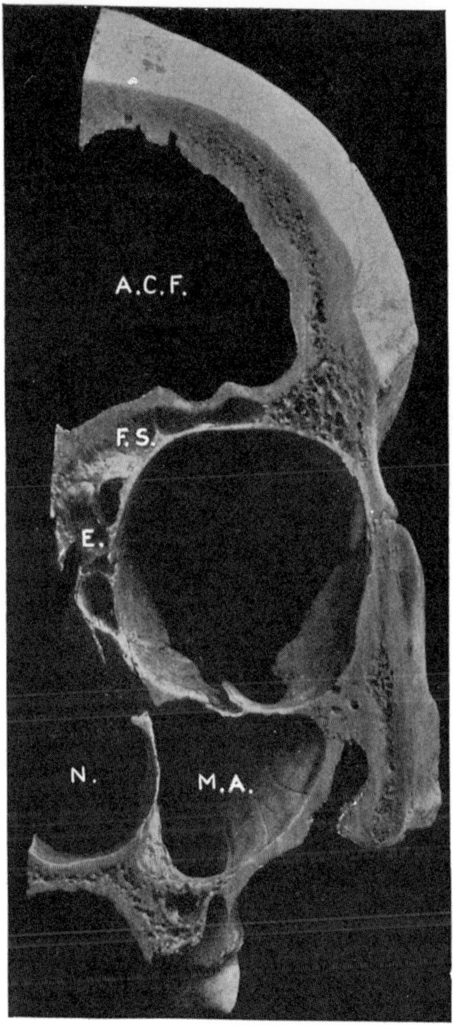

FIG. 20.—**Frontal Section** of Left Half of Skull just behind Orbital Margin and passing through largest part of cavity (greatest orbital height = 37·5 mm.); seen from in front. Natural size. The frontal section removed is seen in Fig. 5.

The floor of the orbit is deeply grooved by the infra-orbital sulcus.

A.C.F. = Anterior cranial fossa.
F.S. = Frontal sinus.
E. = Ethmoidal air-cells.
N. = Nasal cavity.
M.A. = Maxillary sinus (showing the *ostium maxillare*).

lines and roofs over the infra-orbital sulcus, so enclosing the infra-orbital nerve and artery; on the lateral side the orbitalis muscle is incorporated with it (Fig. 42, p. 87). The inferior oblique muscle arises from and lies across the floor anteriorly; the inferior rectus muscle is in close contact with it behind at the apex, but is separated from it in front by the inferior oblique muscle and a mass of orbital fat. Externally, that is inferiorly, the floor is related in nearly the whole of its extent to the maxillary sinus, as seen in the last figure and also in Fig. 21; behind, at the apex, the orbital process of the palatine bone may be hollowed out by an air-cell, as is well shown also in Fig. 21. Abnormal posterior ethmoidal cells in rare instances undermine the maxillary part of the floor along its postero-medial border.

(iv) **The Medial Wall** (inner, nasal, or ethmoidal wall; *paries medialis*).—This is the smallest of the orbital boundaries. Its limits lie broadly between lines parallel with the upper and lower borders of the ethmoidal bone. It is marked off from the roof by the fronto-ethmoidal suture prolonged forwards towards the point nasion at the root of the nose, and from the floor by the line of sutures referred to on p. 43; anteriorly it is bounded by the posterior lacrimal crest of the lacrimal bone. The surface is convex about the centre, and concave in front in harmony with the contour of the eyeball, though these curves are but slightly marked; it lies almost parallel to the mesial plane and to its fellow of the opposite side, as shown in Fig. 17, but curves into the floor below. It is formed by four bones, which are from before backwards: (1) the lower part of the medial angular process of the frontal lying above (2) the lacrimal bone, (3) the orbital plate (*lamina papyracea* or *os planum*) of the ethmoidal, and (4) a small part of the lateral aspect of the body of the sphenoidal bone. The individual bones take part in the formation of this wall to a variable degree, but the largest and most important is the lamina papyracea of the ethmoidal; it is oblong in shape, and articulates with the frontal above and the maxilla below, with the sphenoidal bone behind by the spheno-ethmoidal suture and with the lacrimal in front by the ethmo-lacrimal suture. These last two sutures are both vertically placed, but, like the bones, are subject to much variation. Posteriorly the palatine bone commonly extends upwards on to this wall; sometimes a descending process from the frontal

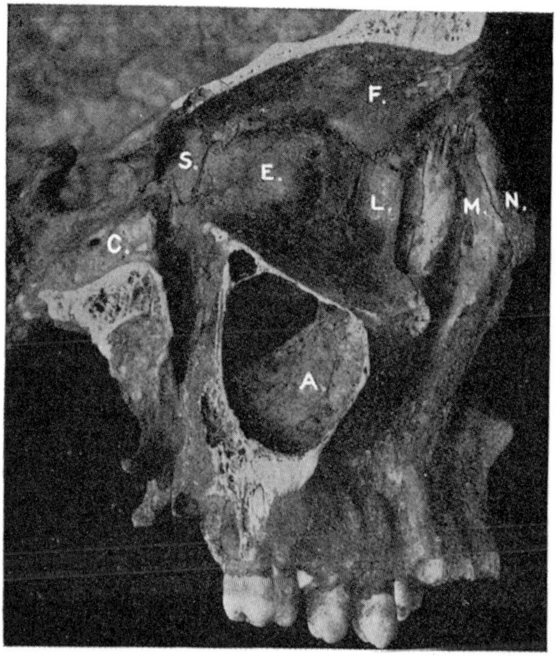

FIG. 21.—Sagittal section through Right Orbit to display the **Medial Wall.** Natural size.

S. = Sphenoidal bone, and is placed just in front of the optic foramen.
E. = Ethmoidal bone; L. = Lacrimal bone; M. = Frontal process of maxilla. Between L. and M. is the fossa for the lacrimal sac, bounded behind by the sharp posterior lacrimal crest, in front by the blunt anterior lacrimal crest.
N. = Nasal bone; F. = Frontal bone.
A. = Maxillary antrum opened into, and at its postero-superior angle the palatinal air cell is seen.
C. = The position of the cavernous sinus at the side of the body of the sphenoidal bone; above it lies the *sella turcica*; the lower limb of the letter C is pointing towards the *foramen rotundum*, through which the section passes.

meets it and completely separates the ethmoidal from the sphenoidal bone. In this region also a part of the sphenoidal turbinated bone has been seen. Anteriorly the ethmoidal surface may be separated from the lacrimal by the junction of prolongations of the frontal and maxilla, an 'orbito-maxillo-frontal suture' being formed by their union (A. Thomson,

1890). Small ethmo-lacrimal ossicles may occupy the upper or lower ends of the ethmo-lacrimal suture (Macalister, 1884). The lamina papyracea may be subdivided into two or more pieces by additional vertically placed sutures, a condition found by Bianchi (quoted by Ledouble) to occur more frequently in the skulls of imbeciles, and confirmed by Greig (1924). Poirier (1912) figures the very rare 'orbital Wormian bone' behind the ethmoidal and between the palatine and sphenoidal bones. It may be noted that in mammals below the primates the ethmoidal bone does not enter into the formation of the orbit, its place being taken by the frontal.

FIG. 22.—Well-developed Lacrimal Bone with a large **hamular process** extending forwards on to the orbital margin as the *pars facialis*. The naso-lacrimal canal in this specimen is of large calibre (7 mm. × 5·5 mm.). Peruvian Indian skull. $\times \frac{1}{2}$. Compare with Fig. 23.

The lateral or orbital surface of the **lacrimal bone** is divided into two parts by a vertical ridge, the posterior lacrimal crest (*crista lacrimalis posterior*); the bone behind the crest is flat and forms the forepart of th orbital wall, the are front is excavated by a gutter, the *sulcus lacrimalis*, wh joins a similar groove on the frontal process of the maxilla to form the fossa for the lacrimal sac (p. 67). The lower end of the posterior lacrimal crest usually terminates in a forwardly directed and hook-like **hamular process** (*hamulus lacrimalis*) which articulates with the lacrimal notch of the maxilla, and defines the opening of the naso-lacrimal canal. Much variation is found in this process: it is more often present in males than in females, and may either form a short spur, as in most cases, or reach the lacrimal tubercle of the anterior crest, or even extend on to the facial aspect of the maxilla as the *pars facialis* (Gegenbaur, 1881), a condition which exists in lower mammals and primates generally, and was found in 4·25 per cent. of 1,000 human skulls by Macalister. In rare instances it forms a separate ossicle (the *os hamulus* or lesser lacrimal bone), as found in 1½ per cent. of subjects by Macalister and Thomson, or it may be separated from the adjacent floor of the orbit by an ossicle (the *ossiculum canalis naso-lacrimalis* of Grüber, 1877).

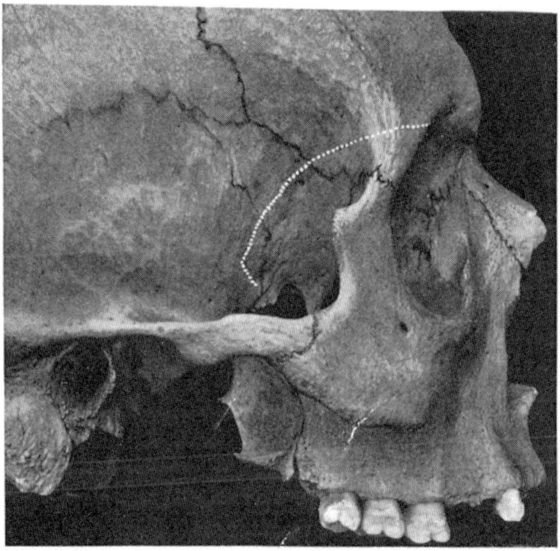

FIG. 23.—**Absence of Lacrimal Bone,** the whole fossa for the lacrimal sac being formed by the maxilla. The outline of the orbit is dotted in. The lateral margin presents a well-developed *processus marginalis* on its temporal border. × ½.

When the hamulus is well developed, the level of the upper opening of the naso-lacrimal canal is raised accordingly. It may be notched or pierced by a branch of the infra-orbital artery on its passage to the lacrimal sac, as seen in the above figure.

The lacrimal is the most fragile of all the bones, and is usually perforate owing to incomplete ossification or to absorption in old age; it is fenestrated in the foetus (Macalister, 1884), and Merkel considers a complete bone to be an anatomical curiosity. Its strongest part is along the line of the posterior lacrimal crest.

Numerous **variations** are to be found in the lacrimal bone, which is evidently undergoing retrogressive changes in man. Either half in front or behind its crest may be undeveloped, the deficiency being made up in the one case by the frontal process of the maxilla, as seen in Fig. 23, and in the other by an extension of the lamina papyracea, though this condition is probably due to an early union of the bones, such fusion with adjacent bones being often found, especially with the frontal process of the maxilla. It may be absent altogether (1·2 per cent. of 2,021 skulls of all races—Ledouble). On the other hand, as just noted, it may develop a hamular process, extending forwards on to the orbital margin. Occasionally the bone is bipartite, indicating an accessory centre of ossification, and finally 'peri-lacrimal ossicles' sometimes occur in relation with it behind, in front, and below, as many as twelve varieties having been described by Flecker (1913), though only one of them, representing a separated hamulus, appears to be derived from the lacrimal bone itself. Kofler (1929) illustrates many such variations.

The **features** of this wall to be noticed are the sutures described above, the ethmoidal foramina, the architecture, and relations. It will be more convenient to describe later in a separate section the fossa for the lacrimal sac (p. 67), which is sometimes considered as a feature of this wall.

The ethmoidal foramina (internal orbital foramina, *foramen ethmoidale anterius vel posterius*).—These small oval orifices lie in the angle between the roof and the medial wall of the orbit, along the line of the fronto-ethmoidal suture. They are two in number, distinguished as anterior and posterior, and lead into corresponding canals, which are formed above and in greater part by the frontal bone (and so would more appropriately be termed fronto-ethmoidal canals), as seen in Fig. 31. They are directed anteriorly and medially, and open into the anterior cranial fossa just above and at the sides of the lamina cribrosa of the ethmoidal bone; in rare instances the posterior foramen is duplicated, or there may be a third foramen intermediate in position, and even four have been noted. Ono (1928) illustrates a fissure in the fronto-ethmoidal suture.

The anterior ethmoidal canal is the larger of the two, and slopes more forward. Its orifice lies 20 mm. from the orbital margin and 30 mm. behind the point nasion (on an average of twenty skulls measured by the writer). It contains the naso-ciliary nerve (here called the anterior ethmoidal) and the anterior ethmoidal artery.

The posterior ethmoidal canal is usually placed from 10 to 15 mm. behind the anterior one, sometimes at a slightly

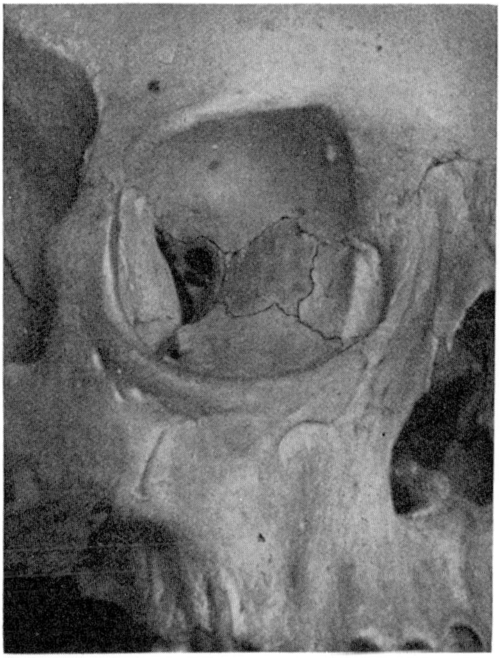

FIG. 24.—Right Orbit to show **features of Medial Wall.** The specimen also illustrates a doubled optic foramen, a *spina trochlearis*, the tubercle of attachment of the *pars lacrimalis* (Horner's) muscle at upper end of posterior lacrimal crest, and the projection of orbital floor over posterior extremity of inferior orbital fissure. Double zygomatico-facial foramina are also present, and the curved edge of the infra-orbital foramen is well marked. Natural size.

lower level and in the spheno-ethmoidal suture. Through it pass the posterior ethmoidal artery, and occasionally a nerve filament (the *nervus spheno-ethmoidalis* of Luschka) derived from the naso-ciliary nerve. Both canals are lined by the periorbita, and in addition possibly transmit lymphatic vessels.

The **architecture.**—The medial wall, save for the small comparatively thick sphenoidal area, is the thinnest of the orbital boundaries; the orbital plate of the ethmoidal bone (os planum) is of such paper-like delicacy (as is well seen in Fig. 17) that it is termed the *lamina papyracea,* but in spite of its thinness (0·2 to 0·4 mm.), which may allow the outlines of the divisions between the underlying air-cells to be seen through it, the bone is tough, and is very rarely the site of senile atrophy. Congenital dehiscences, however, along the ethmo-maxillary suture, or at the site of the orbital Wormian bone, have been noticed by several authors; Zuckerkandl found fifteen cases, and Onodi (1913) saw eighteen, three of which communicated with the frontal sinuses as well as with the ethmoidal cells.

With regard to the **dehiscences** found in various parts of the orbital walls, it may be noted that when due to absorption or atrophy of the bone in old age they are usually situated in the middle of a bony plate and the edges are irregular, whereas when due to lack of development they lie between the sutures and the margin is smooth and regular in outline. Congenital dehiscences are found specially in those parts of the orbital walls which are developed by ossification of the primitive cartilaginous nasal capsule, namely, the lamina papyracea and the orbital wall of the maxilla (Bull). Dehiscences of a third kind are pathological, and may also be seen on the medial wall, where they are caused by extension of disease from the ethmoidal cells. Instances are shown in Fig. 25 and Fig. 124, p. 231.

The **relations of the medial wall.**—Internally, the lining periorbita is continuous through the ethmoidal canals with the dura mater of the anterior cranial fossa. The medial rectus muscle is in close contact with the greater part of this wall, and just behind the posterior lacrimal crest is the attachment of its 'check ligament', and also of the pars lacrimalis (Horner's muscle) of the orbicularis oculi muscle; against the anterior half of the upper part of the wall and lying in the interval between the adjacent borders of the medial rectus and superior oblique muscles are the nasociliary nerve and ophthalmic artery. Externally, the medial wall is the most important from a practical point of view, owing to the fact that it is related in its whole extent to the accessory air-sinuses of the nose, from which disease may invade the orbit; posteriorly a small area is in contact with the sphenoidal sinus, but in the rest of its extent this wall is in direct relation to the ethmoidal cells, which underlie not only the lamina papyracea but the lacrimal bone as well, and

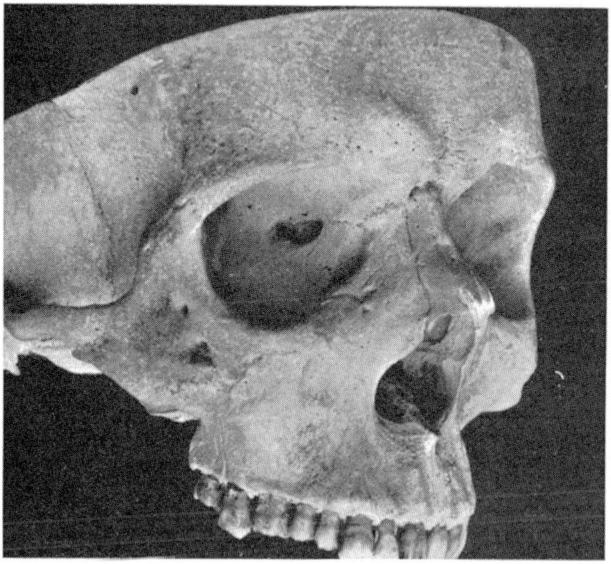

FIG. 25.—Skull showing **Dehiscence** of medial wall in fronto-ethmoidal suture, leading into ethmoidal cells.

may even encroach backwards upon the sphenoidal region. By these air-cells the orbital cavity is separated on its medial aspect from the nasal cavity. These relations are well displayed in Figs. 3, 17, 20, 34, 42, and especially 197, p. 379.

It is fitting at this point to review the accessory sinuses of the nose as a whole before describing the lacrimal fossa and naso-lacrimal duct, to which, in common with the orbital walls, they present such important relations.

4. THE ACCESSORY AIR-SINUSES OF THE NOSE IN THEIR RELATION TO THE ORBIT

From the above account of the relation of the walls, the orbit is seen to be surrounded by air-sinuses (*sinus paranasales*) on three sides, the lateral wall alone being exempt. The roof in its medial and anterior region is usually related to the frontal sinus. The floor is always undermined by the maxillary sinus, behind which there may be present the small palatinal air-cell. The medial wall is generally related to the sphenoidal sinus and always to the ethmoidal cells, which latter may extend over the roof, or beneath the medial edge of the floor, or replace the sphenoidal sinus in position.

The massing of these air-sinuses round the orbit is illustrated in Figs. 3, 17, 20, 34, 42, and also in Fig. 197, p. 379. They may be divided into two groups: an anterior, comprising the maxillary and frontal sinuses together with the bulk of the ethmoidal cells, and related to the forepart of the orbit wherein is lodged the globe of the eye, and a posterior group, consisting of the posterior ethmoidal, the palatinal cell, and the sphenoidal sinus, which are related more closely to the optic nerve. The openings of these sinuses into the nasal cavity are arranged around the middle concha, those of the anterior group lying below its line of origin, those of the posterior group above it (Logan Turner, 1908); on the orbital aspect the two groups may approximately be demarked by a line running downwards and backwards from the anterior ethmoidal foramen to the floor of the orbit (Ranglaret, 1896). All have in their venous system branches which pass into the orbit, and thence through the ophthalmic veins to the cavernous blood sinus; some of their lymph channels possibly take the same course, but this is not certain. Their size and extensions vary not only at different periods of life, but often in the two sides of the same individual; their communications with the nasal fossae, however, are fairly constant.

These air-spaces are all derived as outgrowths of the primitive nasal cavities; they appear in early childhood, increase most actively in size at puberty, and continue to grow until about the age of thirty, with a slight increase in old age through atrophy of their bony walls. They retain communication with the nasal cavity and are lined by a direct continuation of its mucous membrane, here blended so closely with the

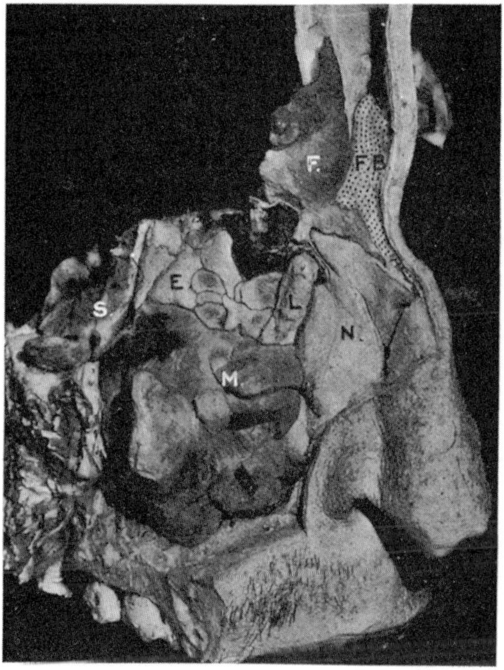

FIG. 26.—The Accessory Air **Sinuses of the Nose.** Prepared by removing the bones piecemeal from around the muco-periosteal lining of the air-spaces after hardening in formalin. (Preparation by Dr. E. J. Curran of Harvard.) $\times \frac{2}{3}$.

The anterior ethmoidal cell in relation to the upper part of the lacrimal sac (L.) has been blackened.

F. = Frontal sinus. F.B. = Frontal bone. N. = Nasal mucous membrane.
L. = Lacrimal sac. M. = Maxillary sinus. S. = Sphenoidal sinus.
E. = Ethmoidal cells.

periosteum as to form a 'muco-periosteum', a fact which explains how readily disease can spread into them from the nose. Their collective capacity is more than twice that of the nasal cavity (Braune), and a good idea of their extent can be obtained from such a dissection as has been photographed in Fig. 26. The method, for which the writer is indebted to

Dr. E. J. Curran of Harvard University, consists in removing the bones piecemeal with chisel and forceps from around and between the sinuses after soaking the specimen for two weeks in strong formalin solution; the lining muco-periosteum is left intact, and will be found sufficiently hardened to retain the shape of the cells. Their form can also be well shown either by wax models (Jackson and Connor, 1917) or by metal casts (Fig. 27).

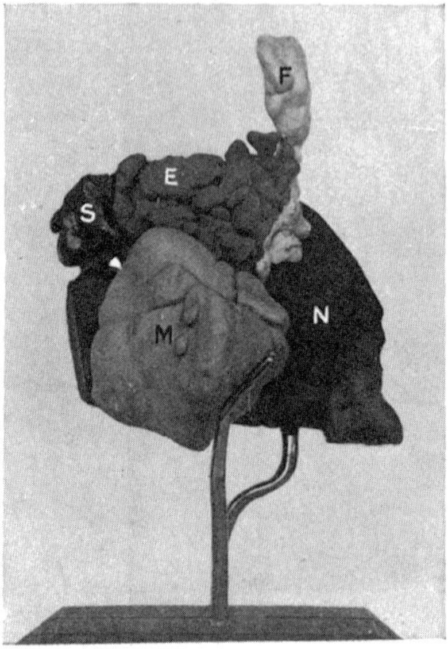

FIG. 27.—Fusible metal **cast of the Accessory Sinuses** of the Nose (made by Sir Auckland Geddes, in the McGill Anatomical Museum).

F. = Frontal sinus. S. = Sphenoidal sinus.
E. = Ethmoidal cells. M. = Maxillary antrum.
N. = Nasal mucous membrane.

The complex arrangement and variable extent of these sinuses, together with the irregular modelling of the lateral nasal wall to which they are related, will be understood better by a consideration of their mode of growth, before proceeding to a brief description of the adult formation.

Development of the air-sinuses.—Before cartilage has been formed in the walls of the primitive nasal cavities, linear outgrowths of the lining epithelium (which is ectodermal in origin) occur in the side and roof of each. These gutters are the future meatuses of the nose, and the ridges left between them eventually form the three *conchae* or turbinated processes, superior, middle, and inferior: three is the number usually described, though Zuckerkandl (1892) found four such processes in 87 per cent. of 267 skulls, a number which agrees with the description given by Sappey (1867), the additional process being found above, or below and hidden by, the one usually designated superior. The superior concha is the shortest, the middle is nearly twice its length, and the inferior extends still further forwards; the first two are processes of the ethmoidal, the last is formed by a separate inferior turbinated bone (*concha nasalis inferior*). Above the superior concha is a narrow space, the 'spheno-ethmoidal recess'.

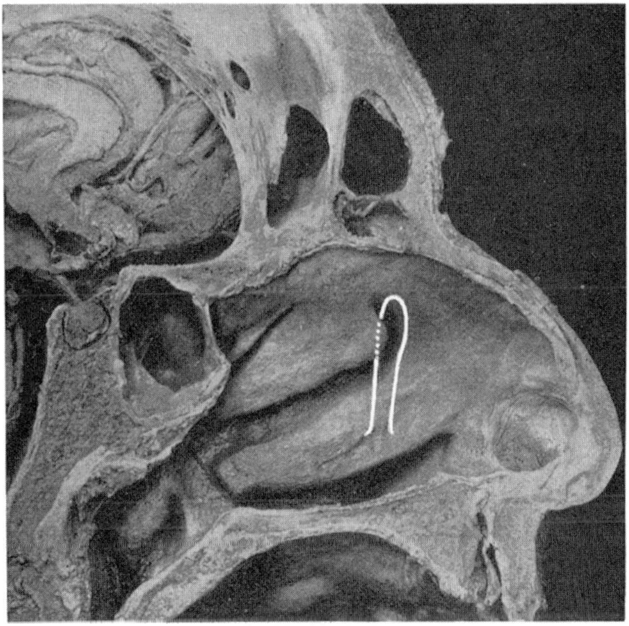

FIG. 28.—Sagittal section of Head of Adult Female. Left half viewed from within to show the left **lateral nasal wall.** See Fig. 29 for guide to parts. The relative positions of the fossa for the lacrimal sac and the naso-lacrimal canal are indicated by the white line. × about $\frac{2}{3}$.

In front of the attached margin of the middle concha there is often a prominence on the frontal process of the maxilla, the *agger nasi*; this prominence (named from the Latin *agger*, a mound or rampart) forms the upper boundary of a shallow depression, the *atrium*, which is an anterior continuation of the middle meatus, and lies above the 'vestibule' or cartilaginous dilatation at the entrance of the nose (Fig. 29). On the lateral wall of the middle meatus a secondary linear depression develops, forming in the adult a curved gutter, the *hiatus semilunaris*, which is bounded below by a sharp crest, the *processus uncinatus*, and overhung by a prominence, the *bulla ethmoidalis*; it leads into a curved and upwardly directed channel, the *infundibulum ethmoidale*. These features on the meatal wall are hidden by the overhanging middle concha, but are well displayed in Fig. 29.

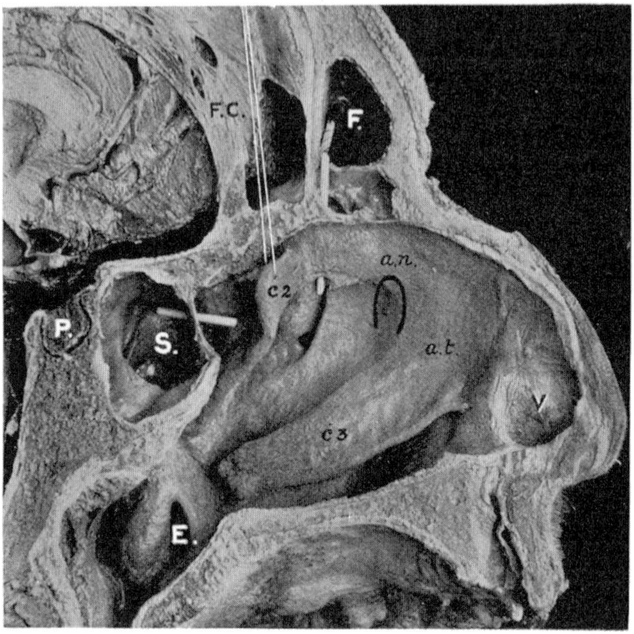

FIG. 29.—The same preparation as in Fig. 28. The middle concha has been pulled upwards to expose the middle meatus. The position of the fossa for the lacrimal sac relative to this wall was ascertained by driving pins through from the opposite side and is outlined in black. Rods have been passed through the opening of the sphenoidal sinus and down the *infundibulum* of the frontal sinus; the latter leads into the *hiatus semilunaris*, which is bounded above by the rounded *bulla ethmoidalis* and below by the *processus uncinatus*; the *ostium maxillare* of the antrum is also seen. × about $\frac{3}{4}$.

Compare with the deeper stage of a similar dissection seen in Fig. 128, p. 236, Fig. 199, p. 384, and the X-ray on p. 73.

P. = *Hypophysis* or pituitary body lying in the *sella turcica*. S. = Sphenoidal sinus. F.C. = *Falx cerebri*. F. = Frontal sinus. *c* 2, 3 = The middle and inferior concha. *a.n.* = *Agger nasi*, here little developed. *a.t.* = *Atrium*, and V. = Vestibule of the nose. E. = Orifice of auditory or Eustachian tube.

The nasal mucous membrane continues to grow outwards in various directions, and its projections develop into the accessory sinuses of the nose. Thus the frontal sinus is formed at the upper end of the infundibulum by a bud of mucous membrane which grows upwards through the ethmoid, and pushes its way between the outer and inner lamellae of the frontal bone to a very variable extent. Usually absent at birth, this outgrowth appears in the first or second year of life, but is nascent until the fifth or sixth; at the end of the seventh it is a definite space, remaining small until puberty, and reaching full development at twenty-five years. The stalk of this outgrowth forms the 'fronto-nasal duct', and either retains its original continuity with the infundibulum or, as in half the number of cases, opens independently into the middle meatus, the infundibulum then ending blindly above, owing to the anterior end of the uncinate process fusing with the forepart of the bulla. The ethmoidal cells bud out into the nasal wall above and below the middle concha, and are present at birth as little *culs-de-sac*. They multiply until each lateral mass of the bone is composed of a number of cells (8–10), the 'ethmoidal labyrinth', an increase in growth which causes the greater width between the eyes in the adult. The *bulla ethmoidalis* is inflated by one or more such cells, and the term, derived from the Latin *bulla*, a bubble, well expresses the formation of these cellular outgrowths. The sphenoidal sinus is primarily formed by a similar budding from the region of the spheno-ethmoidal recess into the body of the bone; present as a round depression in the first to third year, it develops more rapidly after the sixth or seventh, and is complete at twelve years. The maxillary sinus is formed about the third foetal month by an outgrowth from the middle meatus in the region of the future hiatus semilunaris, which bursts through the lateral nasal plate of cartilage and distends the maxillary process of the foetal face. This is the only one of the sinuses that is more than a mere rudiment at birth, at which period it forms a shallow depression about the size of a pea above the germ of the first milk tooth; at the age of ten it is well formed, and during adolescence it expands and fills out the whole of the maxilla, reaching full size after completion of the secondary dentition.

In the adult much variation in the size and extensions and consequently in the relations of these air-spaces, especially of the ethmoidal group, is found, as may well be expected from their mode of development. The ethmoidal group, together with the maxillary sinus, are constantly present, the others may in rare instances be undeveloped even in the adult. The frontal, sphenoidal, and palatine sinuses are homologous with, and may be succeeded or displaced by, the ethmoidal cells, the growth of which, indeed, may be so exuberant as to encroach upon the area usually occupied by any of the others.

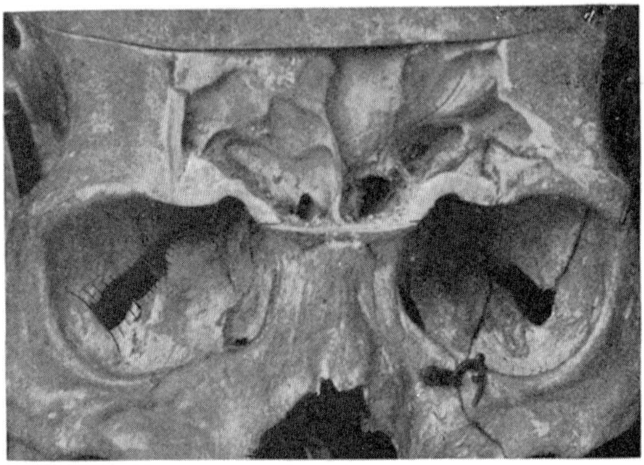

FIG. 30.—The **Frontal Sinuses** exposed by removal of their anterior walls. The irregularity and roughness of the cerebral wall is abnormal, but the preparation shows a median dividing septum and the openings at the back and lower part. (The position of the *orbitalis* muscle of Müller is indicated by the shading across the right inferior orbital fissure.) $\times \frac{2}{3}$.

The **frontal sinuses** (*sinus frontales*) lie above the upper and inner angles of the orbital entrances and behind the supra-orbital ridges. Usually two in number, they are rarely of equal size, the dividing septum being generally displaced to one side of the mid-line; they may be entirely undeveloped, as in the upper two adult bones seen in Fig. 13, p. 31. Logan Turner gives normal dimensions of $1\frac{1}{4}$ inches high, 1 inch wide, and 1 inch deep from before backwards, or 31, 25, 25 mm. respectively, and the average capacity is from 3 to 5 c.c. in males, but much less in females. Schaeffer found the extreme capacity of both to vary from 1 to 45 c.c. The opening (*ostium frontale*) of each lies posteriorly and near the median plane (Fig. 30), and leads through the fronto-nasal duct into the middle meatus of the nose, either separately, or through the infundibulum (as found in 56 per cent. of cases by Schaeffer and as illustrated in Fig. 29); the duct is separated from the lacrimal fossa by ethmoidal cells. Rawling describes the infundibulum as opening into the middle meatus under cover of the middle turbinate on a level with the inner margin of the palpebral fissure. The width of the sinus is its most variable feature, and, as already noted on p. 31, it may extend to a considerable distance over the roof of the orbit (orbital part of the sinus), as shown in the figures overleaf.

The **ethmoidal cells** (*cellulae ethmoidales*) form a mass, the ethmoidal labyrinth, lying between the upper part of the nasal fossa and the orbit, from the latter of which it is separated by a smooth thin plate of bone, the *lamina papyracea*. These cells are variable in number; usually there are eight or ten, with extremes of four to fifteen (Rollet), and averaging the size of a pea, though smaller in front than behind; separated from one another by delicate bony partitions, they open either individually or through one another into the nasal cavity. They may be subdivided into three groups, anterior, middle, and posterior, in accordance with their relative positions, the site of their openings, and the bones completing their walls. The anterior cells, one to three in number, open into the middle meatus, either into the infundibulum or into the recess above (or more rarely beneath) the bulla ethmoidalis, and their walls are completed by the frontal, lacrimal and maxillary bones; they may project into the frontal sinus to form a frontal bulla, or hollow out the anterior extremity of the middle concha and also the agger nasi, and as 'lacrimo-ethmoidal cells' they present important relations to the fossa for the lacrimal sac (see p. 70). Those of the middle and largest group open above and below the bulla ethmoidalis, which they also inflate; their walls are all formed by the ethmoidal bone itself, save where they project, like the frontal sinus, above the orbit between the lamellae of the frontal bone. The posterior cells, unlike the other two groups, open into the superior meatus and above the middle concha; usually three in number, their walls are completed by the ethmoidal, sphenoidal and palatine bones; they may project

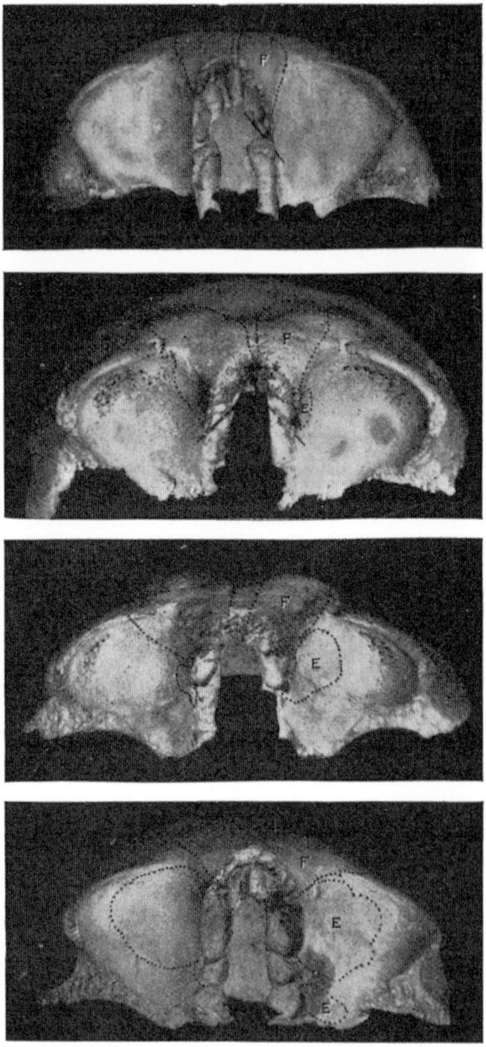

Fig. 31.—Four Frontal Bones showing varying degrees of **extension of frontal sinuses** (F.) and **ethmoidal cells** (E.) over the orbital roof. In the upper two specimens bristles have been passed through the anterior ethmoidal canals. $\times \frac{1}{2}$.

or even open into the sphenoidal sinus or replace the air-cell of the palatine bone, and one or more cells may present important relations to the optic canal, as has been seen on p. 29. The palatinal air-cell hollows out the orbital process of the palatine bone, and so is related to the apex of the floor of the orbit (Fig. 21, p. 49); it usually opens into either the sphenoidal sinus or a posterior ethmoidal cell.

The **sphenoidal sinus** (*sinus sphenoidalis*) is one of a pair of air-spaces hollowed out in the body of the sphenoidal bone; they are generally completely, but rarely equally, divided from one another by a septum, and are frequently encroached upon by ethmoidal cells; an average size is 12 mm. high, 15 mm. wide, 22 mm. deep, with a capacity of 7·5 c.c., but the extent to which they hollow out the basi-sphenoid is most variable, Schaeffer, for example, giving a range of from 0·5 to 30 c.c. each. Cope (1916, 1917) classifies three types according to their degree of development; out of 292 sinuses examined he found that 155 extended backwards beneath the fossa hypophyseos (sella turcica), 72 were confined to the pre-sphenoid in front of the olivary eminence, and 65 were intermediate in type and only slightly encroached upon the basi-sphenoid; examples are seen in Figs. 29, 38, 128, 199. They lie at the back and upper part of the nasal cavities; in the skull, the distance of the midpoint of the anterior wall of the sinus from the point nasion is about 48 mm. or 2 inches (varying from 42 to 60 mm. in 20 cases measured by the writer), and from the lateral margin of the anterior nares is about 70 mm. or $2\frac{3}{4}$ inches (varying from 64 to 74 mm. in 13 preparations). Each communicates by an opening (*apertura sinus sphenoidalis*) situated above the centre of its anterior wall with the corresponding spheno-ethmoidal recess. They are related to the cavernous sinus laterally (where the bony wall has been found deficient by St. Clair Thomson, 1906) and to the hypophysis cerebri or pituitary body superiorly and posteriorly; their important relations to the optic nerves and chiasma are described on pp. 28, 385; and further it is to be noted that the pterygoid or Vidian canal, containing the sympathetic and motor roots of the spheno-palatine or Meckel's ganglion, may form a prominence in the floor, as may also the internal carotid artery on the side wall. By extension into the greater wing of the sphenoid they may even bear relation to the temporal lobe of the brain, or to the foramen ovale and mandibular nerve. They were found totally absent in one case out of an hundred in the McGill collection, which is the percentage generally given in the literature, although Van Gilse (1926) did not find this condition once in a thousand skulls. Variations in size and relation are illustrated in Fig. 32.

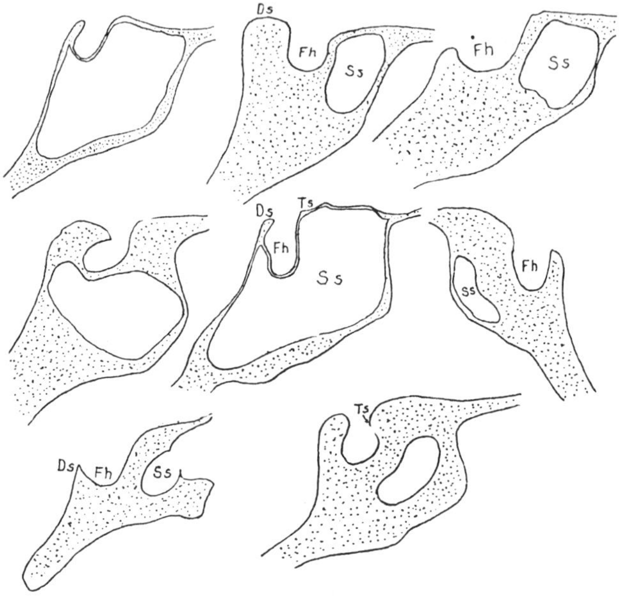

FIG. 32.—Sagittal sections through the region of the **Sphenoidal Sinus** and hypophysial fossa. (The types indicate how variations in morphology may influence the growth of an enlarging hypophysis.)
Ds. = *dorsum sellae*; Ts. = *tuberculum sellae*; Fh. = *fossa hypophyseos*; Ss. = *sinus sphenoidalis* (From Parsons Schaeffer, 1924).

The **maxillary sinus** (*sinus maxillaris*, antrum of Highmore) occupies the body of the maxilla, lying beneath the orbital floor, to the lateral side and extending a little below the level of the nasal cavity; its average dimensions are $1\frac{1}{2}$ inches high, 1 inch wide, $1\frac{1}{4}$ inches deep, with a capacity of 15 c.c. Its opening, the *ostium maxillare*, which is duplicated in nearly half the number of cases (i.e. in 42·4 per cent. of 125, as found by Schaeffer), is situated at the upper part of its nasal wall (Figs. 20, 124), and leads into the middle meatus through the hiatus semilunaris (Fig. 128, p. 236); in those cases where the fronto-nasal duct leads directly into the hiatus semilunaris, this gutter could drain the frontal sinus more or less completely into the antrum, a condition seen in Fig. 29. On the upper half of the forepart of this nasal wall of the antrum there is a thin-walled 'lacrimal protuberance' formed by the naso-lacrimal canal (Fig. 129, p. 237). The upper or orbital wall is thin and ridged in its forepart by the infra-orbital canal and by bony channels containing the superior alveolar or dental nerves. The cavity may thus be partly or completely subdivided by septa, and part of it circumscribed anteriorly to form a 'prelacrimal recess' or 'infra-orbital sinus' (p. 76). The floor is generally related to the second bicuspid and first molar teeth, the sinus usually extending downwards between the fangs of the latter (Rawling). This sinus, like the others, may present abnormal extensions passing into any of the processes of the bone, such as the frontal or the zygomatic (with, but very rarely, a continuation into the zygomatic bone itself, the *sinus malaris vel zygomaticus*), or it may replace the palatine air-cell, or even be prolonged under the medial wall of the orbit. The variations are less common than those of the other air-spaces.

For further details of the anatomy and development of these accessory sinuses the reader should consult the works of Zuckerkandl (1882), Logan Turner (1901), Killian (1903), Rollet (1903), Onodi (1913), Skillern (1913), Schaeffer (1920), and papers by Coffin (1905), Underwood (1910), Schaeffer (1921), Fawcett (1911), and Budge (1930).

5. THE OSSEOUS LACRIMAL PASSAGES

These comprise the lacrimal fossa together with a downward continuation, the naso-lacrimal canal.

(i) **The Fossa for the Lacrimal Sac** (*fossa sacci lacrimalis*) is a broad groove occupying the lower two-thirds of the medial orbital margin, which, as explained on p. 17, divides to enclose it; shallow above, where it usually extends up to the frontal bone, it becomes deeper at its lower end, where it is continued downwards into a bony channel, the naso-lacrimal canal, the two combined resembling the cut end of a quill pen seen sideways (Fig. 39). It is formed by two bones, the frontal process of the maxilla in front and the lacrimal behind; each is grooved by a *sulcus lacrimalis*; the approximation of the two sulci forms the fossa, and the vertically disposed 'lacrimo-maxillary suture' between them divides it into halves. The extent to which the two bones participate in this formation varies considerably. In Fig. 23, p. 51, for example, the maxilla forms even the whole fossa, though the writer has never observed a similar encroachment by the lacrimal bone. Such variations depend on the degree of development of the lacrimal (p. 52); also one would expect the maxillary element to predominate in the heavier type of face and jaw. The fossa is said to be broader in the brachycephalic and narrower in the dolichocephalic type of skull. It is bounded in front and behind by two 'crests' (Fig. 21, p. 49).

The rounded anterior lacrimal crest (*crista lacrimalis anterior*) of the maxilla is inconspicuous above but well marked below, where it becomes directly continuous with the infra-orbital margin; here it may present, as in 17 out of 50 skulls examined by the writer, a 'lacrimal tubercle' which can be felt in life, and from its position may be taken as a guide to the lacrimal sac which lies just behind and above it. Gérard (1907) found that in 28 per cent. of 140 orbits the tubercle projected backwards over the lower part of the fossa as an 'anterior lacrimal spur' (Fig. 33), and so might form an obstruction to catheterization of the lacrimal passages. Attached to the anterior lacrimal crest is the medial palpebral ligament of the eyelids above, with fibres of the orbicularis oculi muscle below.

The posterior lacrimal crest (*crista lacrimalis posterior*), formed by the lacrimal bone (see pp. 50, 51), is much more sharply defined, and may even be prolonged as a thin plate of bone curving forwards over the fossa, as is well seen in foetal skulls towards term or at birth (Fig. 48, p. 101); this formation, indeed, is considered by Goldnamer (1923) to exist always in recent adult skulls, but to be lost in cleaning them. It is thickened above at the site of attachment of the *pars lacrimalis* (Horner's) muscle, as seen in Fig. 24, and at its lower extremity it usually ends in a projection, the hamulus lacrimalis (described above on p. 50), which curves round laterally and so defines the upper opening of the naso-lacrimal canal. Just behind the posterior lacrimal crest is affixed the septum orbitale.

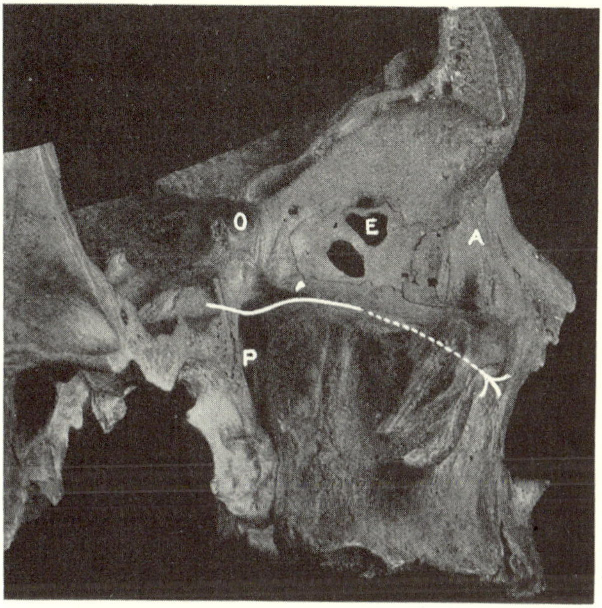

FIG. 33.—Sagittal section through Right Orbit to show **medial wall**. There is a well marked 'lacrimal tubercle' at the base of the anterior lacrimal crest, and a slender hamular process at the base of the posterior lacrimal crest. The orbital roof is doubled throughout by extensions from ethmoidal cells. $\times \frac{3}{4}$.

A. = Anterior lacrimal crest.
E shows an ethmoidal cell broken into; above it is a large anterior ethmoidal foramen.
O. is the optic canal through which the section passes.
P. is placed in the pterygo-palatine or spheno-maxillary fossa.
The white line indicates the course of the infra-orbital nerve.

The fossa has three walls: an anterior, narrow and well marked only in its lower part; a medial, broad and divided vertically by the lacrimo-maxillary suture; and a posterior, variable in width. As regards the medial wall, it is important to note that there is a striking difference in the architecture of the two regions demarked by the suture, the anterior or maxillary half being strong and resistant, the posterior lacrimal half being thin and fragile, as confirmed by the observations of Patton (1923); the subdivision is not always equal, as noted above, for the maxillary portion may encroach upon the other, and in very rare instances even form the whole of the medial wall of the fossa (Fig. 23, p. 51), in which cases not only is the fossa found much shallower and smaller than usual, but the calibre of the naso-lacrimal canal is correspondingly reduced (p. 79), as is seen so well in the casts figured by Zabel (1900), Fig. 39. The posterior wall, always formed by thin bone, may be comparatively broad and conspicuous when the skull is viewed from the front, giving depth to the fossa, or it may be narrow and less prominent, the fossa then being shallow; these differences are most noticeable in the upper region, and depend upon the degree of development of the underlying ethmoidal cells; they are illustrated in the left orbits of the skulls shown in Fig. 50, p. 107. At the upper end of the lacrimo-maxillary suture (i.e. at the point dacryon) there is present in 20 per cent. of cases a foramen which transmits to the ethmoidal cells an arteriole from the nasal branch of the ophthalmic artery, as well as a filament from the infra-trochlear branch of the nasal nerve (Macalister, Ledouble). In 2 per cent. of cases small twigs from the angular and infra-orbital arteries pass through small orifices lower down in the suture. Zuckerkandl, moreover, describes a 'lacrimo-facial vein' into which blood passes from the anterior ethmoidal cells through the lacrimal bone.

The relations of the lacrimal fossa are as follows: on the lateral or orbital aspect it is covered by the lacrimal fascia of the periorbita passing from one crest to the other; across the front of its upper part lies the medial palpebral ligament and skin; whilst behind it in the same region there is found the pars lacrimalis or Horner's muscle; the lower part of the fossa has the skin and orbicularis muscle related to it in front, the septum orbitale behind. These relations are illustrated and referred to further on p. 232 in describing the lacrimal sac.

On the medial or nasal side the fossa is related to ethmoidal air-cells, which intervene to a variable extent between it and the nasal cavity (Fig. 118, p. 223). These 'lacrimo-ethmoidal cells' are usually two in number, superior and inferior, and belong to the anterior group; their walls are completed by the lacrimal and frontal bones and sometimes in addition by the frontal process of the maxilla, and they open into the infundibulum. When well developed they may bound the fossa on its posterior, medial, and anterior aspects. Their precise relations are important in connexion with the above emphasized difference in the thickness of the walls where formed by the lacrimal bone behind, or by the maxilla in front, a difference strongly marked even when the frontal process of the latter also is undermined by cells. In an examination of 100 orbits by the writer, approximately the lower half of the whole fossa was found in every instance directly related to the middle meatus of the nose, which space could always be entered easily through the thin *postero-inferior quadrant* of the fossa formed by the lacrimal bone (a relation found generally by Toti, in all cases by Dieulafé (1905), and Aubaret (1910), and in the proportion of 80 per cent. of 82 skulls by Thorsch (1909), and in 18 out of 29 by

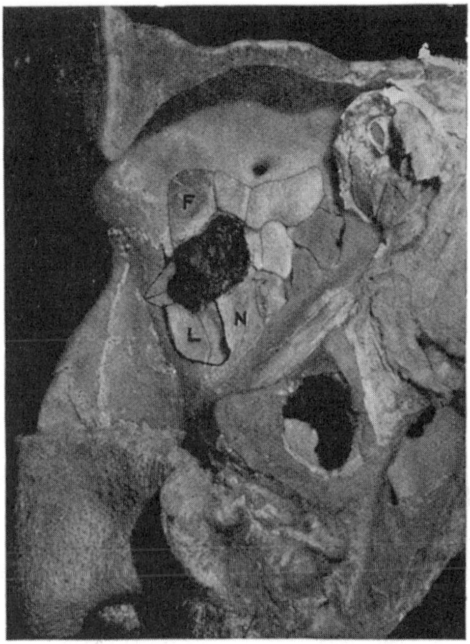

FIG. 34.—Dissection of Medial Wall of Left Orbit to show the relation of the **fossa for the lacrimal sac** (L.) in its upper part to an anterior ethmoidal cell (in black), above which is seen the frontal sinus (F.), and in its lower part to the nasal mucosa (N.). The muco-periosteal lining of the air-cells has been exposed by removal of their bony orbital walls. A wedge-shaped piece has been cut out of the anterior lacrimal crest to show its thickness. Natural size.

Salus). On the other hand, the upper half of the fossa presented relations to either one or two anterior ethmoidal cells, which extended (i) to the posterior wall only in 14 orbits; (ii) as far as the suture in the centre of the medial wall in 32; (iii) completely across the fossa and up to, or even into, the frontal process of the maxilla in 54 (Whitnall, 1911). These variations are shown in the adjacent figure.

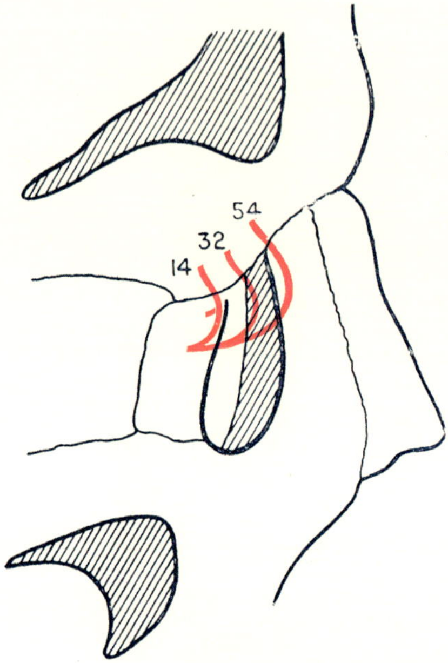

FIG. 35.—To show the varying forward relations of an **Anterior Ethmoidal Cell** to the upper half of the fossa for the lacrimal sac. The figures give the percentages of cases found in 100 orbits examined. The lower limb of the lines corresponds approximately to the free margin of the middle concha. Compare with Fig. 36 opposite.

There was in most cases a single intervening cell situated in the fore end of the processus uncinatus of the ethmoid in the region of the agger nasi (*cellula lacrimalis*, or 'cell of the agger nasi', or the 'cellule fronto-ethmoido-ungueale' of Aubaret), as is well shown in Figs. 26 and 34 above, and in Fig. 128, p. 236, and it should be particularly noticed that this air-cell separated the upper half of the lacrimal fossa from the nasal cavity, and would first be opened into on attempting to enter the cavity by this route, or, on the other hand, in obtaining access to the lacrimal sac by the nasal route. In no case was the fronto-nasal duct found in direct relation to the lacrimal fossa, but there was always a cellular interval between them, as seen in Fig. 26, though Onodi (1913) has seen an enlarged frontal sinus itself extending downwards to the upper end of the fossa. Finally, it should be mentioned that an extension from the upper part of the hiatus semilunaris reaching to the posterior wall of the lacrimal fossa is often present; such an 'ethmo-lacrimal recess' was found by Grünwald (1910) in 56 out of 79 preparations.

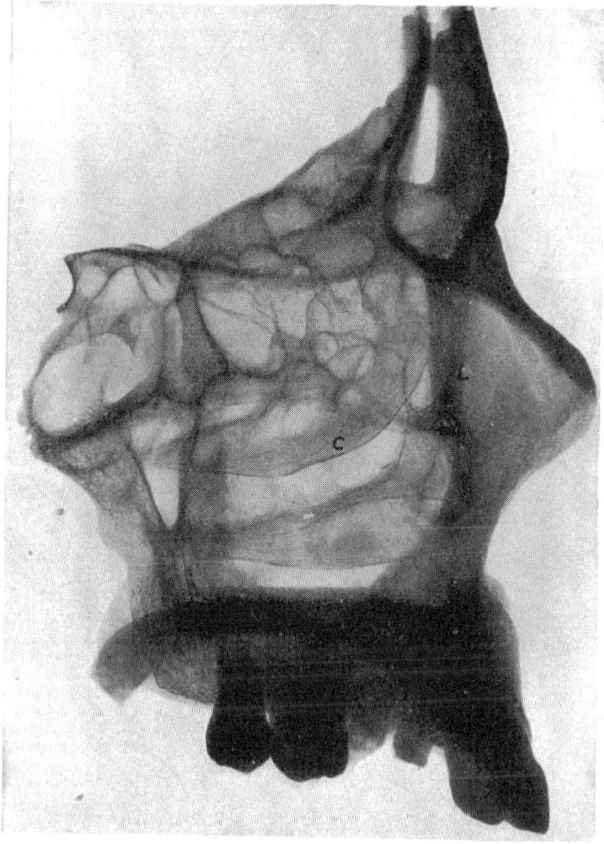

FIG. 36.—Radiograph through the left lateral nasal wall seen from the medial aspect, to show the position of the **fossa for the lacrimal sac** (L.), on the further side, relative to the fore end of the middle concha (C.) on this side. The outline of the naso-lacrimal canal is also well shown. Compare with Fig. 128, p. 236. Natural size. (By Dr. J. D. Morgan.)

As regards the deeper relations to the lateral wall of the nasal cavity, the upper half of the lacrimal fossa lies opposite the fore end of the attached border of the middle concha and its forward continuation, the agger nasi, an air-cell usually intervening, as described above. The lower half of the fossa is directly related to the upper part of the middle meatus, just below the attachment of the middle concha (the remainder of the meatus in vertical height being related to its continuation,

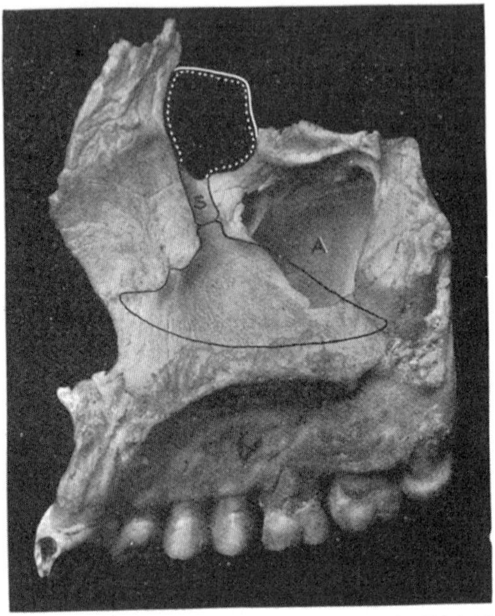

FIG. 37.—**Formation of the naso-lacrimal canal.** Right Maxilla seen from Nasal Side, showing outlines of lacrimal bone, above, and inferior concha or turbinated bone, below, and the formation of the *sulcus lacrimalis* of the maxilla, S., into the naso-lacrimal canal by the junction of processes of these two bones on its medial side. The extent to which the lips of the sulcus normally form the medial wall of the canal is here seen. Compare with Fig. 38 opposite, with the bones in place.

The fore and upper part of the antrum, A., presents a well-defined 'prelacrimal recess'. Natural size.

the naso-lacrimal canal). In Figs. 28, 29, and 199 the position of the fossa relative to this wall was indicated by pins driven through its outlines from the orbital aspect, and in Fig. 36 the exact relation is clearly displayed. The free extremity of the middle concha does not usually project as far forwards as the fossa, since Thorsch (1909) found that an opening made through the fossa into the nose was covered by the middle concha in 14, partly covered in 13, but completely uncovered in 52 out of 79 cases.

(ii) **The Naso-lacrimal Canal** (*canalis naso-lacrimalis*) is a short bony tube leading downwards from the lacrimal fossa to the inferior meatus of the nose, and containing the mem-

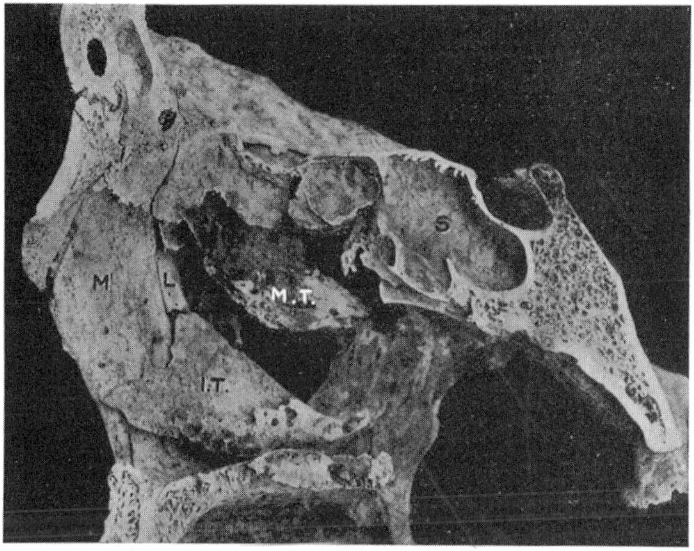

FIG. 38.—The Bones of the Lateral Nasal Wall, showing the descending process of the lacrimal bone, L., and the lacrimal process of the inferior concha or turbinated bone, I.T., which together form the **medial wall of the naso-lacrimal canal.** Compare with Fig. 37. Natural size.

M. = Frontal process of maxilla.
M.T. = Middle concha or turbinated process of ethmoidal bone.
S. = Sphenoidal sinus.

branous naso lacrimal duct. It is formed by three bones, the maxilla, lacrimal and inferior nasal concha (inferior turbinated bone), of which the greater part is contributed by the maxilla. On the forepart of the nasal surface of the body of this last bone is a deep groove, the *sulcus lacrimalis,* the downward continuation of the same sulcus on the posterior aspect of the frontal process which helps to form the lacrimal fossa, and the gap between its lips on the nasal side is completed by the articulation of the descending process of the lacrimal bone from above with the lacrimal process of the inferior turbinated bone (*concha nasalis inferior*) from below, as shown in Figs. 37, 38.

These two processes, of paper-like delicacy, as are the bones to which they belong, complete the medial wall of the canal and shut it off from the nasal cavity; the extent to which they take part in the completion of the canal is variable, and is compensated by the degree of development of the lips of the sulcus lacrimalis. The upper aperture of the canal is situated in the antero-medial angle of the orbital floor, and its lateral margin, the *incisura lacrimalis,* is circumscribed by the hamulus of the lacrimal bone when this process is developed; the inferior oblique muscle of the eyeball arises from the orbital floor just wide of this margin (Fig. 19, p. 43). The lower or nasal opening of the canal lies in the apex of the roof of the inferior meatus and near the junction of its anterior and middle thirds; it is distant from 15 to 20 mm. or $\frac{3}{4}$ inch (average 16·1 mm.) from the fore end of the inferior concha, and from 11 to 23 mm. (average 17·3 mm.) above the floor of the nasal fossa (Swerschewsky, 1910). (This opening of the bony canal is not to be confused with that of the contained membranous duct, which commonly lies at a lower level beneath the mucous membrane lining the side wall of the meatus, and is described on p. 238.)

The naso-lacrimal canal lies between the adjacent foreparts of the maxillary sinus and the middle meatus of the nose, projecting into either or both spaces. As regards the former, it has been noted (p. 67) that the canal usually projects more or less boldly into the inner and forepart of the sinus to form the 'lacrimal protuberance' of Zuckerkandl (Fig. 129, p. 237); this columnar prominence extends from the roof of the sinus, where an abnormal ethmoidal cell may present relations to it, to nearly half-way down the medial wall, that is, as far as the line of attachment of the inferior turbinated bone; the thinness of its wall, about 1 mm., is to be remembered in curetting the sinus; Fein (1912) found a lacrimal protuberance well developed in 9 but absent in 17 out of 37 bones, and noted that the alveolus of the first bicuspid tooth may be prolonged to meet it. Part of the maxillary sinus may be related more intimately to the canal by the formation of a 'prelacrimal recess' (Killian, Onodi) or 'infra-orbital recess' (Zuckerkandl), the upper, inner, and fore corner of the cavity being then circumscribed by the lacrimal protuberance on the medial side, with projecting ridges due to the infra-orbital canal laterally, and the anterior superior alveolar or dental canal inferiorly, the roof of the sinus bounding the space above. Out of 21 maxillae, 5 presented such a recess about as large as a pea, curving round the front of the lacrimal protuberance and separated off from the rest of the sinus by a constricted neck (Whitnall, 1913); its formation and position are well shown in Fig. 37 and are also seen in Fig. 178, p. 332.

On the lateral wall of the nasal cavity, the line of the naso-lacrimal canal, which may be indicated by a protuberance similar to that seen within the sinus, extends downwards with a backward slope from the attached end of the middle concha, as indicated in Fig. 28 and shown in Fig. 36, and bounds the atrium posteriorly. The relations to this wall should be considered in conjunction with those of the lacrimal fossa given above.

The **form, dimension,** and **direction** of the bony lacrimal passages all show considerable variation, owing chiefly to the extent to which the individual bones participate in their formation, in a lesser degree according to the type of skull, and also from the influence of certain intrinsic factors (see p. 238). The form of the passages is shown best by the fusible metal casts taken by Zabel (1900), as seen in Fig. 39, and is indicated, though less accurately, by the plaster moulds of the interior of the sac and duct prepared by Aubaret (1910).

The lacrimal fossa is widest laterally at its base, where it is confluent with the circular opening of the naso-lacrimal canal; it presents an antero-posterior oval section towards its middle and tapers off at its upper end; in the infantile skull it is almost cylindrical throughout, and is extensively overlapped by the posterior lacrimal crest. It is shallower in degree as it is formed in greater extent by the maxilla, and its breadth varies with the shape of the skull, being wider in the brachycephalic and narrower in the dolichocephalic type. The width varies from 4 to 9 mm. (with an average of 6·5 mm. in 122 orbits; Gérard, 1907), and its length averages 16·5 mm. in males and a millimetre shorter in females (Ledouble, 1906). Facial asymmetry explains variation in the two fossae of the same skull, found 10 times in 138 orbits by the latter author.

The naso-lacrimal canal is usually flattened from side to side, but is sometimes cylindrical. Gérard found it narrowest at the orbital opening; Zabel, on the other hand, described a slight constriction towards the middle with dilatations at the upper and lower openings, the planes of which apertures are parallel. The upper or orbital opening is generally oval and much more cleanly cut, especially when defined by the pre-

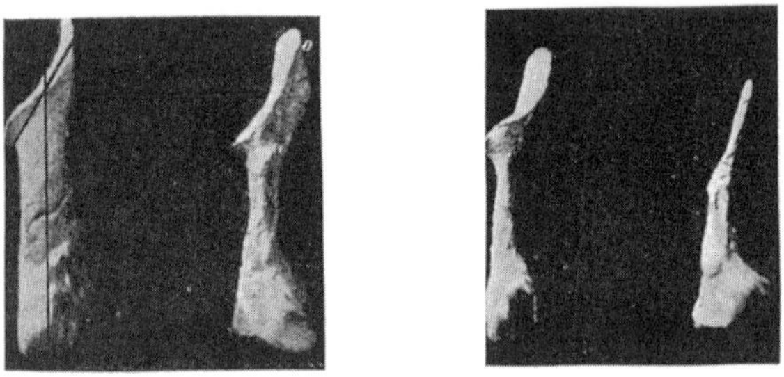

a *b*

FIG. 39.—**Casts of Osseous Lacrimal Passages** by Zabel, modified. Taken from right orbits and seen from anterior aspect. About natural size.

a = From two skulls with normal lacrimal bones, showing variation in calibre of naso-lacrimal canal. The black lines show the inclinations of the long axes of the fossa and canal.
b = From two skulls with badly developed lacrimal bones, showing marked reduction in calibre of canal.

sence of a hamular process, than the wider, almost funnel-shaped inferior or nasal aperture. Kofler (1929) illustrates, amongst other variations, an anomaly of the opening lying entirely in the maxillary process.

The dimensions of the naso-lacrimal canal vary considerably. The length, generally less in dolichocephalic and greater in brachycephalic skulls, is affected by the development of the hamular process at the upper end and by the level of the articulation of the inferior concha at the lower end. From its upper to its lower opening an average length is 12·4 mm. or ½ inch, combining the following extreme ranges: Zabel, from 2·5 mm. to 15 mm.; Merkel, from 10 to 12 mm.; Swerschewsky, from 6 to 15 mm. (average 9·2); Schwalbe, from 12 to 24 mm.;

Power, an average of 15·3 mm. in 287 European skulls (see Fig. 122, p. 228).

The height of the upper or orbital opening of the canal from the level of the nasal floor was found by Power (1886) to average 27·4 mm. in 292 European skulls; Gérard's figure is 23·3 mm., or nearly an inch, with extremes of 25 and 40 mm. in 69 skulls. Theobald (1900) found that the average combined length of the fossa and canal was 30·7 mm. in 286 skulls. The transverse width (the narrower dimension) of the upper orifice of the canal is 4·6 mm., on an average made from the following figures:

(i) Zabel, from 4·3 to 6·8 mm. (sagittal width from 5·9 to 11·9 mm.).

(ii) Onodi, from 3 to 4 mm.

(iii) Merkel, 5 mm.

(iv) Gérard, from 3 to 6 mm.

(v) Power, an average of 3·77 mm. in 205 European skulls (with an average antero-posterior diameter of 4·14 mm.). The right duct was slightly narrower than the left in both dimensions.

(vi) Swerschewsky, from 3 to 6 mm. (average 5·1), and transverse width of lower opening from 4 to 8 mm. (average 7).

(vii) Theobald, 3 to 7 mm. (average 4·11) in 70 adult canals measured by rigid probes; with a difference in the measurements on the two sides in 18 cases.

The above dimensions may be compared with those of the membranous passages given on p. 240, and Onodi's measurements of the duct in children will be found on p. 241.

It is rather remarkable that the calibre of the canal appears to be affected by the degree of development of the lacrimal bone, being narrowed when this bone is rudimentary; the extent to which it is then formed by the maxilla compensates for this deficiency. In an examination of 50 maxillae by the writer, it was found that in 31 the lacrimal sulcus formed about three-quarters of the circumference of the canal, the lips being separated by about 4 mm., and the gaps between completed as usual by processes of the lacrimal bone and inferior concha (Fig. 37); in these cases the canal was always roomy, an antero-posterior diameter of 8 mm. being found near the middle of several, and there was nowhere any narrowing; in 12 bones in which the lips of the sulcus were only one or two millimetres apart, the canal was a little narrower, but without constriction; in 7 bones the lips met and fused (Fig. 40), and the central part of the canal thus formed entirely by maxilla was much narrower, no less than four of these cases presenting a decided constriction of the lumen in this region (Whitnall, 1912). Whilst the naso-lacrimal duct enclosed in such a narrowed canal may have been permeable, it is obvious that a slight cause would lead to its occlusion. In these cases the part contributed to the formation of the canal by the lacrimal bone was naturally reduced; and Zabel has noted that in some skulls where the lacrimal bone is rudimentary or absent not only is the lacrimal fossa shallow and small, but the canal is narrowed transversely (Fig. 39). In 12 complete skulls with rudimentary lacrimal bones seen by the writer, 7 presented shallow fossae, with decided narrowing of the orbital openings of the canals. The constriction can be well seen by holding the skull with the vertex towards one and bringing the orifice of the canal into view by glancing over the supra-orbital margin, as illustrated in the following figure.

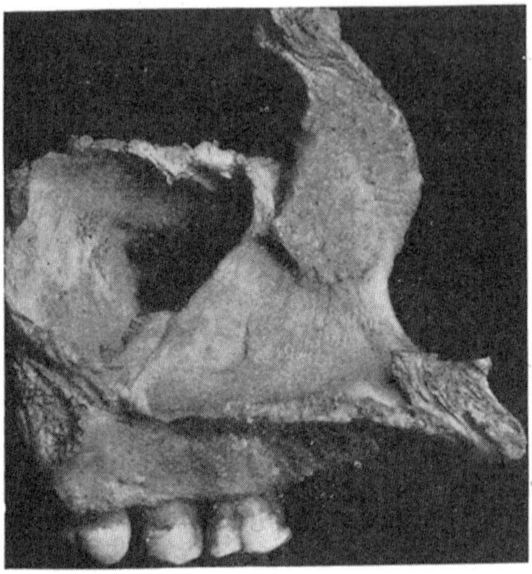

FIG. 40.—Left Maxilla viewed from Nasal Side, showing approximation of lips of *sulcus lacrimalis* to form completely the mid-part of the **naso-lacrimal canal,** with constriction of lumen to 3 mm. Compare with Figs. 19, 37. Natural size.

The direction and inclination of the lacrimal passages stand next to the dimensions in importance from a practical point of view; they are dependent on the configuration of the facial skeleton, being influenced by the breadth of the bridge of the nose and of the nasal aperture, and by the development of the maxillary air-sinus. Seen in profile, both fossa and canal slope slightly backwards as they pass downwards, the long axis of the canal forming an angle of from 15° to 25° (Testut), with

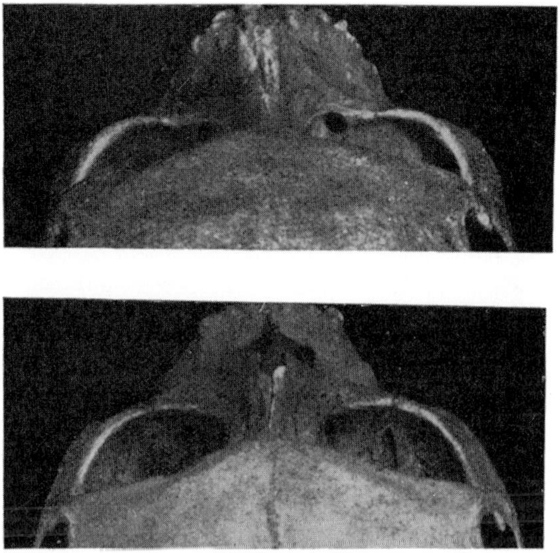

FIG. 41.—Foreparts of two Skulls, viewed from above to show **orbital opening of naso-lacrimal canal,** in the upper specimen of normal calibre, in the lower one narrowed. The latter is the skull shown in Fig. 23, p. 51, with absence of lacrimal bone. × ½.

the frontal plane (Figs. 36, 125, 128). Seen from in front, the inclinations differ, the fossa showing a lateral deviation inferiorly, whilst the canal on the other hand inclines slightly medially towards its lower end; hence the long axis of the fossa and canal together form an angle, or rather a wide parabolic curve, with a concavity directed towards the nose and a summit near the upper orifice of the canal (Fig. 39 *a*). The lateral obliquity of the medial wall of the fossa is more pronounced in leptorhine or narrow-nosed skulls than in the platyrhine or broad-nosed type; it forms an angle with the median sagittal plane open below and on the average of 26° (with extremes of 13° and 37°), according to Picou. The long axis of the canal forms a similar angle, but open above, and of 12°; the two canals therefore converge inferiorly, in agreement with the observations of Schwalbe, Merkel, Zabel, and Picou, though a divergence of direction is described as normal by Henle, Hyrtl, Arlt, and Onodi. Theobald in 275 European skulls found the canals were parallel in 37, sub-parallel in 117, converging below in 121. The naso-lacrimal canal in the negro has been studied by Santos Fernandez (1903 and 1921), who finds it shorter, broader, and straighter than in white races.

In the living subject, the line of the long axis of the fossa is indicated by the upper part of the naso-jugal fold of the skin (Figs. 8 *a*, 9; Fig. 53, p. 118; and Fig. 148, p. 269). According to Arlt (1855), the deviation of the canal, which he considers is a lateral one, is shown by a line drawn from the medial canthus of the eyelids to the lower part of the naso-labial fold. In his words (quoted by Onodi), 'when the interval between the alae nasi, measured at the place where they pass on to the cheek, is equal to the interval between the middle points of the two medial palpebral ligaments, then there is no lateral deviation of the nasal duct. But when the first-mentioned interval is the wider, as is usually the case, then the lateral deviation of the duct is represented by half this difference. In exceptional cases the alae nasi approach each other so closely that the space between them is less than that between the middle points of the palpebral ligaments. When this is the case the lateral deviation is reversed. The deviation can readily be ascertained by applying a straight sound to the base of the ala nasi below and to the middle of the palpebral ligament above.' In life it should be remembered that a probe passed down the lacrimal sac and duct has to accommodate itself to the oppositely inclined long axes of their osseous boundaries, though its passage is no doubt facilitated by the elasticity of the delicate bones forming the medial wall of the channel, which it must touch in the middle of its course. 'It is commonly said that probes passed one into each nasal duct meet above on the glabella; this is certainly not often the case during life' (Fisher, 1904). From this point of view the inclination of the fossa is negligible, since its contained sac is confined laterally by soft and extensible structures, whereas the direction of the duct lying in its osseous canal is of greater importance.

The backward inclination as seen in profile is indicated by a line passing from the medial commissure of the eyelids towards the first molar tooth. The line passes just in front of this tooth, according to Onodi and Testut, but behind it according to Luschka and Merkel. In children the direction is towards the first temporary molar tooth (Onodi).

Of the authors quoted in this section, the work by Onodi (1913) will be found of special interest, in that he discusses the various operations performed on the nasal wall of the lacrimal passages, and illustrates their relations to the surrounding parts by numerous photographs of fresh preparations in the child as well as in the adult.

6. TABLE OF APERTURES OF THE ORBIT

AND THE STRUCTURES WHICH TRAVERSE THEM

The orbit communicates with the **Cranium** by the

1. Optic canal, transmitting
   - the optic nerve,
   - the ophthalmic artery, along which run
   - the sympathetic fibres from the carotid plexus;
   - the orbital prolongations of the cerebral meninges as sheaths round the optic nerve.
2. Superior orbital fissure, transmitting to the middle cranial fossa:
   - the III, IV, VI, and ophthalmic division of V cranial nerves;
   - the ophthalmic vein;
   - the sympathetic, and sometimes also the sensory root of the ciliary ganglion, and some filaments from the cavernous plexus of the sympathetic system;
   - the orbital branch of the middle meningeal artery.
3. Ethmoidal canal, anterior, transmitting the continuation of the naso-ciliary as the internal nasal nerve, and the anterior ethmoidal artery to the anterior cranial fossa and thence to the nasal cavity.
4. Ethmoidal canal, posterior, transmitting the 'nerve of Luschka' and posterior ethmoidal artery (less constant than the anterior vessel) into the posterior ethmoidal cells and nasal fossa; probably also lymphatics through both canals.

With the **Nose** by the

5. Naso-lacrimal canal, transmitting the naso-lacrimal duct.
6. Ethmoidal canals.

With the **Face** by the

7. Inferior orbital fissure (to infra-temporal fossa), transmitting the infra-orbital nerve and blood-vessels;
   the veins connecting the ophthalmic and pterygoid plexuses.
8. Zygomatico-temporal canal (to temporal fossa), transmitting the corresponding branch of the zygomatic or orbital branch of the superior maxillary division of Vth nerve, together with an artery.
9. Zygomatico-facial canal (to cheek), transmitting the corresponding branch of the zygomatic or orbital branch of the superior maxillary division of Vth nerve, together with an artery.
10. Infra-orbital canal (to cheek), transmitting the infra-orbital nerve and vessels.
11. Supra-orbital notch and frontal notch (to frontal region), transmitting branches of the supra-orbital nerve and vessels.

As regards the relations of the orbit to the **teeth**, Magnasco (1928) gives a full account of the bony, vascular and nervous connexions, and divides the possible ocular affections into those occasioned by continuity, such as perivascular abscess, maxillary sinusitis and secondary orbital cellulitis, and those of more distant or general infectious dental origin. He distinguishes in the latter a group of functional disturbances of the fifth nerve (reflex lacrimation, corneal anaesthesia, conjunctival ulcers, herpes), and a group of motor derangements (blepharospasm, photophobia, midriasis, neuritis of the optic nerve). Wilmer (1930) reviews the subject historically, and thinks the 'eye-tooth' relation is based more on popular imagination than on clinical observations; changes in the blood-stream especially affect the uveal tract of the eyeball. Dewey (1920), more particularly concerned with lymph channels, emphasizes the importance of the anterior and medial superior alveolar canals passing between the infra-orbital canal and the front teeth, as noted also on p. 16 and illustrated on p. 8, Fig. 5. Hardy (1917), Ackland and Lang (1923), and Redslob (1929) may also be referred to.

It is of interest to note how much more closely connected

the orbit is with the **brain** than are the other cavities of the face. The orbit communicates with the cranial cavity by several apertures; its lining periorbita is a direct continuation of the dura mater, which, together with the other two cerebral membranes, the pia and arachnoidea, are prolonged around the optic nerve into the orbit; the blood comes from the internal carotid artery just before this vessel terminates in branches which supply the brain, and is drained into the cranial venous sinus system; a venous link is also afforded by the ophthalmo-meningeal vein (p. 313); and, lastly, the chief contents of the orbit are the globe and optic nerve, outgrowths from the brain.

A distant means of connexion between the two orbits is afforded by the cavernous sinuses into which the ophthalmic veins drain, since they are joined together by inter-cavernous sinuses; and possibly also by sympathetic nerve fibres (p. 370).

7. THE PERIORBITA

is the name given to the periosteal lining of the orbital cavity. Like the periosteum elsewhere, with which it is continuous in many places around that cavity, it consists essentially of a sheet of dense fibrous tissue supporting the minute nutrient vessels of the bone; it is not of uniform strength throughout, but is exceedingly thin where it lines the roof and comparatively thick at certain other points. It is easily detachable from the surface of the bones, and therefore can be raised from them by sanguineous or pathological effusions; but it is adherent (i) along the sutures, where in early life it is continuous with the periosteum covering the opposite sides of the bones; (ii) at the fovea trochlearis, where it binds down the trochlea of the superior oblique muscle; (iii) at the fissures and foramina through which it becomes continuous with the lining of the cavities into which they lead; and (iv) at the orbital margin, where it is confluent with the periosteum covering the bones of the face; here it is thickened to form a ridge, the *arcus marginale*, which marks the exact line of the orbital margin, and from which arises the septum orbitale of the eyelids.

By removing the surrounding bones piecemeal in a formalin-hardened preparation, the whole contents are retained in the periorbita as a cast of the orbit; the trochlea is left with the membrane, but the lacrimal sac requires severing from the naso-lacrimal duct; posteriorly the continuity of the periorbita with the dura mater of the cranial cavity is demonstrated (Fig. 197, p. 379).

Internally, save for numerous fine loose strands which connect it with the stroma of the orbital fat, muscle sheaths, and lacrimal gland (the last constituting the so-called *ligamenta glandulae lacrimalis*), the periorbita will be found, on removal of the entire contents of the orbit and so viewing it from within, to line the space smoothly, hiding the sutures, and so continually maintaining the contour of the medial wall as it bridges across the lacrimal fossa to form the *fascia lacrimalis* that the position of the enclosed sac is only indicated by the severed points of its canaliculi (Fig. 124, p. 231). As in the case of the lacrimal fossa, the membrane not only lines but roofs over the infra-orbital sulcus, and in both these places its deeper osteogenetic layer may develop bone to a variable degree, giving an accentuated posterior lacrimal crest in the one case and a more or less complete bony roofing to the infra-orbital sulcus in the other. Further reference is made to the lacrimal fascia on p. 230.

The blood-vessels of the periorbita are derived from various branches of those which traverse the orbit, and it is innervated by the sympathetic system through filaments from the spheno-palatine ganglion (p. 360); Gray (1930) figures a supply from a long ciliary nerve to the orbitalis muscle.

The periorbita presents a point of especial interest in the region of the inferior orbital fissure, where a mass of smooth or involuntary muscle fibres is found incorporated upon its superficial aspect. This muscle, described by H. Müller in 1858, is known as the 'orbital muscle of Müller', or *musculus orbitalis* or muscle of Gegenbauer, or spheno-maxillary muscle, and should not be confused with the involuntary palpebral muscles also discovered by him. The term 'periorbital muscle' is suggested as a better name. It completely fills the inferior orbital fissure, and is narrow, but two or three millimetres deep, in the centre, and fan-shaped at the extremities, where it spreads in a thin layer over the orbital floor; its position and extent are indicated in Fig. 30, p. 62. Anteriorly it is continuous, according to Charpy (1912), with the orbital expansion of the inferior oblique muscle, a relation the writer

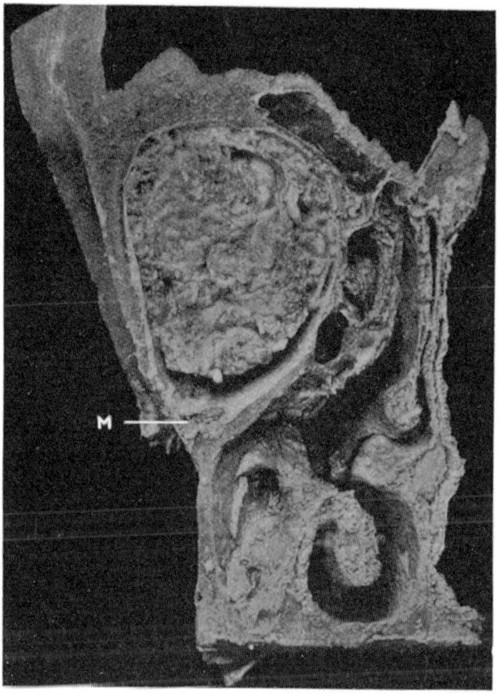

FIG. 42.—Frontal section through the Right Orbit seen from in front, showing the *musculus orbitalis* or '**orbital muscle**' **of Müller,** M., in the inferior orbital fissure, and its connexion with the lining periorbita. Natural size.

has been unable to confirm; posteriorly it can be followed beneath the annulus of Zinn to the lower extremity of the superior orbital fissure, and even up to the wall of the cavernous sinus; whilst inferiorly it is in contact with the fat of the pterygopalatine fossa (Hesser, 1913). Through it there pass only small nerves, and venules from the ophthalmic system to the pterygoid plexus (p. 316); some of its fibres accompany the veins out of the orbit (W. Krauss, 1911), and the inferior ophthalmic vein is sometimes related to it beneath the annulus of Zinn (see Fig. 167, p. 313). Like other vestigial structures, it shows much variation in size.

It is of twofold interest: first, in that it represents in man the larger muscular sheet (*membrana orbitalis musculosa*) which completes the lateral wall of the orbit in lower animals where the bone is absent (p. 39). It exists as a smooth-fibred muscle in all mammals save in the aquatic group, where it is cross-striped as in amphibians, lizards, and birds, and where, according to Groyer (1903), who, like Burkard (1902), has studied it from the point of view of comparative anatomy, it appears to play an important role as an organ accessory to the lacrimal apparatus. Its function in this respect is not clear, and any use it may have in animals beyond keeping tense the orbital wall is to protrude the eyeball, since Müller found that irritation of the cervical sympathetic nerve causes protrusion of the globe in beasts where the muscle is well developed, and in them it is antagonized by a special 'retractor bulbi' muscle. In man it is purely vestigial, and excitation of the nerve does not cause protrusion of the globe (Turner, 1862); moreover, a retractor bulbi muscle is not developed. The second point of interest is a pathological one, an explanation of the **exophthalmos** of Basedow's disease being sought in the relation of the periorbital muscle, firstly to the orbital contents as a whole, and secondly to the venous system. It is hardly possible either that contraction of such a small muscular mass, for the most part deeply confined in the inferior orbital fissure, as seen in the above figure, and elsewhere blended with the closely adherent periorbita, could affect the position of the eyeball by compression of the mobile orbital fat (i.e. that the muscle in its vestigial condition could retain the possible action of its full development), or that occlusion of such venules as traverse the muscle (see p. 313) could affect the volume of the orbital venous system which communicates so freely, not only within itself, but with the extra-orbital system. Festal, indeed, concludes from the effect of his injections that the orbital veins are normally never quite full. Treacher Collins (1921), like Tilley (1926), favours the first view of the cause, that is, compression of the orbital contents as a whole by contraction of this muscle. W. Krauss (1912), who has studied the muscle in the new-born, and Fründ (1912) consider both explanations as sufficient. Vitali (1929) thinks that the muscle must have a functional value in view of its peculiar nerve supply (p. 360), and could assume a certain importance in exophthalmos by compression of the vessels traversing it. Hesser (1913) on the other hand expressly denies that the condition could thus be produced since only in exceptional cases did he find either the inferior or superior ophthalmic vein in a position to be so affected, and he agrees that the free anastomoses between the veins of the orbit and those of the face would obviate the questionable engorgement. It may be of interest to note that other explanations of the condition are: contraction of the peribulbar involuntary musculature (Landström, 1908; see p. 299), an orbital oedema from vaso-motor disturbances (Sättler, 1911); a localized arterial dilatation (which would cause a pulsating exophthalmos: see reference to Krauss on p. 313) or venous congestion (which should entail dilatations of the retinal veins); a hypertrophy of the retro-ocular fatty tissues (Basedow); relaxation and elongation of the ocular muscle from fatty degeneration or atony (Cooper, Egeberg, Lemke, Traube, Bristowe, Dalrymple, &c.); but the most probable cause of the condition is dilatation of the orbital blood-vessels by excitation of the sympathetic nervous system (Fuchs, 1917). The exophthalmos is said to disappear after death, and no unusual gross conditions have been found post-mortem; in a dissection of two orbits taken from a subject who had died from the disease and had presented the well-marked sign, nothing abnormal could be found by the writer.

8. MENSURATION

The dimensions of the orbit considered here are the volume, the size of the entrance, the depth and the length of the walls, and certain angles; measurements of the smaller features have been given in the text.

Since the form of the orbital cavity varies according to race, age, sex, and in individuals, and since the exact points from which the dimensions are taken are neither identical nor definitely specified in all cases, it is not surprising that there is a great lack of agreement in the figures given by different authors. In measuring the orbit, it is of primary importance to fix the points clearly; they should, as Gayat notes, be well marked, easily recognizable in ancient as well as in recent skulls, constantly present at all ages, and the progress of growth should neither accentuate some nor obscure others. It is to be noted also that the presence of the periorbita makes a difference in the measurements, it being stated that the depth, for instance, is diminished by from 3 to 6 mm. after removal of the membrane from the margin; further, it is hardly necessary to point out that great accuracy is required, since a difference of one millimetre in the height converts an orbital index of 86·3 into 88·9 or 83·7, and one millimetre difference in the width converts the index into 84·1 or 88·6, according as the dimension is increased or decreased by that amount.

The **volume** has been obtained by filling the orbit with lead pellets, by which means Gayat (1873) found it to be 22 c.c. in a child of ten, and 29 c.c., with extremes of 25 and 33, in eleven adult skulls; Weiss gives 29·74 c.c. for adults. Relative to the size of the whole skull, it is larger in females.

The dimensions of the orbital **entrance** or base are conveniently obtained by measuring along threads fixed across it; the vertical diameter or height should be parallel to the lateral and medial margins, and the horizontal diameter or width parallel to the upper and lower margins; these two lines are nearly at right angles to one another, but, owing to the greater projection of the supra- and infra-orbital margins, they do not touch where they cross, but are separated by an 'axial distance', which varies from 2 to 10 mm. (Ambialet, 1905). The line of height at its upper extremity falls usually to the lateral side of the supra-orbital notch, but should they coincide the distance is taken to an imaginary line crossing the edges of the groove; inferiorly a point should be chosen which avoids the undue prominence of an infra-orbital tubercle at the suture. The line of width is on a level with the lateral orbital tubercle (p. 13), so being opposite the commissure of the eyelids, and if the margin here presents no distinct edge, one can be indicated by drawing the flat of a lead-pencil down it; the point where it abuts on the medial margin is less easy to define owing to the presence of the fossa for the lacrimal sac (see p. 17). Topinard measured to the upper part of the anterior lacrimal crest, the width thus being affected by

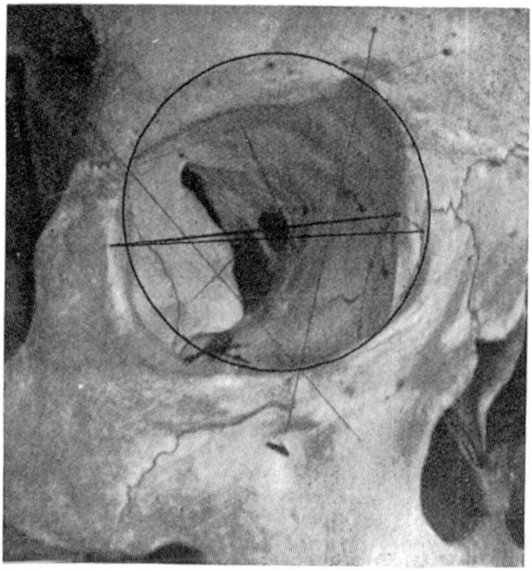

FIG. 43. Right Orbit to illustrate (*a*) the difference in **orbital width** (2·9 mm.) according as whether taken to Flower's lacrimal point (upper horizontal line), or to the point dacryon (lower line) on the medial side. The orbital index in the former case is 86·3, and in the latter 80·3. (*b*) The figure also indicates the directions in which the **primitive circular outline** of the orbital margin has been altered by growth in the surrounding parts. The circle is drawn from a central point equidistant from the supra-orbital notch, the zygomatico-frontal and the zygomatico-maxillary sutures.

the configuration of the frontal process of the maxilla and the nasal bones at the root of the nose; Broca chose the point *dacryon*, where the upper end of the lacrimo-maxillary suture meets the frontal bone, but whilst marking accurately the centre of the orbital margin, it is often obscured by fusion of the bones. Flower marked a 'lacrimal point' further back, where the upper end of the posterior lacrimal crest meets the frontal bone, and although its position is affected by the degree of projection of the underlying ethmoidal cells, yet this variation is shared by the orbital wall and gives the dimension a truer value as indicative of the width of the forepart of the cavity. The exact point taken is important, since a difference of as much as three millimetres may exist between the above three points in the horizontal direction, so affecting the orbital index, as is illustrated in Fig. 43.

The height of the orbital entrance is approximately 35 mm. or $1\frac{3}{8}$ inches, and the width 40 mm. or $1\frac{1}{2}$ inches, but precise figures vary greatly—the height from 30 mm. (Arlt) to 40 mm. (Richet, Luschka), and the width from 36 to 50 mm. as given by the same authors. According to Baker (1900), the figures given by Emmert (1880) are on the whole the most extensive, and were very carefully taken in a series of skulls free from abnormalities; the race, however, is not noted. His figures are:

	Adult Males aged 20–67.	Adult Females aged 23–67.	Children aged 10–17.
Width in mm.	41·6	39·8	34·3
Height in mm.	34·0	33·6	29·2

Other figures are:

	De Wecker.	Arlt.	Legrange.	Richet.	Weiss.
Width in mm.	39	36	40	40–6	39·7
Height in mm.	35	30	35	40	35

The distance from the lateral margin of the one orbit to that of the other (i.e. the extra-orbital width) is given by Emmert as 99·7 mm., or 4 inches, in males, 96 mm. in females, 80·8 mm. in children, and the inter-orbital width at the margins is, according to Broca, 25 mm.; diagonal together with other measurements are given in the table on p. 104.

At birth the height of the orbital entrance is almost equal to the width, but during growth the width predominates, more in man than in woman, and, it is interesting to note, more or less according to race. Broca gave this relation a mathematical value by expressing the proportion of the orbital height to the orbital width, using the formula:

$$\frac{\text{Orbital height} \times 100}{\text{Orbital width}},$$

the result being termed the **orbital index**; for example, in the skull illustrated in the frontispiece:

$$\frac{\text{Marginal height}}{\text{Marginal width}} \quad \frac{33{\cdot}4\times 100}{38{\cdot}7} = 86{\cdot}3 \text{ the orbital index.}$$

Three classes of orbit are obtained by this means:

Megaseme, with an orbital index of 89 and over; this high numerical value is remarkably constant in the yellow races, excepting the Esquimaux, and the opening is usually round (Fig. 50, second skull, p. 107).

Mesoseme, with an index between 89 and 83, found in the white races, where Flower gives a mean figure of 87 for 208 European and 88·4 for 33 English skulls.

Microseme, with an index of 83 and less, characteristic of the black races, where the opening is rectangular (Fig. 44).

Cameron (1920) has devised a composite 'naso-orbito-alveolar index,' according to which he divides modern races into two groups, applying the terms, broadly speaking, Eurasiatic and Negro. The orbital measurement concerned is the height from the level of the inferior orbital margin to the point nasion.

The vertical height is proportionately greater in the yellow and in females of all races. The orbital index is greater in infants than in adults of either sex, and is higher in females than in males. This character is of anthropological interest, and will be found fully considered in the works of Topinard (1885), Broca (1875), and Flower (1907). Duckworth (1904), however, concludes that its range of variation is too great to render accurate information in most instances, since so many different records exist, and it has been accorded quite a secondary place in the list of selected indices. Broca states that the difference due to sex alone amounts to as much as 31 per cent. of the total racial variation.

The **depth** of the orbit and the length of its walls.—The point at the apex from which these dimensions are taken is best considered to lie at the centre of the bridge of bone which separates the optic foramen from the superior orbital fissure; this point is accurate in that it is situated in the middle of the annulus of Zinn from which the recti muscles of the globe arise, and lies between the optic nerve and the bulk of the other nerves which enter the orbit; and it is convenient, since it can be straddled by the forked extremity of a graduated rod along which a cross-piece may be slid against the desired point on the margin and the distance between them read off, as suggested by Gayat. This is the central point of the apex or 'apical point' of the orbit, and it is separated from its fellow of the opposite side by 25 mm. or an inch.

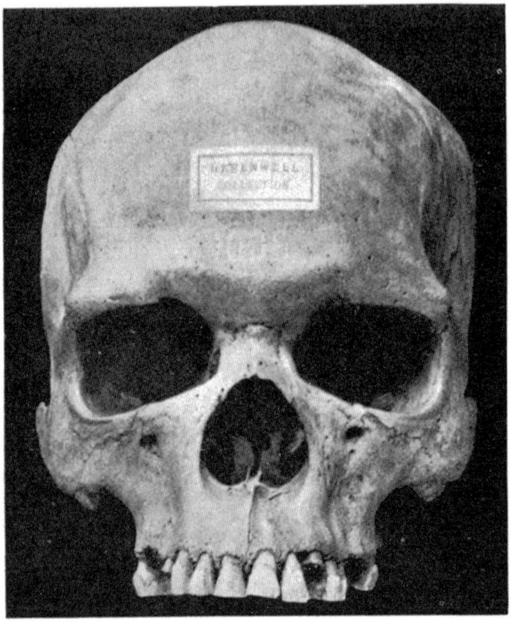

FIG. 44.—Australian Skull, showing **rectangular orbital margin.** Orbital height, 27 mm.; orbital width (to Flower's point), 39 mm. Orbital index, 69·2. If the width is taken to the point dacryon (42 mm.), the index becomes 64·3. The orbital height is unusually low. $\times \frac{1}{2}$.

The orbital depth is usually measured to various points on the margin, though actually such figures give the length of the corresponding walls or the distance of the anterior opening of the optic canal from the margin; theoretically, the true depth is that of the long axis measured to the centre of the axial distance (i.e. to the mean of the distance which separates the centre of the vertical from the centre of the horizontal plane of the entrance), but practically the centre of the vertical plane or marginal height is easier to define, and, moreover, corresponds to the normal position of the most prominent part of the cornea.

The figures given show again much variation: the orbital depth is about 40 mm. or 1½ inches, but the extremes in literature are from 39 mm. (Merkel) to 50 mm. (Richet, Tillaux); Emmert's distances of 39·8 mm. in males, 39·4 mm. in females, 34·75 mm. in children are taken to the centre of breadth of the entrance; Merkel's figure of 39 mm. is taken to where the breadth and height cross; Arlt gives 42 mm.; Gerlach 43 mm.; De Wecker and Lagrange 40–50 mm.; the depth is less in females.

The **walls** are longer on the temporal side and floor than on the nasal side and roof; their lengths from the apical point to the margin (which, it may be noted, are measured in a straight line and do not follow the curves) give also the distance of the optic foramen from the latter. The figures which follow are in millimetres:

Medial wall:

Emmert: 41·4 in males, 40·3 in females, 36 in children (apical and marginal points not stated).

Gayat: in 56 adult skulls, 50·6 from apical point to anterior tip of fovea trochlearis and 47·4 to crest of lacrimal bone.

Roof:

Broca: 50·9, from apical point to supra-orbital margin.

Zanda and Geissler (Gayat), 53.

De Wecker (Testut), 43 from optic foramen to centre of margin; 40 to 41 from optic foramen to supero-medial angle of margin; 43 from optic foramen to supero-lateral angle.

Lateral wall:

Emmert: 46·4 in males, 46 in females, 39·4 in children.

Gayat: 48·2 from apical point to zygomatico-frontal suture.

Floor:

Gayat: 49·1 from apical point to zygomatico-maxillary suture.

De Wecker (Testut): 46 from optic foramen to centre of margin.

Zander and Geissler (Lagrange): 53 from optic foramen to margin; 47 from optic foramen to medial angle; 56 from optic foramen to lateral angle.

The dimensions of the right orbit in the adult male skull

shown in the frontispiece, and used as type in Figs. 10, 15, 43, and 49, are:

Orbital height, 33·4 mm.; orbital width to lacrimal point, 38·7 mm. (to the point dacryon, 41·6).
Orbital index, 86·3.
Lateral wall, 47 mm. long; floor, 53 mm. long; roof, 51·5 mm. long.
Medial wall, length to dacryon, 39 mm.; to anterior lacrimal crest, 45 mm.
Anterior end of inferior orbital fissure to orbital margin, 16 mm.
Anterior end of superior orbital margin to orbital margin, 29 mm.
Inter-orbital width between dacryon on each side, 21 mm.
Inter-orbital width between lacrimal point on each side, 25·3 mm.
Extra-orbital width between lateral orbital margins, 98 mm.

In the female skull illustrated in Fig. 49 the measurements are:

Orbital height, 34 mm.; width to lacrimal point, 35·8 mm.; orbital index, 94·9.
Length of roof from apical point, 45 mm.; floor, 43·5 mm.; lateral wall, 42 mm.
Length of medial wall from apical point to posterior lacrimal crest, 34 mm.; to dacryon, 36 mm.

The **angles** to be described are those of (*a*) the planes of the orbital entrances; (*b*) the walls; (*c*) the long axis of the cavity.

(*a*) The horizontal planes of the two orbital entrances form an angle with one another, the summit of which is placed at the root of the nose and the extremities at each margin just below the lateral angular process of the frontal bone. This 'naso-malar' angle is given by Flower as 131° in 130 Europeans, and is much more open in yellow races where the root of the nose is flattened. Weiss gives about 140°, Lagrange 145°–150°. Merkel gives angles of 144·6° in children, 146·5° in adult females, 147° in adult males. Evatt (1907) has devised a means of plotting the position of the orbital entrances on paper by joining up points on the zygomatic bone just below the lateral marginal suture with the lacrimal points; he thus obtains a 'basic angle', which is another measure of their obliquity, and classifies 'ithybasic,' 'mesobasic,' and 'loxobasic' types accordingly.

(*b*) The angles formed between the lateral and medial walls are from 45·9° to 48·6° (Lagrange), and between the lateral wall of each orbit, from 87·4° to 90·6° (Lagrange), or 89·9° in males, 89·9° in females, 87·4° in children (Emmert). The two medial walls are rarely parallel (only in 11 cases out of 100; Merkel, 1901), but are usually about 3 mm. wider apart behind (see Fig. 136, p. 254).

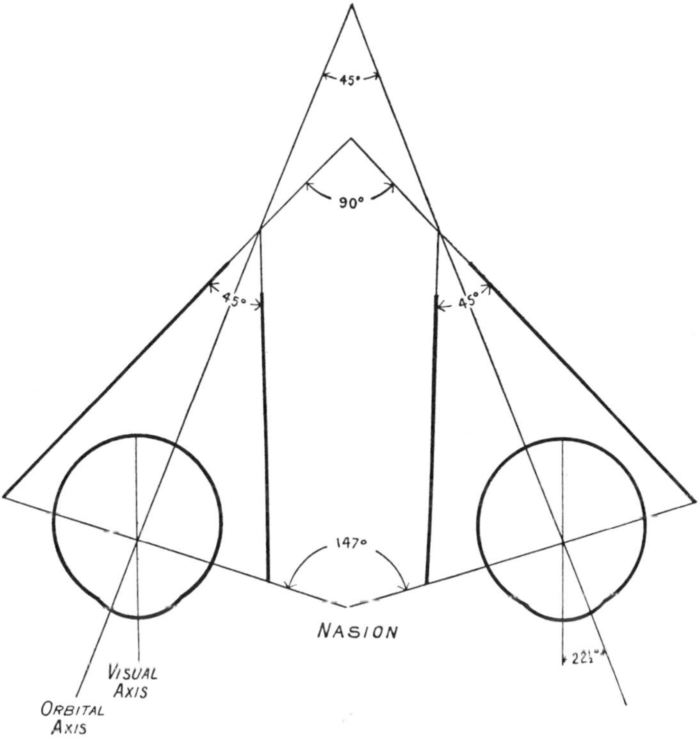

FIG. 45.—Linear Reconstruction of Horizontal Section through the Orbits, to illustrate the angles formed by the walls. Compare with Figs. 17, p. 38, and 136, p. 254. Natural size.

(*c*) The long axis of the orbital cavity is level in infants, but in adults is depressed anteriorly to form an angle of 15° to 20° with the horizontal plane (Fig. 21, p. 49). With its fellow of the opposite side it forms rather less than half a right angle, or more precisely an angle, according to Emmert, of 43·4° in males, 44·7° in females, 42·4° in children; other figures given are 40·6° to 44·7° (Lagrange), 40° (Poirier), 46° (Testut). The orbital and visual axes therefore do not coincide, but differ by 18° (Testut), or by 20·3° to 23·2° (Lagrange), or 23° to 25° (Druault). The distances between the orbital axes at their anterior extremities is given by Emmert as 60 mm. in males, 58·3 mm. in females, 48 mm. in children. These axes and angles are illustrated in Fig. 45 above.

As regards the measurements given by the various authors cited above, it is noteworthy that, according to Adachi (1904), they were made upon only 20 skulls by Merkel and Kallius, 25 by Emmert, 100 European by Weiss. Adachi himself, in a most careful and detailed examination of Japanese orbits, took 94 skulls of that race and 30 European. Ono (1928) also has written a monograph in which detailed measurements of the orbit in 214 Japanese skulls are tabulated; Kuryrova (1930) describes the form in Czech skulls. Goldnamer (1923) tabulates most authors' measurements.

9. DEVELOPMENT AND GROWTH

Towards the end of the third week of foetal life five processes arise from the base of the primitive cerebral capsule, and by the end of the second month they have united to form the facial part of the head. Of these processes, two are placed on each side and are called the maxillary and mandibular, whilst the remaining process, the nasal (or fronto-nasal), is median in position, and is composed of symmetrical right and left halves, each of which subdivides later into lateral and mesial nasal processes. The chief structures developed from the original five processes are the nose from the nasal, the cheek and upper jaw from the maxillary, and the lower jaw from the mandibular (see Figs. 97, 98, p. 188).

The orbital walls are formed superiorly by the capsule of the fore-brain, in which the frontal bone is developed, laterally and inferiorly by the maxillary process, in which the zygomatic bone and maxilla (with the exception of its frontal process) arise, and medially by the lateral nasal process, which gives origin to the frontal process of the maxilla, nasal, lacrimal, and lateral mass of the ethmoidal bones. Posteriorly the orbit is completed by a separately developed part of the sphenoidal bone derived from the base of the skull; between the primitive pre- and orbito-sphenoidal parts of this bone the optic nerve enters the cavity, and gaps left on each side of its greater wing form the orbital fissures. The naso-lacrimal canal is formed between that part of the lateral nasal process which gives rise to the frontal process of the maxilla and the maxillary process. The approximate number and positions of the centres of ossification are given in the adjacent figure.

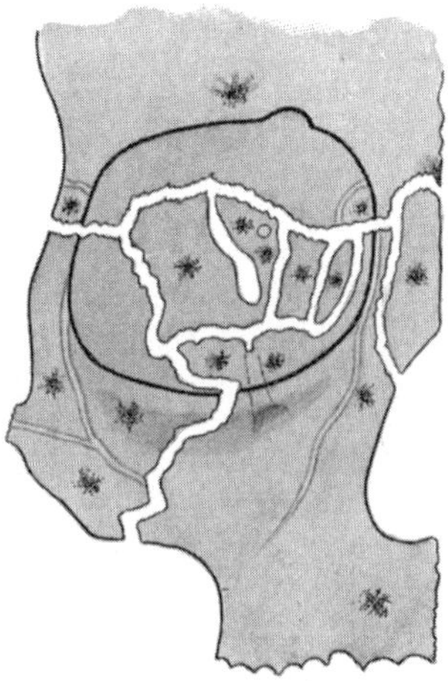

FIG. 46.—Diagram to show the approximate position and number of **centres of ossification** in the bones which form the Orbit. The centres appear between the sixth and eighth weeks of foetal life, and, except for the sphenoidal, have fused into the component bones before birth (6th to 7th month). In very rare cases one of the secondary centres may remain ununited.

Primitively the orbital boundaries must tend to be moulded upon the optic cup, the precursor of the eyeball, since this is formed before the matrices of the bones are laid down; in foetal skulls the margin, for example, is circular up to the age of six or seven months. Later on the various adnexa of the globe cause the orbital shape to be modified in the regions they occupy in the adult. The globe, lacrimal apparatus, and oblique muscles may therefore be considered to be the chief internal factors which determine the form of the orbital cavity; there are also external influences to be noted due to the growth of the surrounding parts, but their effect is more marked after birth. In agreement with the precocious development of the eyeball the dimensions of the orbit are considerable relative to those of the other cavities of the face, and seem to be correlated also to the size of the cranial cavity. Even so, the orbit at first is outpaced in growth by the globe, as seen in the sagittal section of a six-months foetal head in Fig. 47, where the margin is closely applied to, and only extends as far as, the equator of the globe, instead of arching well over and standing away from it as in the adult.

At **birth** the chief features to be noted are as follows: the large size of the orbital entrance, its sharp and strongly developed margin, and the close position of the infra-orbital foramen below it. The infantile margin naturally conforms to that of the primitive form, fitting closely round the globe, and

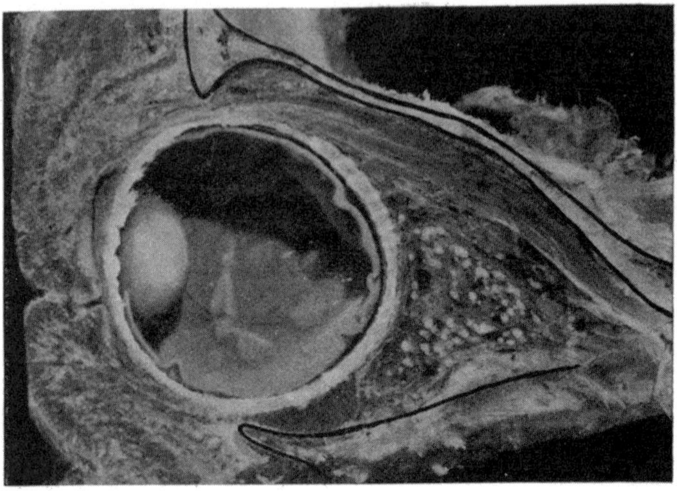

FIG. 47.—Sagittal section through the **Orbit of a 6–7 months foetus,** showing the small extent to which the walls enclose the globe. Note the thickness of the eyelids. ×4.

(The hyaloid artery is faintly seen, but throws a well-marked shadow on to the retina below.)

the development of the surrounding parts has not yet affected it to an appreciable extent; the circular outline tends to be retained, or to recur, during early life up to puberty, after which time the changes occurring in the secondary period of skull growth affect this region to a much more marked degree than the cavity itself. The height and breadth of the entrance at birth are almost equal. The top of the nasal aperture bears the same relation to the orbit at all ages, but its lower border is nearly on a level with the infra-orbital margin at birth, and is much below it in the adult. The orbit bears nearly the same proportion to the cranium at all ages, but at birth the marginal height is about half that of the face, and in the adult rather less than one-third; as the face grows, the vertical diameter of the orbit increases rapidly, so that, according to Merkel, at five years the entrance lacks only two or three millimetres of the adult height, which it gains in the next two years. The bones at birth are still widely separated, and the walls are completed by membrane; the roof is proportionately large, and flat or even convex; the medial wall is low and not definitely separated from the floor, which is likewise small; the greater wing of the sphenoidal bone forms a relatively larger part of the lateral boundary than it does in the adult. The fossa for the lacrimal gland is a deep conical recess, that for the lacrimal sac is cylindrical, and faces anteriorly instead of laterally owing to the large development of the posterior lacrimal crest. The long axis of the cavity is horizontal, instead of being inclined downwards as in the adult; a frontal section behind the margin presents a long oval outline, with a blunt pole directed supero-laterally and a pointed pole directed infero-medially, whereas in the adult a similar section is either circular or square.

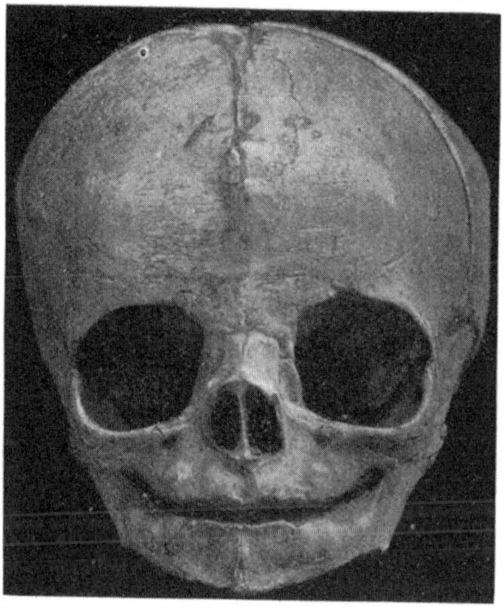

FIG. 48.—The **Skull at Birth.** Note the broad posterior wall of the fossa for the lacrimal sac and the accentuation of the posterior lacrimal crest. × $\frac{2}{3}$.

In the **child** the features noticed at birth are gradually modified by the growth of the adjacent parts. The orbits lie between the bones of the calvaria and those of the face, as shown in Fig. 2; the former are influenced by the expansion of the brain, the latter by the development of the nasal cavity with its accessory air-sinuses and especially of the teeth, the face being fundamentally a masticatory apparatus. Hence there are regions of great developmental activity above, between, and below the orbits, which react upon their form. The margin becomes more oval, but retains in early years, and always in the female, its sharp edge; the directions in which the primitive round opening becomes thus modified can be seen in the frontal plane by drawing a circle round it from the centre, as in Fig. 43. The development of the bridge of the nose carries both margins forward on their adjacent sides. That the shape of the anterior nasal orifice has an influence upon the contour of the medial part of the infra-orbital margin, and so is correlated with the shape of the orbital entrance, will be apparent on a comparison of Figs. 6 and 44 with Fig. 49. On the other hand, the rectangular shape of such orbits as are seen in Fig. 44 suggests the compression of the more delicate facial bones between the calvaria and the mandible through the action of the muscles of mastication in the strong-jawed type of skull. The growth of the nasal fossae and ethmoidal cells, together with the development of that part of the frontal lobe of the brain which rests upon the lamina cribrosa of the ethmoidal bone, increases the width between, and the height of, the medial orbital walls.

A further point of interest is the determination of the period at which growth may be considered to have ceased. The practical importance of this lies in regard to the changes which may follow enucleation of the eyeball. The orbit conforms to the law of adaptation of the organ to the function which it is called upon to fulfil, and if the globe be removed before the orbit is fully formed, growth of the latter is checked. E. Thomson (1900) found in experiments upon rabbits that a uniform deficiency in orbital size amounting to 10 per cent. of all dimensions followed enucleation of the globe. Merkel states that in early childhood the changes occur in as soon as ten weeks after operation, the result being that in the adult the marginal height is reduced, the upper wall is flattened, and the lower slightly raised, and the cavity is narrowed in all its diameters, and more or less facial asymmetry is found according to the stage of growth which had been reached.

The full dimensions of the orbit are variously stated to be attained as early as at three years, or at from five to seven, or not until eighteen or twenty. From the following measurements of a series of skulls from birth to the adult condition, it appears that at three years the dimensions fall considerably short of those found in the adult, at six there is still a difference, whilst at twelve certain measurements approach those of the adult, but others are not fully attained until puberty. Although the eye, like the brain and the ear, attains its full structure and dimensions much earlier than the rest of the body, one would expect to find that the full size of the orbit is not reached until the end of the first and most active period of cranial growth, namely, until puberty.

See Table overleaf. The reader may also refer to the various dimensions given by Koenigstein (1896); as reproduced by Goldnamer (1923), they agree almost exactly with those of the writer.

At **puberty** the sexual differences become marked, the bones in the male being stouter, the supra-orbital ridges more prominent, and the marginal edge, especially superiorly, more rounded. The development of these ridges, and of the frontal sinus above and the maxillary sinus below, throws the margin forward in these two regions.

In the **aged** the margin is flatter, the entrance broader and especially higher than in the adult, owing to atrophy of the bones, a factor which causes also a loss of substance in the walls, chiefly in the lateral; the sutures are obliterated.

10. VARIATIONS, ASYMMETRY, AND DEFORMITIES

The chief variations in the orbit as a whole are found according to the sex and type of skull.

Sexual Differences.—In the female the bones forming the orbit, as elsewhere in the skeleton, show a greater delicacy of construction; there is a tendency towards the retention of the early circular margin, de Wecker finding perfectly round openings in female skulls only. This opening is larger relative to the face, being broader by 1·5 mm. and

Age.	Margin.		Intra-orbital Height.	Trochlear Height.	Depth.				Marginal Obliques.
	Height.	Width.			Roof.	Floor.	Lateral Wall.	Medial Wall.	
6½ months f.	15	16	17	10	23	19·5	20	..	19 by 15
8 ,, f.	17·5	22	..	14	23	10	9	10	23 ,, 19·5
Birth	18	21	22	14	32	25	27·5	22·5	22 ,, 19
1½ years	30	32	33	22	46	40	38	36·5	33 ,, 31
3 ,,	28	32	30·5	23	42	38	37·5	33·5	35 ,, 27
4½ ,,	28	29	30	20	41	41·5	36·5	37	33 ,, 30·5
5 ,,	30	33	33	26	43	42·5	41	35·5	35 ,, 30
6 ,,	29	32	32	25	47	47	44	38·5	33 ,, 39
7 ,,	30·5	33	35	25	46	46	42·5	39	36 ,, 32
12 ,,	33	34	34	27	45	39	38·5	34	40 ,, 34
16 ,,	37	38	38	26	51	43	45	39	42 ,, 35
19 ,,	34	36	38	29	51	49	46	40	40 ,, 33·5
22 ,,	36	37	38	31	48	45·5	43	39	41 ,, 35
Adult M.	33	39	39	29	49·5	52·5	47	39	42 ,, 36
Adult F.	34	36	38	26	45	43·5	42	34	34 ,, 39

NOTES

(i) The measurements are to the nearest half-millimetres, and were taken in right orbits only.
(ii) The 'intra-orbital height' is the greatest height of the cavity, just within the margin.
(iii) The 'trochlear height' is taken as explained on p. 106, and shown in Fig. 50.
(iv) The marginal width is taken medially to Flower's 'lacrimal point' (where the upper end of the posterior lacrimal crest meets the frontal bone).
(v) The 'marginal obliques' are from the infero-lateral to supero-medial × supero-lateral to infero-medial angles.

higher by 0·5 or 0·7 mm., and the orbital index is greater than in the male; the edge is sharper, especially in the supra-

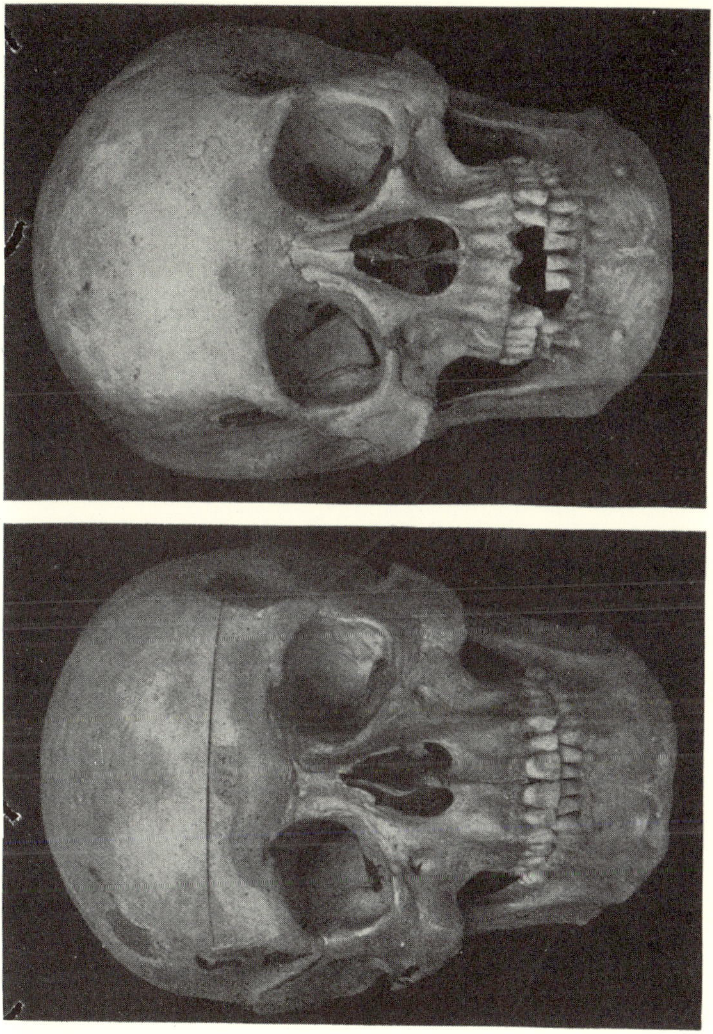

Fig. 49.—Male and Female Skulls, showing **sexual differences** in the orbital margins. Note in the male the influence on the orbital contour of the comparatively stronger framework of the facial bones supporting the teeth, correlated with greater upward pressure of the lower jaw. × about $\frac{1}{2}$.

orbital region where the ridges are less developed or even absent, and the lateral angular process is more slender and pointed than in the male, features which are characteristic differences between the sexes. The female orbit is more elongated and its cavity relatively larger (Merkel).

In the male the margin as a whole is more strongly developed, so affording greater protection to the eyeball, and is correlated with the more powerfully developed jaws. The walls are longer, the upper and lower are more curved owing to the greater prominence of their margins; the margin is less flat and the axial distance is consequently longer; the absolute depth of the cavity is greater, but relative to the size of the entrance and length of the walls it is less.

Type of Skull.—It is said that in the long or dolichocephalic skull the margin is more circular, the naso-malar angle and also the orbital index less, but the cavity deeper than in the short or brachycephalic type, where the margin is oval. The shape of the orbit has been considered to have an influence upon the power of vision, deep orbits being associated with myopic eyes and shallow ones with hypermetropic, but this is denied by Roy (1919). Also Stilling (1888) stated that short-sightedness was commonly associated with a low orbital entrance, the explanation suggested being that in such microsemic skulls the superior oblique muscle was placed at a lower level from the orbital floor than usual, and by compressing the globe caused it to become elongated in an antero-posterior direction with consequent myopia. His theory was supported by Seggel (1890) and Ask (1906), but disputed by Schmidt-Rimpler (1889), who found nearly the same orbital index in myopic and emmetropic subjects, and Hamburger (1904), who observed that on experimental raising of the intra-ocular pressure by injection of the globe in the cadaver there is no furrow caused by the tendon of the muscle. Moreover, the position of the pulley of the superior oblique muscle, that is, the point from which its action in compressing the globe against the inferior oblique muscle would originate, does not appear to be correlated to either the height, shape, or index of the orbital entrance; indeed the distance between the fovea trochlearis, where the pulley is attached, and the site of origin of the inferior oblique muscle, both of which points lie in the same vertical plane and deep to the actual margin, was found remarkably constant in 30 skulls of different type and race, and much less variable (28–30 mm.) than the height of the orbit either at the margin (30·3–38·5 mm.) or at the roomiest part of the cavity (35–43 mm.). This 'inter-oblique distance' or 'trochlear height' of about 29 mm. may be compared with the vertical height of the globe, 23·5 mm. (Whitnall, 1913).

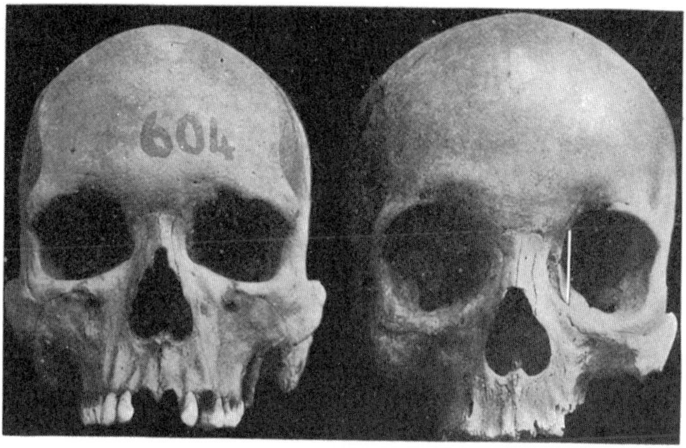

FIG. 50.—Two Skulls, to illustrate criticism on **Stilling's theory** of influence of orbital height upon shape of globe. × about ½.

Difference between marginal orbital heights (31 mm. in first skull, 38 mm. in second) . . . . . . = 7 mm.
Difference between greatest intra-marginal heights . . = 2·5 mm.
Difference between 'inter-oblique distances' (as shown by line in second figure, which was 28 mm. in each skull) . = 0.

Birnbaum (1915), by taking exophthalmometric measurements of normal eyes, found that prominent eyes coincide with large and oval orbital openings, whilst the globes are deeply seated when the openings are small and round. (See also p. 258.)

Asymmetry.—There is usually a slight difference in the shape and position of the orbits and their contents, due, according to Elliot Smith (1907), to an unequal development of the frontal areas of the brain. This inequality is only part of a general asymmetry which is a characteristic difference between man, especially in the higher races, and apes. Ambialet found that the right or left orbit indifferently is displaced, and Dwight states that one orbit and cheek are commonly anterior to the other, whilst the right is usually placed higher. The margin of the right orbit is said to be weaker and sharper than that of the left. Emmert found that the absolute depth of the left orbit is less than that of the right. Tubby (1919) draws attention to the asymmetrical growth of the skull, especially marked in the face and orbits, secondary to congenital torticollis and leading to the condition known as 'ocular torticollis' (Stephenson, 1913); in a right-sided torticollis the superciliary ridges are not level, and the right orbital aperture is lower and often smaller than the left. Also Lowman (1918) and Lloyd Mills (1919) have written on the effects of faulty skeletal alinement upon the eyes, such as exophoria, and remediable by appropriate orthopaedic treatment of the primary cause. The latter writers found, indeed, that about three-quarters of fifty eye cases met with in regular orthopaedic practice showed marked muscular imbalance of the eyes. The effects of such imbalance upon the facial expression are noted by Stevens (1892).

Deformities, due either to artificial causes, as in some Indian skulls where compression of the brain causes a slight protrusion of the apex of the lateral wall of the orbit, or to pathological processes, may be found. Of the latter, the condition known as Oxycephaly ('Tower skull', Turmschädel, &c.) is interesting. It is a deformity of the skull, probably of developmental origin, and due to premature fusion of certain sutures of the vault and base. Characterized by a high-pointed dome, it is nearly always accompanied by exophthalmos, from shortening of the orbital cavity, and gradual impairment of vision from increased intracranial pressure and traction on the optic nerve. The premature synostosis is held by some writers to affect the sagittal suture of the vault, by others the transverse ones, or it may be present in the base only. The condition may be recognizable at birth, with heredity as a factor and accompanied by deformities of the extremities or be delayed until childhood and confined to the skull.

First described by Von Graefe in 1866, about an hundred cases have since been described in the literature. Power (1894), Dock (1919), Swanzy (1900, with bibliography to date), Dorfmann (1908), Brav (1912), Holloway (1915), Sharpe (1916), Goldenburg (1918) may be referred to. Later papers of note are by Sutherland (1922), Davis (1925), Mirimanoff (1925), and especially Greig (1924, 1926) with full bibliography.

PART II

THE EYELIDS

1. The Eyebrows.
2. The Eyelids.
3. The Conjunctiva.
4. The Lacrimal Apparatus.

1. THE EYEBROWS

THE eyebrows (*supercilia*) are formed by muscle fibres and thick skin, surmounted by strong hairs; they are not facial but cranial appendages, being a modified part of the hairy scalp, the continuity with which is indicated both structurally and also by an accentuation of down or fine hair on the region between the temple and the end of the eyebrow. Each has a head, a body, and a tail or lateral extremity, and in extent is limited by the hairs which it bears (Fig. 53). The heads are usually separated by an 'inter-superciliary region' corresponding to the *glabella* (from the Latin *glaber*, smooth, hairless). The exact position is between the supra-orbital ridge and the orbital margin, but they are a little less curved than the latter, and in abnormal cases lie at a higher level. They overlie the frontal sinuses, and on a deeper plane the second frontal convolutions of the brain; the tail lies upon the lateral angular process and extends to the zygomatico-frontal suture; they are traversed by the supra-orbital vessels and nerves. They serve to divert the sweat of the brow from running into the eyes, and shade them from strong light, and they add greatly to the expression of the face.

In each eyebrow four principal layers can be identified in a mid-vertical section. These are:

(i) The **skin,** intimately united to a dense superficial fascia.
(ii) A **muscle** layer, the fibres of which end in the skin.
(iii) A **cellulo-adipose** layer.
(iv) The **galea aponeurotica** (epicranial aponeurosis).

So constituted by the close union of these layers, the eyebrow glides freely over the periosteum of the bone, from which it is separated by a loose cellular 'sub-epicranial tissue', an arrangement that is the exact counterpart of the scalp proper (Fig. 51).

(i) The **skin** is thicker than in the surrounding regions, and is rich in sebaceous and sweat glands and large hair follicles, between which are numerous striated muscle fibres. The hairs are of interest as being the first to develop in the embryo, appearing at three months (Contino, 1907); generally they

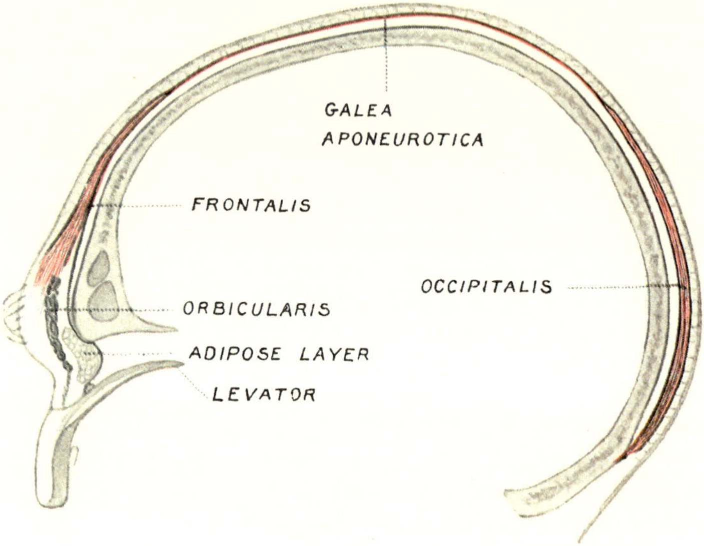

FIG. 51.—The Structure of the **Eyebrow** and its connexion with the **Scalp.**

are directed obliquely and laterally, but those of the eyebrow are often inclined forwards and upwards, owing to the action of the corrugator supercilii muscle; there may therefore be a break in the stream of the hairs, the more frequent occurrence of which in the left eyebrow (153 out of 200) is ascribed by Kidd (1904) to the greater strain placed upon the left muscle in our practice of reading from left to right. The occasional presence of a tuft of longer hairs near the middle recalls the tactile hairs or 'supra-orbital vibrissae' seen in many mammals, and the well-marked sensitiveness to touch of the hairs of the eyebrow is doubtless due to the presence amidst their roots of large sensory bulbs, such as are present in lower animals and have been found in the human foetus by Friedenthal (Keith, 1913). The hairs are more numerous, larger, and more irregularly planted in males than in females, and are less developed in northern than in southern races, in which latter they usually meet in the mid-line. Some of them become markedly long at the middle period of life. The colour is generally the same as that of the hair of the scalp, but may be lighter in tint; it may differ on the two sides, and instances of piebald or wholly white (in albinos), and, indeed, of absence of the eyebrows are recorded. The superficial fascia is composed of dense connective tissue containing fat lobules in its meshes, and it intimately unites the skin to the muscle layer.

(ii) The **muscular layer** is comprised by three orders of fasciculi: the vertically directed fibres of the frontalis, the concentrically arranged fasciculi of the orbicularis oculi, and the obliquely placed corrugator supercilii. These fibres intermingle and are inseparable at their insertion into the skin, where they can be traced through the dense superficial fascia to end between the sweat glands and hair follicles.

The Frontales muscles are the anterior bellies of the *epicranius* or *occipito-frontalis*. This muscular and aponeurotic sheet covers the whole cranium, and is differentiated into two flat posterior bellies, the occipitales muscles, attached to the superior curved line of the occipital bone behind, an intervening tendinous portion, the *galea aponeurotica*, and two anterior bellies, the frontales muscles. The latter have no bony attachments, but blending together at their medial borders extend across the full width of the forehead, and at the supra-orbital margin each intermingles with the corresponding orbicularis oculi to be inserted into the skin of the eyebrow. The medial parts of the muscles between the eyebrows appear to extend downwards to gain attachment to the nasal bones, but these fibres really constitute an independent muscle, the *procerus* or *pyramidalis nasi*, as shown by Duchenne, who termed it the 'muscle of aggression'; it is wrongly classed as part of the frontalis, since comparative anatomy shows that it was derived from the muscle which raises the lip (Barker); it has, moreover, an innervation from the infra-orbital branch of the facial nerve distinct from that of the frontalis (Frohse, 1895), and it contracts synergetically with the corrugator and orbicularis in the presence of strong light. It may be absent. The action is described on p. 184.

The Orbicularis Oculi (*orbicularis palpebrarum*) muscle is a nearly circular sheet of fasciculi placed over each eye, covering the eyelids and orbital margin. The muscle is described at length in connexion with the eyelids, and is figured on p. 125; here it is sufficient to note that it is the upper fibres of the peripheral or orbital part that enter into the composition of the eyebrow, where they mingle with, though lying mainly superficial to, the frontalis. Elsewhere round the orbital circumference the skin can readily be dissected from the orbicularis to display its extent and contour, but in the eyebrow the blending with the frontalis is so intricate and its connexions with the skin so intimate that neither muscle can be isolated as a separate entity.

The Corrugator Supercilii muscle lies deep to the above-mentioned muscles: it is seen upon reflecting the whole eyebrow downwards from the bone in the form of a definite and coarser band of muscle fibres arising from the most medial part of the supra-orbital arcade (the site being often marked on the bone), and running laterally for 2 or 3 cm. to blend with the overlying orbicularis; it is inserted into the skin of the medial half of the eyebrow, the site being marked by a depression when the muscle is contracted (Fig. 96, p. 182). Sometimes distinguished as a separate muscle, the 'depressor of the head of the eyebrow' (Arlt, 1863), it is more generally regarded as a special fasciculus of the orbicularis (Ruge, Henle, Merkel, Virchow, Kidd), though its origin is quite separate, and it can definitely be isolated for most of its course, as seen in the figure opposite. Kudo (1919), however, in a study of the facial muscles of the Japanese, finds that it is frequently inseparable, as indeed Macalister stated it to be in Europeans. The corrugator is differentiated late in the human embryo, after all the other muscles of the face (Futamura, 1906); it may be absent, in which case the eyebrows remain level on lowering the brows and a frown does not appear. Its action is described on p. 183.

(iii) The **cellulo-adipose layer** underlies the muscles, and facilitates the movement of the eyebrow over the orbital margin; it is fairly loose, and contains fat in its meshes; it is abundant in the lower part of the eyebrow, where Charpy (1909) describes it as forming *le coussinet adipeux du sourcil*, an elongated fatty mass 1 cm. high and 5 mm. thick, placed over the supra-orbital margin and extending a little downwards into the eyelid, but it does not always appear so definite a structure as he describes.

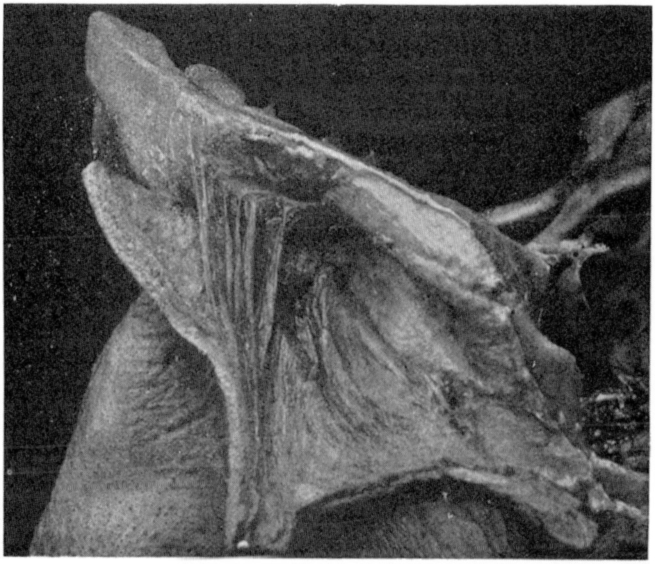

FIG. 52.—Dissection to expose the **Corrugator Supercilii Muscle.** Left orbit seen from the side. The skin has been turned down from the supra-orbital margin carrying the muscle with it; the origin of the muscle from the bone medial to the supra-orbital notch and separate from the rest of the orbicularis is clearly seen. Natural size. Compare with Fig. 96, p. 182.

(iv) The **Galea Aponeurotica** (epicranial aponeurosis): the aponeurotic central portion of this structure splits anteriorly to enclose the frontalis muscle; the superficial layer is very thin and passes with the muscle fibres to the skin of the eyebrow, whilst the deeper and more definite layer passes beneath the muscles and is firmly attached to the supra-orbital margin (Fig. 51).

These four layers—skin, muscle, connective tissue, and

aponeurosis—then constitute the eyebrow proper, and their close union is exactly similar to what is found in the scalp proper, of which this region is, as noted above, merely a specialized portion. In stripping the scalp off the cranial vault the reflection can be carried forwards to the supra-orbital margin, taking the eyebrow with it, and the plane of separation is in the 'sub-epicranial' layer of delicate connective tissue lying beneath and facilitating movement of the scalp over the periosteally covered bone; it is in this space that a traumatic haematoma, suppuration, or emphysema of the frontal sinus extends, and such sub-aponeurotic effusions are prevented from passing down into the eyelids by the above-mentioned attachment of the deeper layer of the aponeurosis to the supra-orbital margin. Charpy (1911) has also demonstrated by means of gelatine injections that the barrier formed by the dense eyebrow mass prevents fluids from extending upwards from the eyelids onto the brow (Fig. 70, p. 143).

The blood-vessels of the superciliary region will be described in detail together with those of the eyelids (p. 166); here it may briefly be noted that there are two chief arteries in this region, the supra-orbital, which passes directly upwards from the supra-orbital notch about an inch from the mid-plane and lies deep to the eyebrow against the bone, and the superficial temporal, which runs parallel with and on the temporal side of the tail of the eyebrow, though some distance from it. The supra-orbital vein does not accompany the artery of the same name upwards over the brow, but runs along the lower border of the eyebrow and forms a marginal arcade (*vena marginalis*), lying deep to the orbicularis. There is, however, a frontal vein which passes directly upwards beneath the head of the eyebrow (p. 169).

The lymphatic vessels drain chiefly into the parotid nodes, but a small number from the head of the eyebrow join those of the medial half of the eyelids and pass directly into the sub-maxillary nodes (see Fig. 92, p. 175).

The muscles of the eyebrows are innervated by the facial nerve, in which respect they present certain points of clinical interest which, together with their actions, will be considered later (pp. 183, 186). The sensory supply to the skin is derived chiefly from the supra-orbital nerve, which accompanies the artery in its course, and, like it, would be severed by a deep horizontal incision along the eyebrow. The head of the eyebrow receives filaments from the external nasal nerve, the tail from the lacrimal nerve; each part may have a further supply, the head by ascending branches of the infra-orbital nerve, the tail from the zygomatico-temporal nerves; all of these are sensory branches of the trigeminal nerve.

2. THE EYELIDS

The eyelids (*palpebrae,* from the Latin *palpare,* to stroke) are two musculo-membranous folds of modified skin placed in front of the orbit, which they shut in anteriorly; they are closely applied to the globe of the eye, for which they act as organs of protection, and they also serve as barriers to the orbital contents between it and the orbital margin. It may be noted, moreover, that they fulfil the role of resting the retina and brain from the rays of light which would otherwise create cortical pictures and maintain cerebral activity; it is only when the eyelids are closed that the visual area of the brain obtains rest. They are distinguished as superior and inferior. The former is more extensive and more mobile than the latter, covering the greater part of the anterior surface of the globe, and being provided with a special elevator muscle. In each eyelid there are to be noted two faces, anterior and posterior, two borders, adherent and free, and two extremities, medial and lateral, uniting to form two corresponding commissures (*commissura palpebrarum*). They are opposed by their free borders or margins, which limit between them a transverse opening, the palpebral fissure or *rima palpebrarum*; the borders of this opening join together at each side at an acute angle to form the medial and lateral *canthi* or 'angles' of the eye (*angulus medialis vel lateralis oculi*). (It is preferable to apply the term 'canthus' to the angle of the opening between the lids, and the name 'commissure' to the actual conjoint ends of the lids themselves.)

They are modified folds of skin, in the thickness of which

is developed a thin sheet of muscle fibres, the orbicularis oculi; the deeper skin layer becomes converted into a mucous membrane, which lines the lids and is reflected on to the anterior part of the globe, and is called the 'conjunctiva'. The membrane is kept moist by the secretion of the lacrimal gland, which is an outgrowth from it, and excess fluid is drained away through a series of channels, the lacrimal passages. The normal skin elements of the free edges of these folds, that is, of the lid margins, become greatly developed for protective and sensory purposes; the hairs are developed as 'cilia' or eyelashes, the sebaceous glands become huge in size and are called 'tarsal or meibomian glands', and the connective tissue supporting them is condensed to form the 'tarsal plates'.

The description of the eyelids will be in the following order:

(1) The Surface Anatomy,
(2) Structure and Muscles,
(3) Vessels,
(4) Nerve-supply,
(5) Movements and Action,
(6) Development, and
(7) Abnormalities.

The conjunctiva will be described in a separate section, as also the lacrimal apparatus.

(1) **The Surface Anatomy.**—The eyelids pass normally into the neighbouring parts of the face without obvious line of demarcation, but their limits are clearly outlined in pathological conditions such as oedema, since the extravasation spreads beneath the skin in the loose connective tissue devoid of fat which represents the superficial fascia of the lids, whereas in the surrounding region of the cheek or eyebrow the corresponding tissue is dense and fatty and connected to the periosteum; thus defined, the upper lid extends to the lower edge of the eyebrow and coincides with the supra-orbital margin, but in the lower lid a swelling can be seen to extend beyond the bony margin as low as the naso-jugal and malar skin-folds.

The form of the lids is essentially correlated to the volume and position of the globe, though also partly dependent upon the shape of the orbital entrance. In infants and subjects of delicate skin the prominence of the cornea is observable through the closed lids. The margin of the upper tarsal plate can readily be felt with the finger.

The **colour** of the skin is generally in harmony with that of the face, but the redness of the cheek does not extend beyond the skin-folds noted above as limiting the extent of the lower lid region, nor usually above the level of the lateral canthus; the lower lid is generally darker, but the tint varies according to health and local conditions. The 'palpebral areola,' which may be black or white, and is so noticeable in fatigue, is attributed by Waldeyer to a reflex of the lymph from the connective tissue spaces beneath the skin, but it is also caused by venous stasis, the vessels showing up through the delicate skin; it is usually most marked in the lateral part of the lower lid region, where, it may be noted, there are no cutaneous arteries of any size. A permanent darker tint commonly noticeable in the medial part of the lower lid or sometimes at the lateral commissure is also explained by venous stasis, but the migration of definite pigment cells (chromatophores) to this region has been described (Dor, 1900), and in many cases there is no doubt as to the presence of pigment in the skin (p. 130); anomalies of such pigmentation are described by Adrogue (1929).

The main **folds** or wrinkles of the skin in and around the eyelids are of topographical interest. They may be folds of movement or structure. In each eyelid there is a horizontally curved 'palpebral' or orbito-palpebral infold (*sulcus orbito palpebralis*), distinguished as superior and inferior according to the lid, and dividing it into tarsal and orbital areas: the tarsal region overlies the tarsal plate and the globe, the orbital is related to the space between the latter and the bony margin. Sappey states that these folds exactly superimpose the lines of reflection of the conjunctiva onto the globe (the 'fornices'), but according to Charpy they indicate the limit between the smooth and folded parts of that membrane. The superior palpebral fold is the deeper; it lies 2 or 3 mm. above the highest point of the cutaneous insertion of the levator palpebrae superioris, the contraction of which muscle causes it to become deeply recessed; the orbicularis oculi muscle is thinnest along this line. The skin of the orbital region of the lid above it often sags forward in middle age when it has lost its elasticity, and forms a fold ('Deckfalte' or 'pli de recouvrement' of foreign authors) which covers the tarsal region even as low down as the eyelashes; it is most marked on the lateral

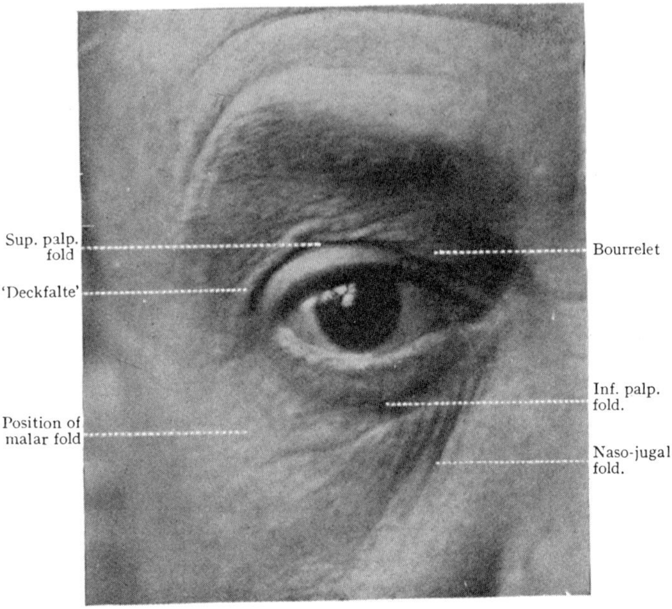

FIG. 53.—The Eyelids and **Palpebral Folds.** Natural size.

side, and may even form a kind of lateral epicanthus in old subjects (*ptosis adiposa*); the fold is obliterated on drawing the eyebrow up. In the medial corner of this region, particularly in the aged when the orbicularis muscle has become weak, there commonly presents a small swelling, the 'bourrelet sénile', or 'palpebral hernia', caused by a hernial protrusion of the orbital fat; such a swelling can be produced or accentuated by pressing the eyeball backwards and towards the nose, the lobe of fat which forms it is shown in Fig. 54. The writer has seen cases where the skin over the lump appeared yellow from the underlying fat.

The volume of the lacrimal gland has no influence on the configuration of the lateral part of the upper eyelid, though it lies immediately behind it. The inferior palpebral fold is well marked in infancy, but may be slight or absent in the adult.

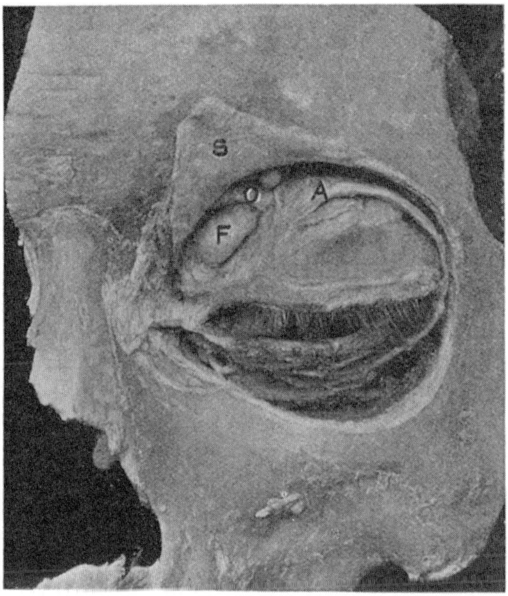

FIG. 54.—The **palpebral** ***hernia adiposa***. Dissection of Left Orbit, seen from in front after removal of the skin, *orbicularis oculi* muscle, and *septum orbitale*; a part of the septum (S.) has been left and turned up. The orbital fat (F.) is seen protruding below the reflected tendon of the superior oblique muscle (O.), and between the orbital margin and the medial extremity of the aponeurosis (A.) of the *levator palpebrae superioris* muscle. The swelling or 'bourrelet' caused in life by this hernial protrusion of the orbital fat is shown in Fig. 53. Above the tendon of the superior oblique is seen a loop formed by the terminal part of the ophthalmic artery. Natural size.

Below the lower lid two wrinkles are usually present on the cheek: a 'naso-jugal' fold running downwards and laterally from the medial commissure, and always well marked, and a 'malar' descending from the opposite side, but often ill-defined; these tend to become continuous with one another, forming the 'Wangenlidfurche' of Arlt or 'sillon palpébro-génien' of Charpy. They are not only folds of movement, seen when the cheek is raised, for example as in laughter, but are structural in that they indicate the differentiation between the loose tissue beneath the skin of the lower lid region and the denser fatty superficial fascia of the cheek; hence they are accentuated by effusion into the lower lid. They also become more marked in ocular fatigue or in the aged when the orbicularis oculi muscle has lost its tone. The naso-jugal fold, moreover, corresponds to a fascial interstice fixed to the bone between the orbicularis oculi and angular head of the quad-

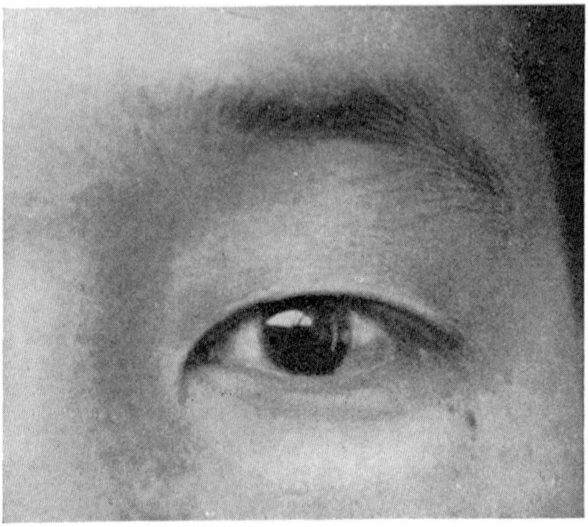

FIG. 55.—Mongolian Eye, showing the ***epicanthus*** or Mongolian skin-fold.

ratus labii superioris (levator labii superioris alaeque nasi) muscles, and along its line there run the facial vein and artery, while its upper part overlies the lacrimal sac (Fig. 57, p. 125).

Another fold is known as the *epicanthus* or Mongolian fold; it is formed by the skin alone of the medial part of the orbital region of the upper lid, and it curves downwards, covering the medial canthus (as the name implies) and joining the naso-jugal furrow of the lower lid. It is generally considered to be characteristic of the Mongolian eye, in which it has been recently studied by Bloch (1911), Forster (1919), and Gifford (1928), who denies that it is a feature of race; for it is present in all races in foetal life (Keith), and is seen in infants, disappearing gradually as the bridge of the nose develops: Metchnikoff (1874), indeed, considered it to be only the persistence of a character peculiar to this period of life and dependent upon the flatness of the nasal bones, as Gifford agrees. Duckworth (1904), however, notes its absence in negroid races, where these bones are even flatter and smaller than those of Mongols, though Testut states that it occurs sporadically in Hottentots; further, Deniker has noted its absence in the young but presence in the adult Kalmouk. An additional feature of such eyelids (but seen in others) is a kind of turgidity, of the upper one especially, which has the effect of turning in the palpebral margins. Epicanthus has been found associated with congenital ptosis (Pockley, 1919).

Other minor skin-folds will be found described in a special article by Charpy (1910).

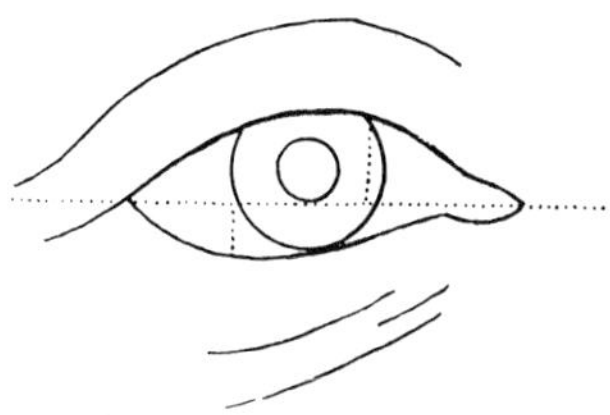

FIG. 56.—The Palpebral Opening, showing the **curves of the lid margins.** Compare with Fig. 53.

The **palpebral opening** (*rima* or *fissura palpebrarum, rictus oculi*) is limited above and below by the free margins of the lids, and at its extremities by their commissures. When the eye is wide open the curves of the margins are not symmetrical, that of the upper lid being more pronounced than the lower, and its greatest height being nearer the medial angle, whereas that of the lower lid is nearer the lateral angle, as shown in Fig. 56; the radius of the upper curve is 16·5 mm., that of the lower 22 mm. (Merkel). The long axis of the opening is not quite horizontal, the lateral extremity being about 4 mm. higher than the medial; in some of the Mongoloid races an obliquity of 50° from the horizontal may be found, although Gifford (1928) considers that the common idea of this being a racial feature is exaggerated, the slope being emphasized by the presence of the epicanthus and the narrowness of the fissure, the latter being really characteristic of such eyes. The opening is given by Luschka, Merkel, &c., as 30 mm. long by 15 mm. high in the normal adult, but less dimensions (28 × 12–14) are more usually found; it is smaller in females and rounder in children, and much individual variation is seen; it is larger in equatorial than in northern races. When the eye is open, the upper lid covers the corneal edge by 3 to 4 mm., and the lower lid just misses its lower margin (Fig. 53). It exposes nearly one-fifth of the surface of the globe, and, it should be noted, represents the mouth of the conjunctival sac. The lateral angle is more acute than the medial, and lies in close contact with the globe; the medial angle extends towards the nose, and is from 5 to 7 mm. away from the globe, being separated by the caruncula and the plica semilunaris. The narrowing of the palpebral opening in both directions on exposure of the eye to glaring light is well displayed in Fig. 96, p. 182. When the lids are closed the opening becomes a fissure, which is not quite straight but forms an elongated and flattened **S**-shaped curve; in relation to the globe it corresponds to the inferior border of the cornea, which is therefore entirely covered by the upper lid when the eye is shut.

The **free margin** of each lid is divided at about the junction of the medial $\frac{1}{6}$th with the lateral $\frac{5}{6}$ths of its length into two definite parts by a small tubercle, the *papilla lacrimalis*, which presents on its summit an orifice, the *punctum lacrimale*. The papilla of the lower lid is directed upwards and slightly inwards, and appears more prominent, whereas that of the upper lid is hidden and the contour of the whole upper margin is uninterrupted when viewed from the front, as seen in Fig. 53; the papillae are best seen by eversion of the lid edge. The medial or 'lacrimal' part of the lid margin (so called from its containing buried in its thickness the lacrimal canaliculus) extends from the papilla to the medial canthus, and is about 5 to 8 mm. long; its surface is rounded, smooth, and devoid of eyelashes; it unites with its fellow of the opposite lid in a semi-elliptical curve, so describing an area, the *lacus lacrimalis*, the floor of which presents a small reddish mass, the *caruncula lacrimalis*, partly lying on a thin fold of mucous membrane, the *plica semilunaris conjunctivae*. The lateral, bulbar, or 'ciliary' part of the lid margin extends from the papilla to the lateral canthus; it is thicker, sharper cut, and more closely related to the globe than the lacrimal part, and it bears the eyelashes. Further details are given on p. 159.

The **deep relations** of the lids: the lateral angle of the palpebral opening lies 10 mm. below the line of the zygomatico-frontal suture and from 5 to 7 mm. distant from the orbital margin; the medial angle lies a little below the fronto-maxillary suture (Fig. 9, p. 21). On dissecting away the superficial parts of the lids (orbicularis oculi muscle and septum orbitale), a view of the contents of the orbit is obtained such as is shown in Fig. 59. The anterior surface of the globe is covered above by the levator muscle expansion and superior tarsal plate, and below by the inferior tarsal plate and orbital fascia; the space between the globe and orbital walls is occupied by masses of fat, removed in the above-mentioned figure to show the position of more important structures. The expansion of the levator extends behind the whole width of the upper lid, the lacrimal gland lies behind its lateral quarter, the pulley and reflected tendon of the superior oblique muscle behind its medial part; the origin of the inferior oblique is situated behind the medial quarter of the lower lid, but the muscle itself recedes deeply away from the mid-point; the lacrimal sac lies in its fossa behind the medial commissure. If the lids, therefore, be divided into quadrants by a vertical line crossing the centre of a horizontal one drawn between the commissures, important structures are found to lie behind each quadrant, except the infero-lateral one, where there is only a mass of orbital fat, and here the finger can be pushed deepest into the orbit between its bony margin and the eyeball.

(2) **The Structure and Muscles of the Eyelids.**—There may be distinguished in each lid four layers or planes of tissue, which are from before backwards:

(i) **Cutaneous.**
(ii) **Muscular** (the orbicularis oculi muscle).
(iii) **Fibrous** (the tarsal plate and the septum orbitale).
(iv) **Conjunctival** (the membrane lining the lids and reflected from them onto the globe).

The tarsal plates form the skeleton of the lids; at each commissure their extremities unite to form medial and lateral palpebral or tarsal ligaments, by means of which they are attached to the corresponding orbital margins at about their mid-points, the medial ligament splitting to embrace the lacrimal fossa. The septum orbitale is a thin fibrous membrane arising from the whole circumference of the orbital margin and extending inwards towards the peripheral margins of the tarsal plates, thus completing the plane of this central fibrous layer. The orbicularis oculi sheet of muscle fibres is laid over these two structures, the septum orbitale appearing as its deep fascia. The conjunctival membrane lines the deep surface of the tarsal plates and is reflected from them onto the anterior surface of the globe, so uniting the eyelids to it, and the fold of reflection is called the *fornix conjunctivae.* Further, the upper eyelid contains and is rendered more complex by the insertions of a special muscle, the levator palpebrae superioris, which forms an additional structural layer intervening between the orbicularis and the tarsal plate; it passes from within the orbit over the summit of the tarsal plate, carrying before it and uniting with the lower dependent edge of the septum orbitale, and its terminal fibres reach the skin by traversing the orbicularis muscle; hence the levator cuts the fibrous layer of the upper lid in two, whereas in the lower lid the septum orbitale reaches the tarsal plate. Finally, in each lid there is a thin sheet of smooth muscle fibres, the palpebral involuntary muscle, attached to the margin of the tarsal plate, and passing thence to the levator muscle in the upper lid and towards the inferior rectus muscle of the globe in the lower lid (cf. p. 145).

This complex architecture of the lids will best be understood by following the steps of a **dissection** from the superficial aspect, as represented in Figs. 57, 58, 59, 60. The skin can easily be raised from the underlying orbicularis, and in doing so it will be seen that on the lateral side and below the orbit the only attachment between the two is by means of connective-tissue septa which pass between the fasciculi of the muscle, whilst in the eyebrow region and below the medial angle of the lower lid cutaneous insertions of actual muscle fibres have to be cut; moreover, the skin may be found adherent to the exposed lower edge of the medial palpebral ligament at the medial commissure. The marked difference between its thinness and delicacy over the eyelids and its thickness and coarseness in the eyebrow and cheek regions is appreciated on carrying the dissection beyond the limits of the lids.

The orbicularis can next be removed with care in one sheet, commencing the reflection on the lateral side (see Fig. 64); in the circum-orbital region the muscle will be found connected

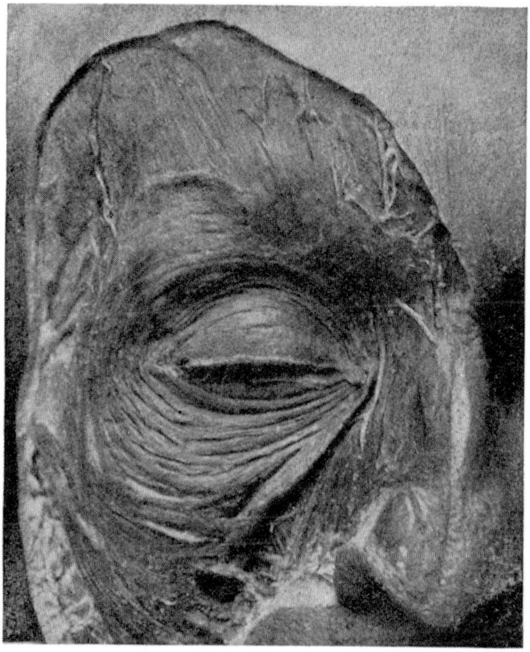

FIG. 57.—**Dissection of the Eyelids:** first stage. The skin only has been removed, showing the *orbicularis oculi* sheet of muscle fibres, which should be compared with Fig. 62 for names of constituent parts. Note the angular vein, running in the cleft between the *orbicularis oculi* and the *levator labii superioris alaeque nasi* or *caput angulare* of the quadratus muscle, along the line of the naso-jugal skin-fold. Natural size.

to the underlying bony surface by a dense felting of connective tissue; over the septal region the reflection is difficult owing to the close adherence of the underlying septum orbitale, which is so closely attached to the deep surface of the muscle as to resemble a fascial lining; in the tarsal part of the lid the muscle is more easily lifted. Near the head of the eyebrow the corrugator supercilii muscle will require severing at its origin from the bone, and from this point onwards towards and below the medial angle of the eye a broad line of muscle fibres has to be cut, marking the osseous attachment of the orbicularis itself. In reflecting the muscle sheet downwards from the superior tarsal plate a succession of fine con-

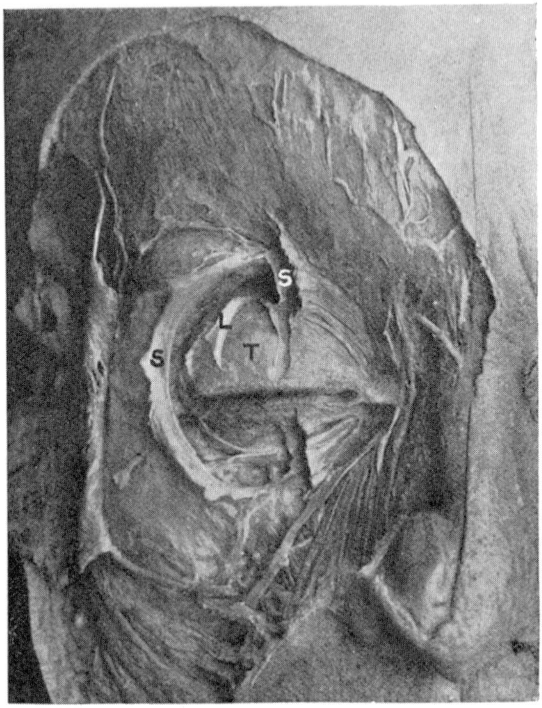

FIG. 58.—**Dissection of the Eyelids:** second stage. The lateral half of the *orbicularis oculi* has been turned back over the temple, the medial half is left *in situ*; the lateral half of the underlying *septum orbitale* (S.) has likewise been turned aside, and the part left shows its connexion with the aponeurosis of the *levator palpebrae superioris* (L.); the latter has been split and turned aside to expose the tarsal plate (T.).

nective-tissue strands are severed; these are the terminal aponeurotic fibres of insertion of the levator, which radiate forwards and traverse the fasciculi of the orbicularis to reach the skin. When cut, they collapse to form a definite layer of fascia, which has to be raised to expose the superficial surface of the tarsal plate (Fig. 72, p. 145). On lifting this fascial layer further upwards, it will be seen to carry the septum orbitale with and in front of it, and the fact is thus demonstrated that the cutaneous insertion of the levator intervenes between the septum and the upper margin of the tarsal plate, and therefore the two last are not continuous with one another,

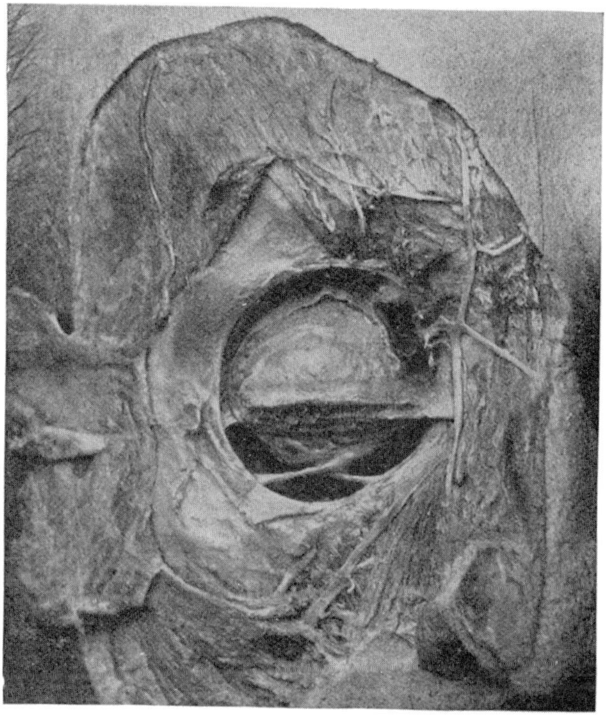

FIG. 59.—**Dissection of the Eyelids:** third stage. The *orbicularis oculi* and the *septum orbitale* have been completely removed, and the fore edge of the aponeurosis of the levator cut away to expose the tarsal plate; the orbital fat has been cleared away. The preparation shows the supra-orbital and supra-trochlear nerves (compare Fig. 9), the pulley of the superior oblique muscle, the anastomosis between the ophthalmic and angular veins, the inferior oblique muscle with its so-called 'check ligament' (the only instance of this structure the writer has met), and the lacrimal gland subdivided into two parts by the lateral horn of the aponeurosis of the levator.

as is sometimes stated. After severing and lifting the septum orbitale, the fascial layer of collapsed levator fibres can be traced backwards, and seen to arise from the front edge of a glistening fibrous band transversely arched across the orbit over the globe; this is the expansion or 'aponeurosis' of the levator, in the posterior edge of which the muscle fibres end (Fig. 71, p. 144).

It will further be noticed that the upper border of the exposed tarsal plate is not free, but is continued upwards and

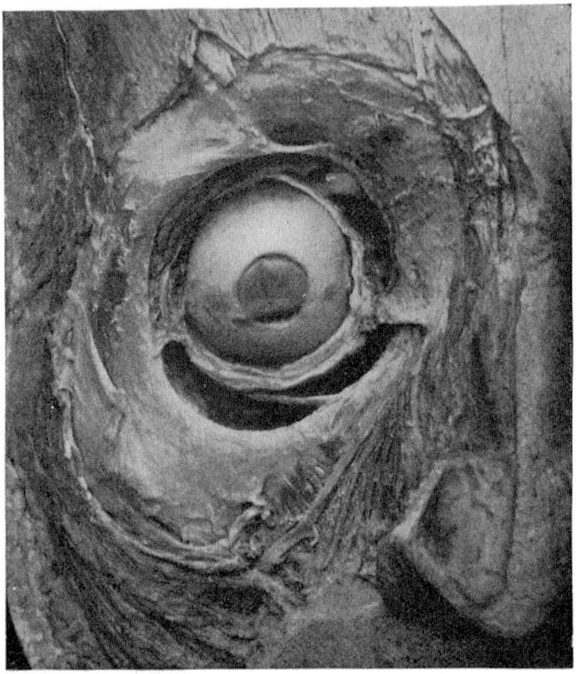

FIG. 60.—**Dissection of the Eyelids:** fourth stage. The globe is exposed by removal of the tarsal plates, and shows below it 'the suspensory ligament' of Lockwood. The medial palpebral ligament is well displayed: also the two parts of the lacrimal gland, divided by the cut edge of the aponeurosis, and below it the *retinaculum oculi laterale* of Hesser.

backwards into the recesses of the orbit above the globe by means of a pale reddish membrane; this is the superior palpebral involuntary muscle, and in a vertical section it can be seen to be attached posteriorly to the muscular belly of the levator and to lie immediately beneath its aponeurosis (Fig. 67, p. 139).

No further demonstration of layers is possible; the tarsal plate cannot be dissected away from the underlying conjunctiva, though the superior palpebral muscle may with difficulty be isolated from it. After removal of the septum orbitale, the orbital fat surrounding the globe is exposed behind both lids; the lacrimal gland is seen to lie behind the

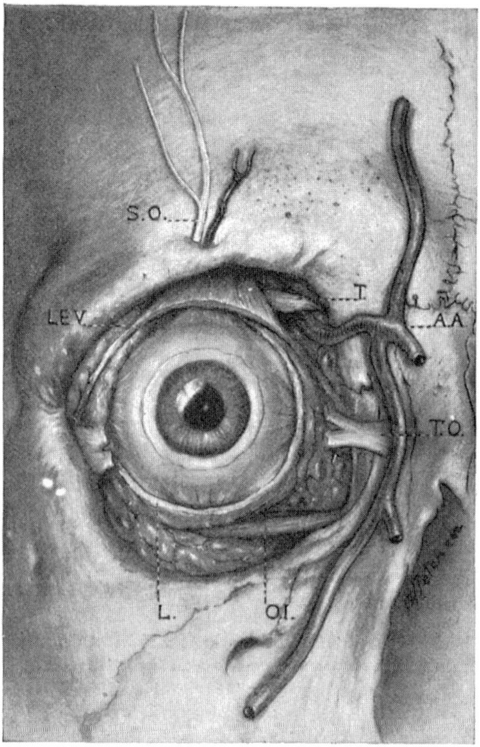

FIG. 61.—Drawing of Dissection of Orbit from in front.

L.E.V. = cut edge of aponeurosis of levator; S.O. = supra-orbital vessels and nerve; T. = tendon of superior oblique; A.A. = angular artery; T.O. = tendo oculi medialis; O.I. = inferior oblique; L. = suspensory ligament of Lockwood.

lateral corner of the upper lid, and the pulley and reflected tendon of the superior oblique muscle behind its medial corner; the origin of the inferior oblique muscle is seen to lie behind the medial half of the lower lid, just wide of and below the lacrimal sac.

These various structures will now be considered in detail. The **skin** is extremely thin, and contains no fat in its deeper layer; the hairs are represented by fine down, or may be so short as not to project beyond the follicles; the dermal papillae, sweat and sebaceous glands are also less developed than elsewhere; pigment cells are found in the stratum mucosum, especially near the medial commissure (see also

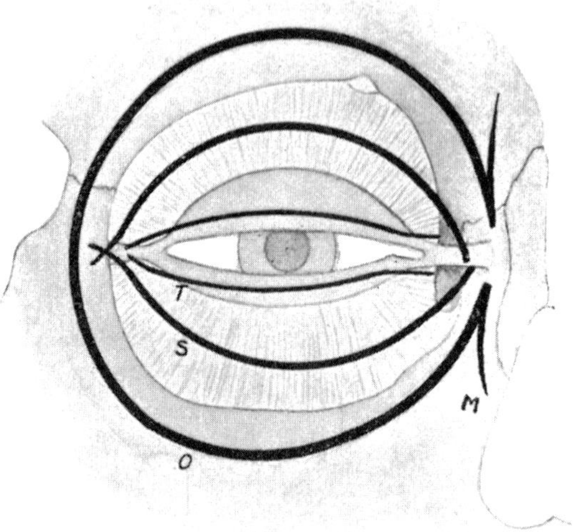

FIG. 62.—Scheme of Right **Orbicularis Oculi Muscle** to show the direction and insertion of its main fasciculi. Compare with Fig. 57.

O. = The *pars orbitalis*, covering the orbital margin.
S. = Pre-septal portion } of the *pars palpebralis* covering the eyelids.
T. = Pre-tarsal portion }
M. = *Pars malaris*.

Note the medial ends of the pre-tarsal fibres passing behind the fossa for the lacrimal sac and constituting the *pars lacrimalis* or Horner's muscle, and the lateral ends of the pre-septal portion which interlace over the orbital margin to form the 'lateral palpebral raphe'. After Merkel, modified.

p. 117), and mast and plasma cells are normally present (Parsons, 1905). The skin is highly elastic, and loosely attached by fine connective-tissue strands to the underlying muscle; this tissue represents the 'superficial fascia' of other parts of the body, but differs in being devoid of fat (though Adachi has found it normally present in the eyelids of the

Japanese); its laxity readily allows the skin to be pinched up or raised by effusions, but it is absent at the commissures where the skin is more adherent, so that effusions in the one lid will not pass readily into the other.

The **orbicularis oculi** (*orbicularis palpebrarum*) muscle is a thin sheet of fibres covering the eyelids and circum-orbital region (Fig. 57); it is somewhat oval in shape, the long axis being horizontally disposed and corresponding to the palpebral opening, round which its well-marked fasciculi are concentrically arranged. Although it forms one continuous sheet, it can arbitrarily be divided into two main portions, an orbital overlying the orbital margins, and a palpebral lying in the eyelids. This distinction, whilst not absolute as regards the attachment of the fibres or their course, is, on the whole, justifiable on both anatomical and physiological grounds, since the two portions show important differences in their relations, attachments, structure, and action.

(i) The orbital portion (*pars orbitalis* or *m. oculo-buccalis* of Burkitt, 1926) is much broader than the palpebral; it extends onto the region of the eyebrow, of which it has already been described as a constituent, the temple, where it lies over the anterior part of the temporal muscle, and the cheek, where it covers the origin of the lip elevator muscles and the infra-orbital and zygomatico facial nerves. The bony attachment of the fibres is to the medial orbital margin and side of the nose, along a curved line extending from the supra-orbital notch to the infra-orbital foramen; fibres also arise from the medial palpebral ligament, which interrupts this line of bony origin (Fig. 65, p. 136).

The fasciculi arising from the line above the ligament sweep uninterruptedly round the orbital circumference, and are fixed again to the bone below the ligament, so forming complete ellipses on the lateral side; in their course they give off cutaneous fibres only to the eyebrow and medial part of the lower lid, elsewhere being connected to the skin merely by connective tissue. In the new-born child the superficial orbicular fibres are not interrupted even on the medial side, but pass continuously over the ligament, and it is usual to find one broad fasciculus passing right over the ligament in the adult, as seen in Fig. 57. The most outlying fibres above and below the orbit have been named the *musculus superciliaris* (Merkel, 1887) and the *pars malaris* (Henle) respectively (Fig. 62); a few of the latter fibres pass into the skin below the medial commissure, and can cause a fine wrinkling in this region, but as a whole these aberrant fasciculi are of no special importance. The pars orbitalis is composed of thicker and coarser fibres than the pars palpebralis, and is redder in colour; its action is auxiliary to that of the latter, and it enters into play in forced closure of the lids, in bright light, and in excessive accommodation of the lens of the eye (p. 186).

(ii) The palpebral portion (*pars palpebralis* or *m. malaris*) is the essential part of the muscle, and forms an integral part of the eyelids; it is composed of two half ellipses, one in each lid; the fibres arise from the middle of the medial orbital margin and sweep across the lids, to terminate at the lateral commissure, where they cross to form the 'lateral palpebral raphe' (*raphe palpebrarum lateralis*). This raphe is formed purely by the interlacing of the muscle fibres, strengthened on its deep aspect by the septum orbitale, and it does not represent the lateral termination of the tarsal plates, as is erroneously described (see p. 153). The pars palpebralis covers both the tarsal plate and the septum orbitale, and consequently 'pre-tarsal' and 'pre-septal' fasciculi may be distinguished; the junction of these two divisions is the thinnest part of the muscle sheet, and it coincides with the superior and inferior palpebral skin-folds.

(*a*) The 'pre-septal' or more peripheral fasciculi arise from the medial palpebral ligament, chiefly from its superficial and partly from its deep aspect, but the lower edge of the ligament is usually left uncovered by muscle fibres (Figs. 57, 58); they also arise from a short line on the bone above and below it; some of the fibres which arise from the deep surface of the ligament have been described by Gerlach (1880) as forming an 'anterior lacrimal muscle', from their close relation to the lacrimal sac, and in microscopical sections their connexion with the lacrimal fascia covering the sac is well seen (Fig. 127, p. 234).

(*b*) The 'pre-tarsal' fibres (*pars tarsalis*), on the other hand, as can well be seen in a dissection of the posterior aspect of the lids (Fig. 63, and Fig. 126, p. 233), form at the medial end of the latter a thick muscular mass about 7 mm. broad by 4 mm. thick, which passes with the deeper part of the medial palpebral ligament behind the lacrimal fossa to be attached to the upper part of the posterior lacrimal crest; the

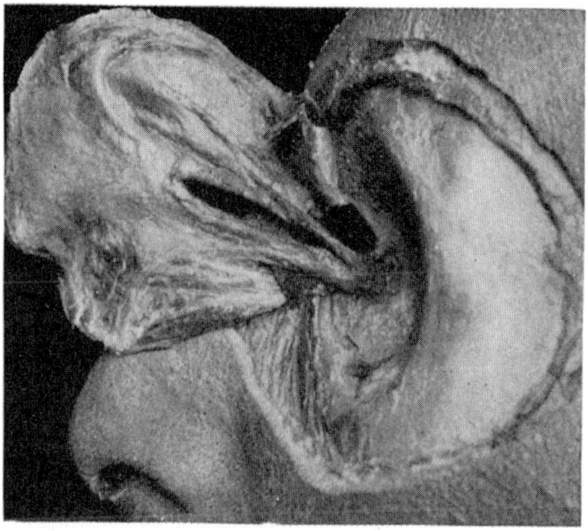

FIG. 63.—The **Orbicularis Oculi Muscle** dissected away from the lateral side and swung medially, to show the direct continuity of its *pars lacrimalis* with the pre-tarsal or *pars tarsalis* fibres which run along the lid margins. The relation to the upper part of the lacrimal sac, which has been exposed by cutting through the lacrimal fascia, is shown, but is better displayed in Fig. 126, p. 233. × $\frac{2}{3}$.

importance and strength of the attachment is shown by the frequent presence in the skull of a thickening or tubercle at this point (Fig. 65), but the area of attachment often encroaches forwards on to the lacrimal fascia or periosteal roof of the fossa in which lies the lacrimal sac. It is this well-defined mass of fibres that has been described as a separate muscle under various names, most commonly as the 'muscle of Horner' (1824), but chronologically more correctly as that of Duverney (1749), who well named it the *tensor tarsi* muscle, or as the posterior lacrimal muscle, or *pars lacrimalis* of the orbicularis. A few fibres are short and end upon the tarsal plate, others are related to the lacrimal canaliculi, which are buried in its mass, as seen in Fig. 127, p. 234, but it should be realized that the above names merely denote the conjoint medial ends of two continuous muscular bands which sweep across the lids from one side to the other and are placed over the tarsal plates, hence being more properly called the pre-tarsal portion or *pars tarsalis* of the orbicularis. Some fibres of the upper bundle cross over to join those of the lower, which latter in their entirety form a specially robust bundle. The practical importance of this part of the muscle lies in its relation to the lacrimal sac, which is considered on p. 234, and on this account the name 'pars lacrimalis' is retained.

Thus the pars palpebralis has a superficial origin from the medial palpebral ligament and adjacent bone, from which fibres sweep over the septal region of the lid, and a deep origin from the posterior lacrimal crest, the fibres from which pass across the tarsal plates. Both sets of fibres interlace in the lateral palpebral raphe.

There is a further subdivision of the orbicularis which calls for mention; this is a small fasciculus of extremely thin fibres, amongst the finest, indeed, of the whole body, which runs close along the free margin of each lid, separated off from the rest of the muscle by and lying behind the follicles of the eyelashes; it is known as the marginal part of the orbicularis (*pars marginalis* or *pars ciliaris*) or ciliary bundle of Riolan, and a few of its fibres, seen in microscopical section to lie behind the openings of the ducts of the tarsal glands, are called the 'subtarsal portion' (Klodt, 1893, Virchow, 1908); they are seen in Fig. 84. The fibres of the pars palpebralis are paler and finer than those of the pars orbitalis; in action they are not only voluntary, but also involuntary (p. 182). The nerve-supply is given on p. 177.

To sum up, the orbicularis oculi muscle sheet may be subdivided into:

(i) **Pars orbitalis,** covering the orbital margin;
(ii) **Pars palpebralis,** covering the eyelids;

and the latter can be differentiated into:

Pars septalis, covering the septum orbitale;
Pars tarsalis, covering the tarsal plates, and including:

Pars lacrimalis (of Horner), behind the lacrimal fossa;
Pars ciliaris along the lid margins;
Pars subtarsalis behind the tarsal glands.

Considerable **Variations** are found in the thickness and extent of this muscle, as might be expected from the fact that all the facial muscles arise from the splitting and differentiation of a common sheet. It is thicker in the negroid and yellow races than in white (Chudzinski); an admirable description of the whole facial musculature in the Australian aborigine is given by Burkitt and Lightoller (1926–27) Accessory fasciculi may be present, forming connexions with neighbouring muscles of the face, especially those of the nose, and it may even be united to its fellow of the opposite side by a *transversus glabellae* muscle (Ruge) and an upward extension of the platysma has been seen to reach the lower eyelid. There may be a complete separation of the palpebral from the orbital portion. On the other hand, cases of absence of the orbital or a rudimentary condition of the palpebral portion have been noted. Such variations are of interest, since they indicate the developmental origin of the orbicularis as a specialized part of a continuous platysmal sheet of muscle fibres; they are recorded in detail in the works of Macalister (1875) and Ledouble (1897). See also p. 150.

FIG. 64.—The **Septum Orbitale** exposed by reflection of the *orbicularis oculi*, which has been turned medially. Note the supra-orbital nerve piercing it above, and the terminal cutaneous fibres of the lacrimal nerve emerging just below the lateral angular process of the frontal bone.

It may be added that the reader interested will find the facial musculature and expression dealt with from the phylogenetic, developmental, and racial aspects in a book by Huber (1931).

The **septum orbitale** (the 'broad ligaments' of the eyelids or of the tarsal plates; superior and inferior palpebral ligaments; orbito-tarsal ligament; palpebral fascia or tarsal membrane; *fascia tarso-orbitalis* or *orbito-palpebralis*) is a thin membrane of connective and elastic tissue, which extends from the entire orbital margin towards, but not reaching, the palpebral opening. Whilst the septum lies in the same morphological plane as the tarsal plates, it is essential to

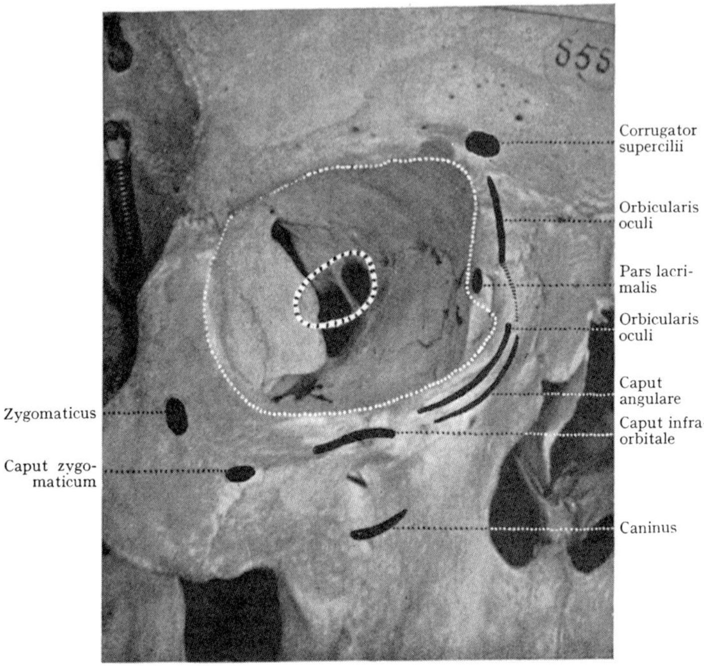

FIG. 65.—Markings of **Muscle Attachments** in and around the Orbit. The centre ring represents the fixation of the Annulus of Zinn; the white dotted marginal line indicates the origin of the *septum orbitale*.

consider it not as a fixed vertically disposed diaphragm, but as a supple floating membrane, changing its contours with the movements of the lids. In fact, its disposition and relations are best realized by regarding it as the *deep fascia* lining the palpebral part of the orbicularis oculi muscle, and in gross dissection of the lids, the skin, muscle, and septum together form one combined layer covering the space between the globe and the orbital margin; it is recognizable as a definite structural layer only by careful dissection, but merits its name in that it can form a barrier to the passage of extravasations from the orbit to the lids and *vice versa* (see pp. 143, 302).

Along its line of origin from the bone it is directly continuous with the periosteum, which is here thickened and raised along the exact line of the orbital margin to form a sharply defined ridge, the 'marginal arcade' or *arcus marginale* (Charpy, 1911). In tracing the septum round the orbital margin, as may be followed in Fig. 65, it is found specially well marked on the lateral side; it then springs from a periosteal bridge across the supra-orbital notch (Fig. 60), and passes in front of the fovea trochlearis along a rough line on the bone to descend behind the area of attachment of the pars lacrimalis muscle (for which the septum acts as a thin posterior covering); it then skirts the lower border of the muscle and crosses the middle of the lacrimal fossa to gain the anterior lacrimal crest and pass on to the inferior orbital margin. The pars lacrimalis muscle and the lacrimal sac thus lie outside the orbit, of which the septum exactly defines the anterior boundary.

In the direction of the palpebral opening the attachments of the septum differ in each lid. In the lower lid it blends with the anterior face of the tarsal plate, conjointly with a fascial expansion from the inferior rectus muscle, which corresponds morphologically to the involuntary part of the levator in the upper eyelid (Fig. 73, p. 147); the union forms a barrier to the orbital fat beneath the globe. In the upper lid it has already been shown (p. 127) to be separated entirely from the plate by the terminal fibres of the levator muscle and carried forwards with them to the skin; at its junction with the levator, however, some of its fibres are reflected backwards towards the aponeurosis of that muscle, so that a smooth concave barrier is presented to the orbital contents above the globe, the concavity containing a well-defined long roll of fat with tapering ends (Figs. 68, 73). Its dependence as a curtain from the supra-orbital margin and its connexion with the levator can be demonstrated by such a preparation as is shown in the following figure, where the bony margin has been sawn and turned aside, carrying the septum with it.

The septum presents local differences in thickness; it is

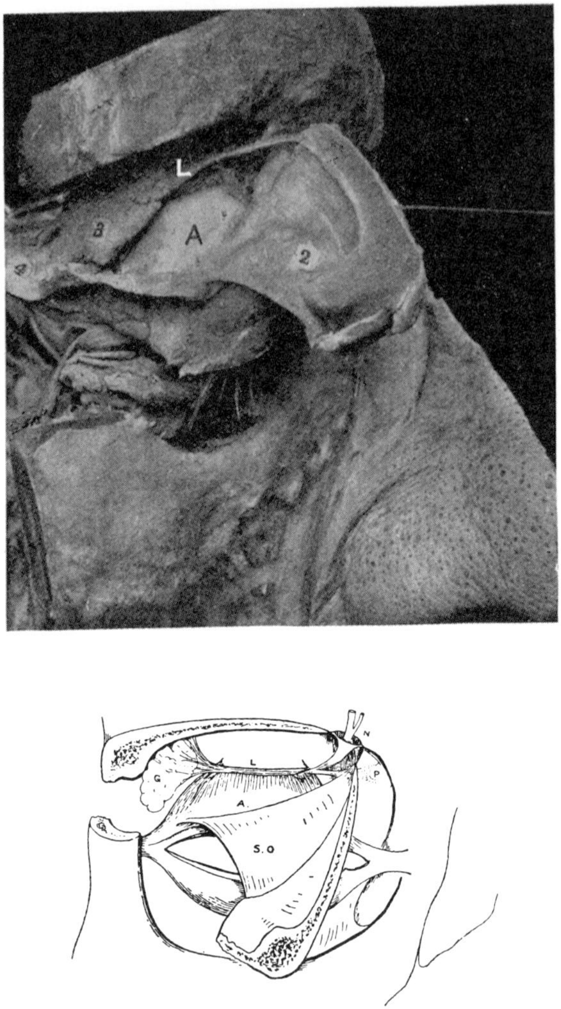

FIG. 66.—Dissection to show **Septum Orbitale.** After removal of the skin and *orbicularis oculi* muscle, the bone forming the supra-orbital margin has been cut and turned down with the attached *septum orbitale* (S.O. and 2), exposing the palpebral aponeurosis (A.) of the levator. Above this is seen lying the transversely placed ligament (L.) formed by its sheath; its expansion which passes forward to bridge the supra-orbital notch (N.), and attachment to the pulley (P.) of the superior oblique on the medial side, and connexions with the lacrimal gland (G. and 3) on the lateral side, are shown. Compare with Fig. 74, p. 149.

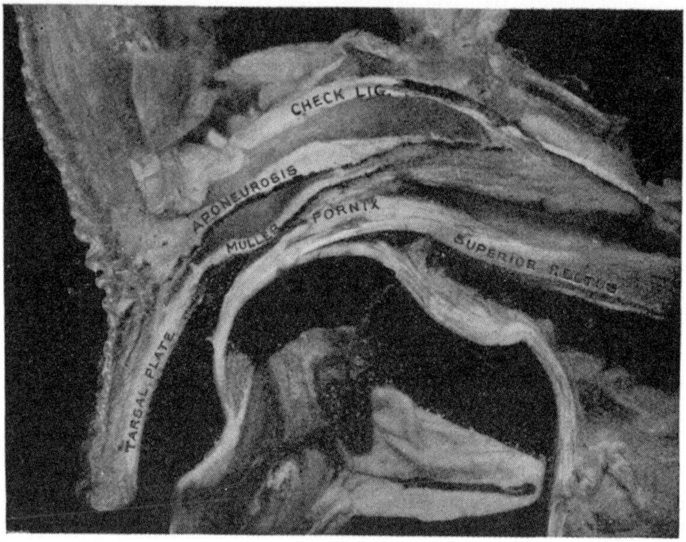

FIG. 67.—Sagittal section through Upper Eyelid and Globe, to show connexions of **Levator Palpebrae Superioris** muscle. ×2.

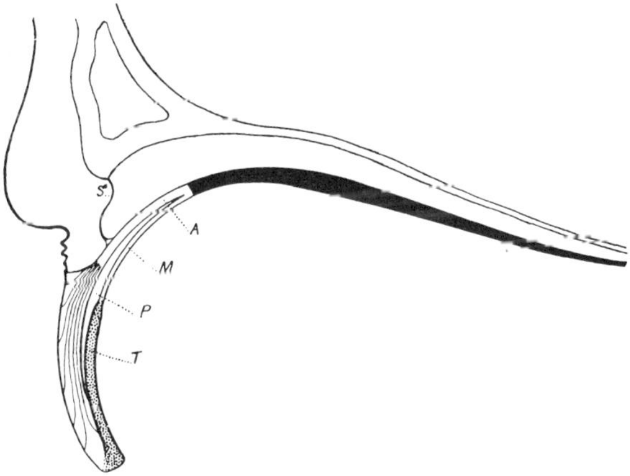

FIG. 68.—Diagram of the **Levator Palpebrae Superioris,** showing its palpebral connexions.

A. = Aponeurosis.
M. = Superior palpebral involuntary muscle of Müller.
T. = Tarsal plate.
P. = Pre-tarsal space.
S. = *Septum orbitale.*

best developed, like the marginal arcade from which it springs, on the lateral side, where it stretches between the orbital margin and the angle of the eye and reinforces the raphe formed by the overlying muscle fibres of the orbicularis; it is thin in the lower lid and most delicate in the medial region; it is perforated by the vessels and nerves passing to or from the orbital cavity, the orifices lying along its peripheral margin (Fig. 59, p. 127).

The **levator palpebrae superioris**, the special elevator muscle of the upper eyelid, lies within the orbit, and only its terminal part enters into the lid. It is thin, flat, and of the shape of an isosceles triangle. It arises at the apex of the orbit and passes forwards towards the base, where its fleshy belly terminates in an expanded tendon called the 'aponeurosis'; the fore edge of this band breaks up into fine connective tissue fibres, which radiate upwards and downwards and are inserted into the skin of the lid. There are therefore three parts of the muscle to be seen in a longitudinal section: the fleshy belly, the aponeurosis, and the terminal fibres (Fig. 68). Moreover, lying in close contact beneath the aponeurosis is a thin lamella of pale smooth muscle fibres, which appears to be a direct continuation of the fleshy belly, but is inserted directly into the upper margin of the tarsal plate. This is the superior palpebral involuntary 'muscle of Müller', and constitutes a secondary insertion of the levator.

The levator arises by a short narrow tendon from the lesser wing of the sphenoidal bone, just above and a little lateral to the optic foramen; it is separated from this orifice by the origin of the superior rectus muscle, with the tendon of which it is blended (Fig. 142, p. 262); shortly after its origin it is crossed obliquely by the trochlear or fourth nerve, which passes medially to enter the adjacent superior oblique muscle. As it passes forwards the levator inclines laterally and lies close beneath the roof of the orbit, the frontal nerve and the supra-orbital artery lying upon it; the nerve crosses it diagonally from the lateral border behind to the medial border in front, and then divides into supra-orbital and supratrochlear nerves, and the vessel lies over its anterior half only. In its whole course the levator lies upon the superior rectus muscle, but does not cover it entirely, the lateral border of the latter being exposed and free; the medial borders of the two muscles coincide and are adherent by their thin fascial sheaths; they are also connected by means of the nerve to the superior rectus, which after piercing that muscle supplies also the levator. These relations are shown in Fig. 69. A bursa is said by Motais to exist between the two muscles, but none could be found in the writer's series of preparations.

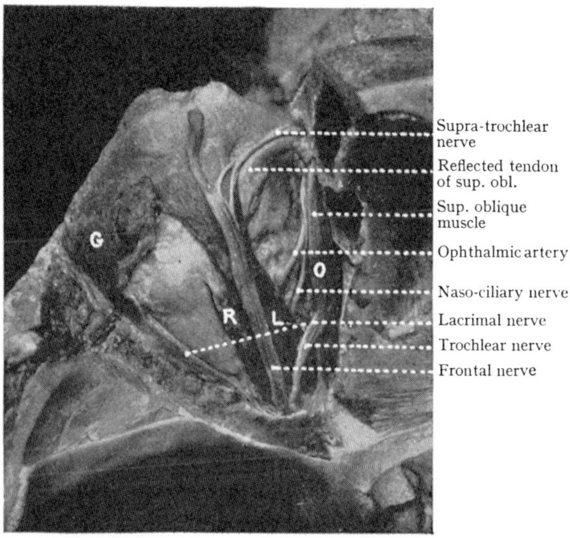

FIG. 69.—Dissection of Left Orbit, to show the ***levator palpebrae superioris*** (L.) and superior oblique (O.) muscles, and the nerves in relation. R. = Superior rectus muscle; G. = lacrimal gland. Compare with Fig. 71. Natural size.

The palpebral *aponeurosis* or expanded tendon of the levator is a glistening white band of dense connective tissue about 7 mm. in depth from before backwards and 0·5 mm. thick; it lies just behind the supra-orbital margin, and arches over the globe from side to side. Its anterior edge splits up into terminal cutaneous fibres; its extremities become narrower but remain tendinous, and are attached to the medial and lateral orbital margins at their mid-points, and may be distinguished as lateral and medial 'horns'. The attachments of the aponeurosis, which, it will be realized, are the actual insertions of the levator, are therefore both cutaneous and osseous. The former are best studied in vertical section, the latter by dissection from above; both are seen in a dissection of the lids from the front.

(i) The **cutaneous insertion** is into the skin of the whole of the pre-tarsal part of the lid by means of vertically radiating fibres of delicate connective tissue; this is the primary and essential attachment of the muscle. In reaching the skin the fibres necessarily traverse the horizontally disposed fasciculi of the orbicularis oculi muscle, as can clearly be seen in microscopical sections (Fig. 79, p. 156). They sweep over the bare superficial face of the tarsal plate in their downward course, and it is the condensation of these fibres that forms the definite layer of tissue which has to be lifted before the plate is exposed in a methodical dissection of the lid, as shown in Figs. 71, 72; they are also raised in splitting the lid along the midline of its edge. The importance of realizing this arrangement in operations on the lids for ptosis, trichiasis, or plastic operations is emphasized by Wolff (1905). The uppermost fibres meet and fuse with the septum orbitale; the lowermost fibres are inserted into the lower third of the face of the tarsal plate and mingle with the bulbs of the cilia (Fig. 68). It will be seen that the centrally situated aponeurosis of the levator with its radiating fibres divides the lid into two potential spaces, a 'pre-tarsal' confined between the tarsal plate behind and the levator in front, and a 'pre-septal' which lies in front of the levator and the septum orbitale, and is limited above by the fatty cushion of the eyebrow (Fig. 68). The pre-tarsal space contains between its boundaries laterally the palpebral lobe of the lacrimal gland and terminal branches (lateral palpebral) of its artery, and within its upper part there runs from side to side the superior tarsal arterial arcade.

These spaces can be injected in the fresh subject by means of coloured gelatine (Clermont, 1909; Charpy, 1911); by using a small amount of fluid the two spaces can be injected separately, but on pushing the injection further the masses become confluent and the fluid will next pass between the fasciculi of the orbicularis and lie beneath the skin, so that the whole lid becomes infiltrated, the muscle lying in the middle of the mass, as shown in Fig. 70. The injection will not readily pass backwards into the cavity of the orbit, owing to the barrier afforded by the septum orbitale, which is rarely penetrated even by abundant injections, though a gradual transudation and percolation through the nervous and vascular orifices of that membrane can take place; extension on to the temple or cheek is prevented by the felting of connective tissue which lies between the orbicularis oculi and the bone in these regions. (See also p. 302.)

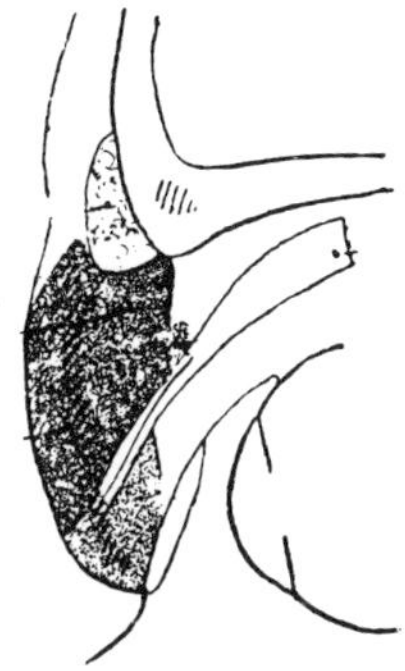

FIG. 70.—Diagram showing superficial **injection of Eyelid** (Charpy, 1911). The injection, introduced sub-cutaneously in front of the *orbicularis oculi,* has penetrated between the fasciculi of this muscle so that it becomes buried in the mass. Part of the injection in lighter tint has transuded beneath the levator below and lies on the tarsal plate, reaching down to the margin of the eyelid. The sketch shows the extent of the swelling in a 'black-eye' or 'poached-egg eye'.

(ii) The **osseous or orbital attachments** of the aponeurosis are effected by means of its medial and lateral horns to the corresponding orbital margins at their mid-points and opposite the commissures of the eyelids. The lateral horn is much more strongly developed, and in the form of a conspicuous ligament, 3 to 4 mm. broad, cuts into and subdivides the lacrimal gland, which is, as it were, folded round it (Fig. 71, and see also Fig. 111, p. 210); below the gland the lateral horn is inserted into the orbital tubercle of the zygomatic bone, covering the insertion of the true lateral ligament of the tarsal plates (just as the terminal cutaneous fibres of the aponeurosis pass in front of the plate); it is not uncommon to find a slender continuation of the horn passing downwards round the lateral angle of the eye into the lower lid, where it joins the fascia beneath the globe, as is seen in Figs. 54 and 58. On the medial side the aponeurosis loses abruptly its tendinous nature as it passes over and comes into close contact with the reflected tendon of the superior oblique muscle; the opposed surfaces are smooth and shining, but no definite bursal formation can be identified. From this point the medial horn turns downwards at an angle, and can be traced only with difficulty in the form of loose strands to fuse with the medial palpebral ligament, and so reach the bone. Thus from the greater strength and more definite develop-

FIG. 71.—Dissection of the Left Orbit from above and in front (the same preparation as Fig. 69, viewed from in front) to show the **aponeurosis** of the *levator palpebrae superioris muscle* and the **lacrimal gland.** The pulley of the superior oblique muscle and its tendon are also seen. Note the cutaneous branch of the lacrimal nerve emerging below the lateral horn of the aponeurosis. Natural size.

ment of the lateral horn, and the fact that the fleshy fibres of the muscle curve laterally and extend further forwards in that direction, whilst the medial border of the muscle is parallel to the mesial plane, the main osseous insertion is the lateral one into the orbital tubercle. Retraction of the muscle is consequently more restrained on this side, and the greater excursion of movement on the other may explain the accentuation of the marginal curve of the upper lid in its medial part noted above on p. 121.

(iii). The **palpebral involuntary muscles** or 'tarsal muscles' (*musculi palpebrales*, or *tarsales* of Merkel, or orbito-palpebral muscles of Sappey) were first described by Müller in 1859; they are not to be confused with the involuntary 'orbitalis' muscle, also described by him, which lies in the inferior orbital fissure and is connected with the periorbita (p. 86). They are found in each lid in varying degrees of development, but usually form a thin sheet of pale red fibres connected with the peripheral margin of the tarsal plate and lying in front of the conjunctiva. That in the upper lid, namely the superior palpebral involuntary muscle, is the better developed structure, and is about 1 cm. broad antero-posteriorly; its position immediately beneath the aponeurosis of the levator has already been noted, but so closely does it lie in contact with it that it may not be recognized as a distinct structure, though the two can be easily separated and their different connexions made out in a vertical section. The fibres arise from between the striated fibres of the levator by elastic tendons, and have a similar mode of fixation to the upper margin of the tarsal plate, a connexion to be seen in such a microscopical section as is shown in Fig. 80, p. 156, and well displayed in Fig. 72, in which the levator fibres have been turned upwards to expose the tarsal plate. For the most part the fibres run in a vertical direction, and are mingled with connective tissue and fat cells; groups of nerve cells analogous to ganglion cells are also present (Groyer, 1905). In the lower lid the

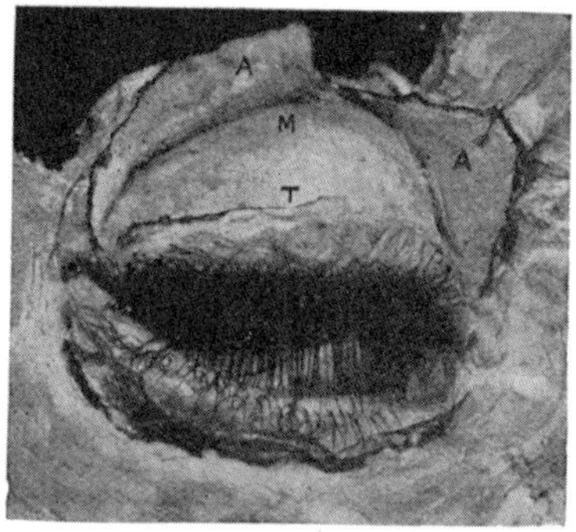

FIG. 72.—Dissection of Right Orbit and Eyelid from in front. The aponeurosis of the *levator palpebrae superioris* muscle (A.) has been reflected upwards from the face of the tarsal plate (T.), to the upper margin of which is attached the **superior palpebral involuntary muscle** (M.). × about $1\frac{1}{4}$.

inferior palpebral muscle is a much less definite structure and is not easily recognizable by dissection, though the fibres can be traced microscopically. It arises from the fascial sheath of the inferior rectus muscle or from the expansion which this sheath gives to the inferior oblique, and it divides into two lamellae at the conjunctival fornix, one of which ends in the bulbar conjunctiva, whilst the other enters the lower lid, where it does not reach as far as the tarsal plate (Virchow, 1908), as represented in Fig. 73.

The morphology of this palpebral involuntary musculature is interesting. In the aquatic mammals, as for example the seal and the dolphin, each of the four recti muscles of the globe divides anteriorly into two lamellae, one of which, the inner, is inserted onto the globe and so constitutes the rectus proper, whilst the other, the outer, enters the eyelid as a 'palpebral muscle'; both lamellae are composed of striated muscle fibres, and both receive the same innervation. In other mammals the outer palpebral lamellae derived from the inferior, medial, and lateral recti are wholly replaced by smooth muscle fibres, and in man are represented by the inferior palpebral muscle and also by the scattered involuntary fibres found in the lateral expansions of the sheaths of the last two muscles, fibres which are sometimes dignified by the name of medial and lateral 'palpebral muscles'. In consequence of the greater development of the upper lid, the outer or palpebral lamella derived from the superior rectus remains almost entirely striated and forms the levator muscle, but a deeper part of the same palpebral lamella, in harmony with the fate of the whole of the corresponding layer of the other muscles, becomes converted into smooth fibres, and is called the 'superior palpebral muscle' (Groyer, 1903, 1905, and 1906). This involuntary musculature as a whole forms an incomplete ring around the globe, and is intimately connected with the fascia around that organ (Tenon's capsule), under which heading further reference is made to it on p. 298. It is innervated by the sympathetic nervous system, possibly by fibres from the cavernous plexus, which enter the third nerve and run along its branch of supply to the levator. Here the superior palpebral muscle calls for notice, since it indirectly forms one of the insertions of the levator palpebrae superioris.

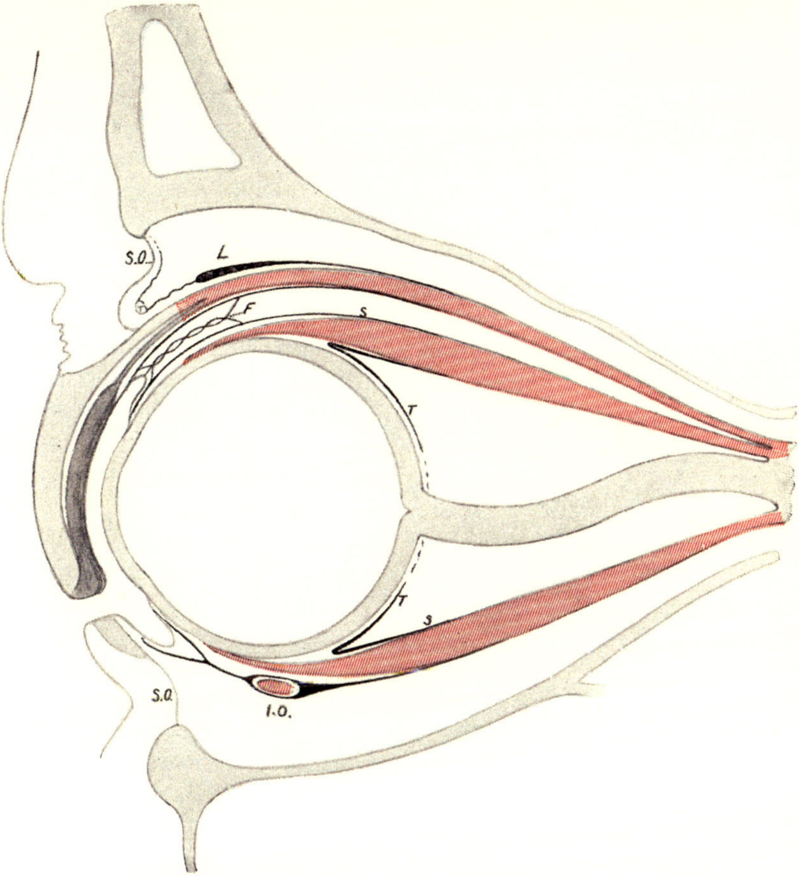

FIG. 73.—Scheme of the **Fascia of the Orbit** (the muscle sheaths and Tenon's capsule) in sagittal section. Compare with dissections shown in Figs. 67, 178.

S.O. = *Septum orbitale.*
F. = Conjoint fascial sheaths of the levator and superior rectus attached to the conjunctival fornix.
S. = Muscle sheaths.
T. = Tenon's capsule.
L. = Thickened part of sheath of levator.
I.O. = Inferior oblique with the 'suspensory ligament' of Lockwood posterior to it (see Fig. 178, p. 332).

(iv) There remains a last co-called insertion of the muscle to be described; this is the conjunctival attachment, which is effected not by the muscle itself, but through the agency of its **fascial sheath**. As is the case in all the ocular muscles, the fascial sheath of the levator is so thin and transparent around its posterior half that the fleshy fibres appear quite devoid of covering; anteriorly, however, the fascia becomes gradually thicker, until just behind the aponeurosis the whole muscle is enveloped in a thick opaque sheath, which fuses underneath with the similar covering of the superior rectus; the common mass of tissue so formed fills up the angle of divergence between the two muscles as they pass to their several insertions, the one on to the lid, the other to the globe. Anteriorly this mass abuts on and is fixed to the conjunctiva along the line of its reflection from the lid onto the globe (superior fornix), and though no muscle fibres pass to it from the levator, the attachment is described as an additional 'insertion' of that muscle (see Figs. 67, 73). It is by means of this fascial attachment that the conjunctiva is drawn up and backwards when the globe is rotated upwards, and harmoniously with it the upper eyelid is elevated; it would be as accurate to describe it as being an additional 'insertion' of the superior rectus muscle.

A further point of interest is presented by this sheath of the levator: its superficial part covering the upper aspect of the muscle just behind the aponeurosis is condensed to form a definite ligamentous band, which stretches transversely across to reach the orbital walls on either side, parallel with but attached at a higher level than the aponeurosis; medially the main connexion is to the trochlea, but behind this some fibres pass to the bone, and a well-marked slip passes forwards to bridge the supra-orbital notch; laterally the band is connected with the stroma of the lacrimal gland and cuts into it, as does the lateral horn of the aponeurosis, and beyond the gland it reaches the lateral orbital margin; it lies for the most part free above the aponeurosis, but stout strands of connective tissue may unite them (Figs. 66, 67, 74). In front of this ligamentous condensation the sheath becomes abruptly so thin that it appears to end in a free border, but it can be traced forwards as a very delicate layer up to the supra-orbital border (see Fig. 73). The band is well developed in the foetus. When traction is made upon the levator from behind this band becomes tense and thus forms a 'check ligament' to the muscle, preventing its overaction, a function it can perform by reason of its situation and attachments better than the aponeurosis (the horns of which are fixed at a lower level), which is generally considered to act in that capacity (Whitnall, 1910). The action of the levator is therefore checked by the attachments of its fascial sheath, as is the case in all the other ocular muscles.

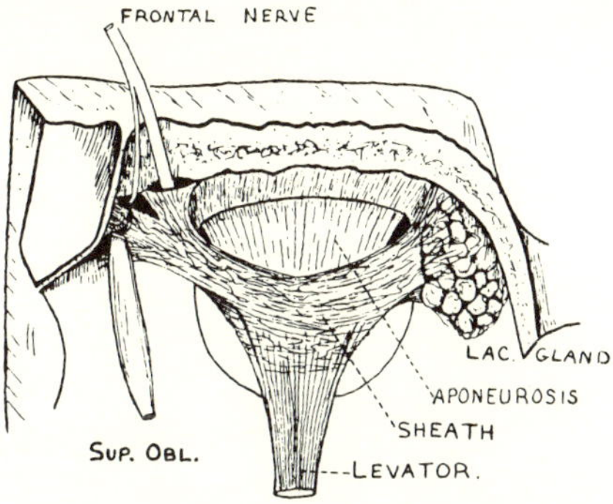

FIG. 74.—Dissection of the *Levator Palpebrae Superioris* from above in right orbit. The muscle is drawn backwards and downwards to make tense the **transverse ligament** formed by its sheath. The lateral attachments of this ligament are seen: on the medial side to the bone below the superior oblique muscle, to its pulley, and by a thin slip of fascia to the margins of the supra-orbital notch; on the lateral side to the stroma of the lacrimal gland and also by direct insertion onto the bone.

To recapitulate: the anterior attachments of the levator palpebrae superioris muscle are as follows:

Primary.—(i) **Cutaneous,** by means of its terminal fibres to the skin of the eyelid; (ii) **Tarsal,** by means of (*a*) some of its terminal fibres, (*b*) the superior palpebral involuntary muscle.

Secondary.—(iii) **Osseous,** by the horns of its aponeurosis; (iv) **Conjunctival,** by means of its fascial sheath fused with that of the superior rectus muscle. The action is described on p. 184.

Three ABNORMALITIES of the levator have been described, explainable by incomplete separation from a common muscle mass:

(i) The commonest is a slip passing from the medial border of the muscle to the pulley of the superior oblique muscle. Such a fasciculus was described by Budge (1859) as the *musculus tensor trochleae*, and appears to be identical with muscles described by Vesalius, Molinette, Kolmus, and Sandifort, the *comes obliqui superioris* of Albinus, and the *gracillimus orbitis* of Bochdalek. Budge found it present in fifteen out of twenty cases examined, but Virchow considers it much more rare, as indeed it was in the writer's observations; an instance is recorded by Rush and Schaeffer (1917). (See also p. 286.)

(ii) Similar fasciculi may be present passing from the lateral border of the muscle to the lacrimal gland. Such strands could have a much more effective action as a *retractor glandulae lacrimalis* than could the medial offshoots as a *tensor trochleae*, since the gland is movable, whilst the trochlea is fixed. In the example figured below, traction on the levator was found to pull the gland backwards, owing to the attachment of these fasciculi. Both these offshoots, be it noted, replace the normal connexions of the fascial sheath of the levator to the trochlea and lacrimal gland.

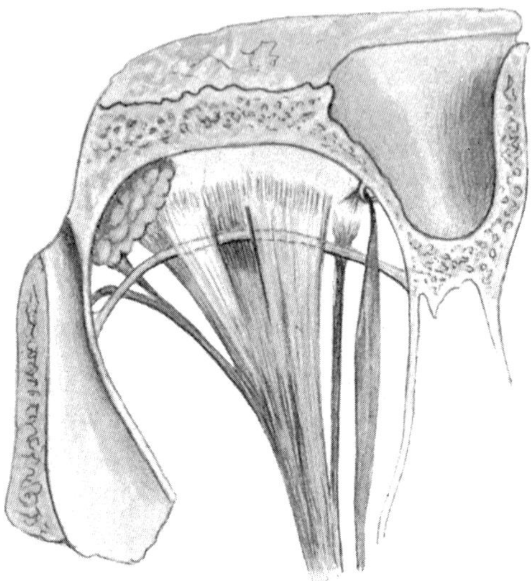

FIG. 75.—Sketch of **abnormal Levator Palpebrae Superioris** Muscle, seen in left orbit after removal of roof. A *transversus orbitis* muscle is represented by the transversely disposed fasciculus; a *tensor trochleae* muscle is seen passing to the trochlea from the medial edge of the levator, and on the lateral side fasciculi pass to the lacrimal gland.

(iii) As a *musculus transversus orbitis*, Bochdalek (1868) described certain fleshy bundles passing from the fore and upper part of the os planum of the ethmoidal bone across the orbit to the lateral wall, passing over but closely connected to the levator, and giving off various attachments to neighbouring fascia. Macalister and Ledouble thought it to be a backwardly displaced slip of the orbicularis oculi. Perna (1905) also described such a muscle slip, and considered it, like the levator itself, to represent the remains of the primitive muscular membrane which surrounds the globe in lower vertebrates. An instance of its occurrence in both orbits has recently been reported by Hueber (1918).

The specimen drawn in Fig. 75 is interesting, in that it presents all three of the above abnormalities. It is suggested that the *musculus transversus orbitis* is normally represented by the transversely disposed check ligament formed by the superficial part of the levator muscle sheath, which was not present in this case (Whitnall, 1921). From the occasional persistence of such variations the levator is evidently derived from a much more extensive muscle sheet, and probably corresponds to the palpebralis muscle of lower animals (see above, p. 146). Wiedersheim (1895) places it in the list of organs showing retrogressive characters, modified, but still performing clearly recognizable functions.

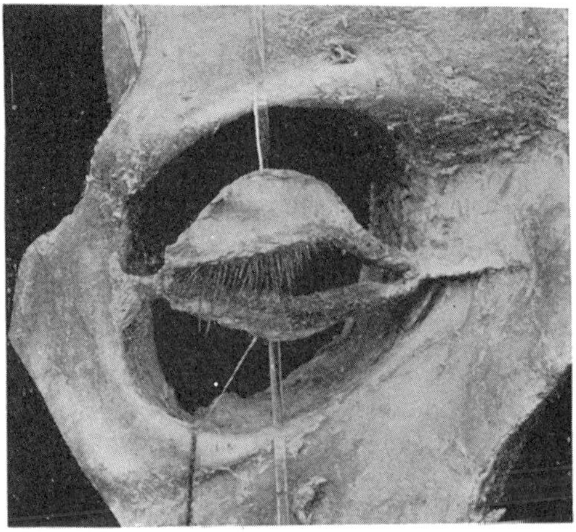

FIG. 76.—Dissection of the **Tarsal Plates** and the Lateral and Medial Palpebral or Tarsal Ligaments. Note the relation of the latter to the upper part of the lacrimal sac. Natural size.

The **tarsal plates** (*tarsus superior vel inferior*, tarsal cartilages; Greek, ταρσός, a flat plate or surface) are two plates of dense connective tissue situated one in each eyelid, and hence distinguished as superior and inferior. Their shapes, the upper that of a D laid on its side, the lower an elongated band, are shown in Fig. 76 above.

They have two borders: a horizontally straight or free one, coextensive with the ciliary margin of the eyelid up to the punctum lacrimale and about 25 to 30 mm. long, and an attached or orbital one, which describes a well-marked curve in the superior plate but is nearly straight in the inferior. The upper plate is from 10 to 12 mm. broad in its middle, the lower one is about 5 mm. wide throughout; they are thickest at the ciliary border and 1 mm. thick in the centre, but taper off towards the attached borders. The curved upper outline of the superior plate can be felt and sometimes seen through the skin; attached to it is the superior palpebral involuntary muscle; it is covered in front by the terminal fibres of the levator, some of which are fixed to its lower third. The inferior plate is covered by the orbicularis; its lower edge is united to the septum orbitale and to an expansion from the fascial sheath of the inferior rectus muscle, in which lies the inferior palpebral involuntary muscle. The plates are moulded upon the anterior surface of the globe, from which they are separated by the closely adherent conjunctiva; hence they are bent with a posterior concavity, a configuration shared by their free margins, and the closeness with which they fit the globe can be appreciated on making them spring from its surface by approximating the thumb and finger-tips placed at opposite ends of the lids. Their extremities meet at each end in the commissures of the lids, and are prolonged to form the medial and lateral **palpebral (or tarsal) ligaments,** by means of which they are attached to the middle of the corresponding margins of the orbit.

The medial palpebral or tarsal ligament (*ligamentum palpebrale mediale, tendo oculi,* internal palpebral ligament, or orbicular tendon) is formed by the junction of direct continuations of the two tarsal plates on the medial side. It is a broad band of connective tissue which is attached to the upper half of the anterior lacrimal crest, but it also spreads out over the frontal process of the maxilla in front of this ridge to a varying extent, even as far as the naso-maxillary suture, as is seen in Fig. 76; it therefore lies in front of the lacrimal fossa, but is partly adherent to the lacrimal fascia which roofs over that space. Superiorly it is thin and blends with the periosteum, so that definition of an upper margin is artificial; it is the weakness of this upper part that allows the fundus of the underlying lacrimal sac to bulge forwards in pathological conditions; inferiorly it is thicker and terminates in a free margin, which is prominent owing to the generally oblique position of the ligament from above downwards and forwards, and is conspicuous since it is bare of muscle fibres (Figs. 58, 148). This edge shows through the skin as a prominent white band on drawing the lids laterally. Fibres of the orbicularis oculi muscle arise from nearly the whole of the anterior surface of the ligament, with the exception of the free edge, and from the deep surface where it is not adherent to the bone or the lacrimal fascia. The relations to the lacrimal fossa are illustrated by Fig. 76, whence it will be realized that the ligament is placed in front of the *upper half only* of the lacrimal sac, but covers that over completely, so that there is no formation of a narrow band crossing its middle as is sometimes figured. Superficially the ligament is crossed by the angular artery and vein, the latter lying nearer to the angle of the eye and about 8 mm. from it (Fig. 90). Almost immediately after the ligament arises from the tarsal plates it divides into two lamellae, an anterior or direct portion, described above as passing in front of the lacrimal fossa, and a posterior or 'reflected' portion, which passes behind it to be attached to the posterior lacrimal crest; this posterior lamella is very thin and of quite secondary importance; it appears in dissections merely as a fascia clothing the front of the pars lacrimalis muscle, being intimately connected with it as it passes behind the fossa (Fig. 156, p. 289). In one full-term foetus, however, this posterior lamella was found better developed than the anterior; possibly the latter becomes the stronger as the orbicularis comes into use. Evans (1925) describes the tendo oculi in different races.

The lateral palpebral or tarsal ligament (*ligamentum palpebrale laterale, tendo oculi externus, ligamentum canthi externum,* external palpebral ligament, or lateral palpebral raphe): the lateral extremities of the tarsal plates form a conjoint tendon as in the case of the medial ends, but this consists only of a single and direct lamella; it is fixed to the orbital tubercle of the zygomatic bone just within the lateral orbital margin at about its mid-point. There is some confusion in the interpretation of this attachment and in the terms applied to it. Commonly the fixation is stated to be by means of the lateral palpebral *raphe,* which is formed by the interlacing ends of the palpebral fibres of the orbicularis, strengthened by the underlying septum orbitale, and passing superficial to the orbital rim; on cutting through the raphe, however, the plates, though a little slackened, will be found still attached

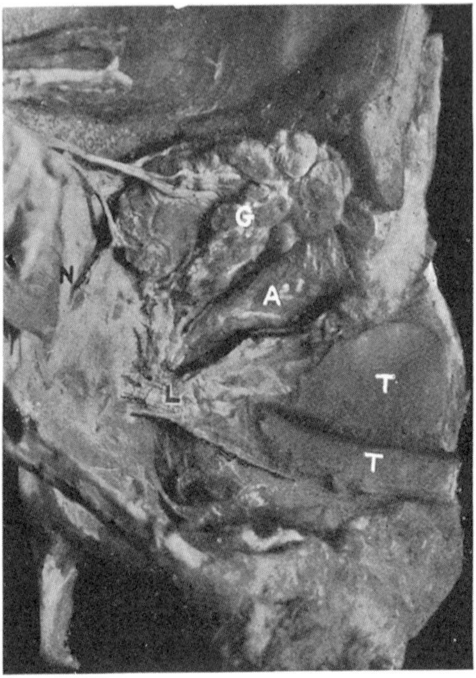

FIG. 77.—Dissection of antero-lateral Region of Left Orbit, seen from the medial aspect, to show the lateral halves of the tarsal plates (T.) and the true **lateral palpebral or tarsal ligament** (L.), attached to the orbital tubercle of the zygomatic bone. The preparation also shows the lacrimal gland (G.) and the anastomosis between its nerve and the zygomatico-temporal nerve (N.); A. is the aponeurosis of the levator. The orifices of the tarsal glands can be seen along the lower margin of the superior tarsal plate. × 2.

to the bone by a deeper-lying and stronger band of fibres, well displayed from the inner surface, as in the above figure. This is the true lateral palpebral *ligament*, and it is separated from the more superficially placed raphe by a narrow but definite cleft, in which there may be found a few lobules prolonged downwards from the lacrimal gland. The ligament is also closely covered at its attachment to the tubercle by the lateral horn of the aponeurosis of the levator, just as the tarsal plate itself above is covered by the same muscle expansion. Hence in this lateral region the order of structures from the surface backwards is skin, orbicularis and septum orbitale in one layer (the raphe), lateral horn of aponeurosis, and the lateral palpebral ligament (Fig. 156, p. 289).

It will be seen that by the recessed position of the lateral palpebral ligament on the one side and the deeply placed attachment of the pars tarsalis muscle on the other the tarsal plates are kept closely curved against the globe in their whole length, an approximation that could not be effected to so complete a degree were they fixed to the bone merely by the raphe laterally and the medial palpebral ligament medially. Moreover, the contractile muscle fibres ensure this contact between tarsal plates and globe under all conditions, whether the latter is recessed or protruded or whether its scleral or more prominent corneal region be presented to the lids in its various movements (Fig. 78).

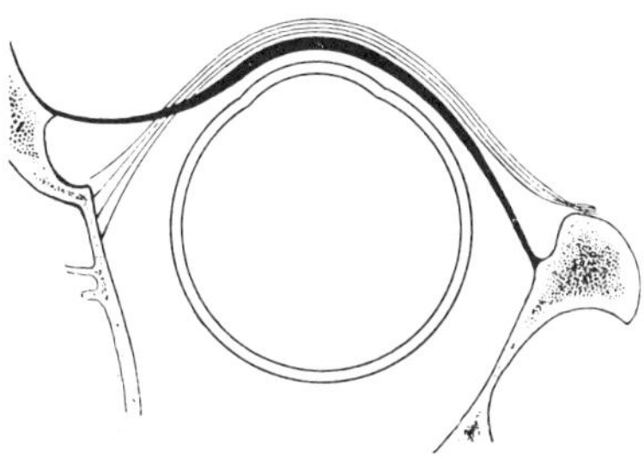

FIG. 78.—Scheme to show relation of **attachments of tarsal plate** (in solid black) and of pre-tarsal fibres of *orbicularis oculi* muscle (the lines) to the medial and lateral orbital margins.

In structure the tarsal plates are formed by dense connective tissue with a certain amount of elastic fibres; they are so firm as to account for their being sometimes called 'tarsal cartilages', though C. Krause pointed out in 1842 that there are no cartilaginous cells present. They are not solid plates, but are hollowed out by a series of large glands (tarsal or meibomian); indeed the most striking feature seen in microscopical sections is the enormous size of these structures, which excavate the plates to such an extent that there is more glandular formation than connective tissue substance (Figs. 80 and 81).

The connective tissue appears simply as a dense supporting stroma or 'peri-meibomian capsule', concentrated in front

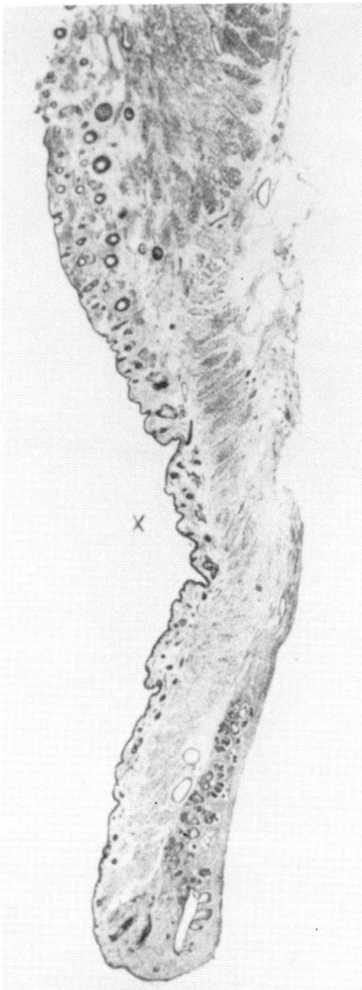

FIG. 79.

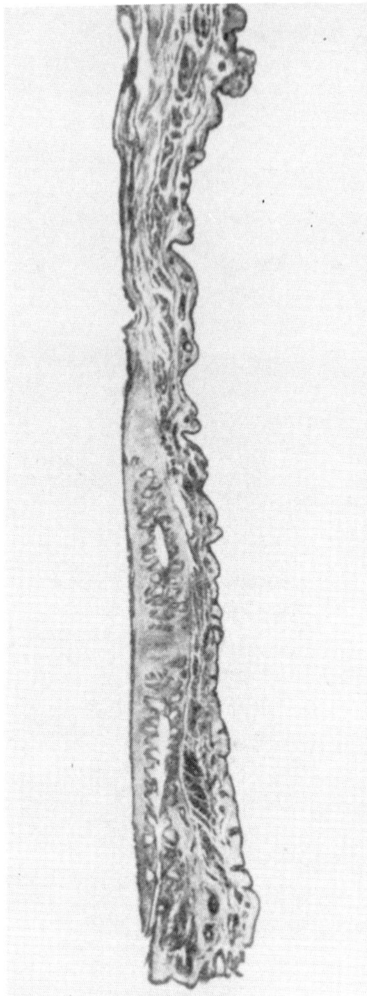

FIG. 80.

FIG. 79.—Longitudinal section of **Upper Eyelid** and Eyebrow, showing various parts and degrees of coarseness of *orbicularis oculi* muscle, and the terminal fibres of the aponeurosis of the levator traversing its fasciculi to reach the skin. The tarsal plate is small. X. is the region of the superior palpebral skin-fold.

FIG. 80.—Longitudinal section of **Upper Eyelid,** showing a large tarsal plate and an almost complete section along a tarsal gland. The superior palpebral involuntary muscle appears as a dark line leading upwards from the upper margin of the tarsal plate.

and behind the glands and connected by transverse strands passing between their acini. This stroma is, moreover, permeated by a system of lacunae and lymph spaces (Parsons, 1904), and is transversed by a network of nerves and blood-vessels supplying the glands. The elastic tissue is concentrated round the glandular acini, where smooth muscle fibres are also stated to be present, and some fat cells are found in the borders of the plates, especially in the lateral convex border of the upper one (Virchow, 1905). Towards the free

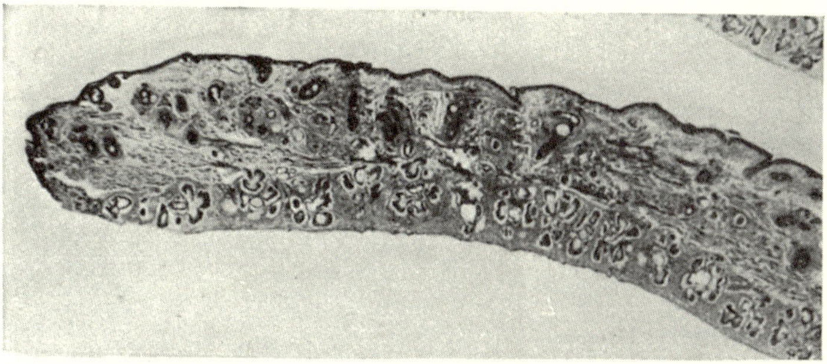

FIG. 81.—Transverse section of the Upper Eyelid to show the **tarsal glands,** and the extent to which they excavate the tarsal plate (on the lower half of the section).

borders of the lids the connective tissue of the plates spreads out anteriorly to intermingle with and support the bulbs of the cilia, so forming a thick marginal basis or 'ciliary mass' (Figs. 79, 86). Their deep surfaces are lined by the closely adherent palpebral part of the conjunctiva.

The tarsal plates stiffen the lids, and by their density render them more efficient organs of protection to the eye from injury or strong light. Phylogenetically they are dermic tissue condensed under the influence of the constant tension of the lids against the globe. By some authors they are considered to be correlated with the presence of the tarsal glands and the greater mobility of the eyelids (Eggeling, 1904), since they do not develop clearly until the secretory activity of the glands is manifest, but others disagree with these conclusions on morphological grounds, finding that their presence is not conditioned by that of the glands, since the eyelids of mammals contain either both glands and plates or only one of these according to species (Argaud, 1913).

The **tarsal** or **meibomian glands** (*glandulae tarsales*) were described by Meibomius in 1666, but had already been figured by Casserius in 1609. They are long, narrow, and nearly straight, though the upper ends may be bent over, and are regularly arranged in a single row perpendicularly to the free margin of the lid; they occupy almost the entire thickness of the tarsal plate, as seen in Fig. 81, sometimes forming a series of ridges on its anterior surface, and extend its whole height, consequently being tallest in its middle and shortest at the extremities. They are relatively longer in the infant.

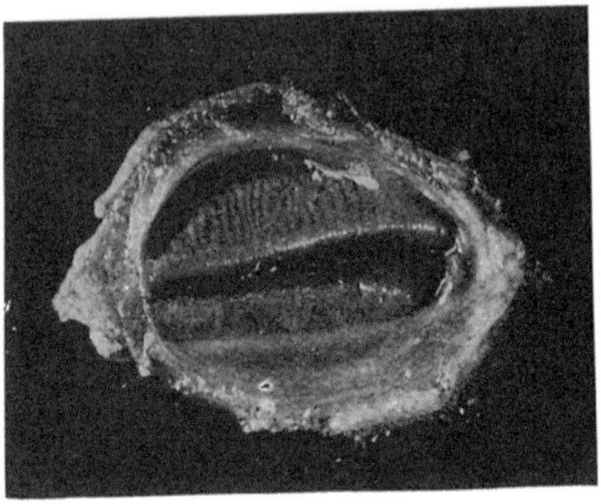

FIG. 82.—The Eyelids seen from behind after treatment with a saturated solution of urea, showing the **tarsal glands** of the tarsal plates. The puncta lacrimalia are also well displayed.

Each gland is composed of a single central canal closed at its upper end and opening below on to the palpebral margin immediately in front of its posterior edge, where the regularly arranged row of minute orifices of these glands can be seen (Fig. 77). Around nearly the whole length of the canal are grouped the acini, which open into it by lateral ducts; the acini are single or composite, and number thirty to forty in the longest gland (Sappey, 1867); they are surrounded by a lymph space lined by endothelium (Parsons, 1904), and the ducts are lined by cubical epithelial cells which become stratified at their mouths. They are sebaceous glands, and the secretion is composed of finely granular fat, the 'sebum palpebrale', which by its greasy nature lubricates the edges of the lids, preventing them from sticking together and also the tears from overflowing; they are absent from the lacrimal part of the lid margin where the tears, unless unusually abundant, first begin to trickle down. Owing to the yellow colour of the secretion the glands can be seen through the conjunctiva on everting the lid. The relation of the horizontally disposed fibres of the pars ciliaris of the orbicularis oculi muscle to the terminal canals of these glands is seen in Figs. 84, 86; they lie chiefly in front, partly behind them, and some minute bundles are even interlobular, interciliary, and retro-tarsal in position.

The tarsal glands represent the ordinary sebaceous glands of a primitive secondary range of eyelashes which have disappeared in man, a fact that explains the presence of a double row of cilia and the absence of these glands in cases of congenital distichiasis (Brailey, 1906; Stephenson, 1901), and the aberrant lashes found in a cyst on the posterior face of the tarsal plate (Szily). (The term 'districhiasis', also derived from the Greek, means the condition of a growth of two hairs from one follicle.)

The free **palpebral margin** is a region of such practical importance as to merit a special description. It is divided along its length into two regions, ciliary and lacrimal, the point of division being marked by the lacrimal papilla. The lacrimal region is devoid of eyelashes, and is smooth and rounded; in the substance of the lid beneath it runs the lacrimal canaliculus. The ciliary region bears the eyelashes, and is flat and about 2 mm. broad; it therefore presents two lips or *limbi*, an anterior, slightly rounded and continuous with the skin, and a posterior, sharply cut and continuous with the conjunctiva and lying against the globe; in the aged the edges lose their sharpness owing to the loss of elasticity of the fibres composing them. When the lids are closed a longitudinal capillary space (*rivus lacrimalis*) is said by some authors to exist between their posterior lips and the globe. The flat surface of the margin between the lips is termed the 'intermarginal region' (Terson, 1903), and it may present along its middle a slight groove or a fine grey or pigmented line, called the 'intermarginal groove', or *sulcus intermarginalis* of Graefe, best seen in the upper eyelid. This line subdivides the flat margin along its length into a ciliary part in front, from the surface of which spring the cilia, and a tarsal part behind, marked by the range of orifices of the tarsal glands; by a reference to the structure of the lids it will be seen that the lid can be split along this line into an anterior layer containing the orbicularis and ciliary follicles, and a posterior containing the tarsal plate (Fig. 83). The

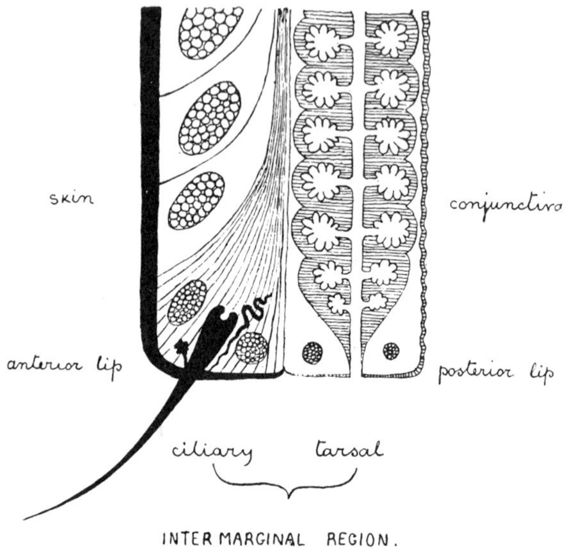

FIG. 83.—Scheme to illustrate the Structure of the Free **Palpebral Margin.**

conjunctival membrane extends as far as the glandular orifices, the rest of the border anteriorly being cutaneous. Structurally, the lid margin consists largely of dense connective tissue directly derived from the splayed-out free border of the tarsal plate: this ciliary mass appears triangular in vertical section, and embedded in it are the follicles of the eyelashes, together with their particular glands, and the fasciculi of the pars ciliaris muscle (Fig. 84).

The **cilia,** or eyelashes, are short, thick, curved hairs, usually deeper in colour than those of the head, but not blanching with age; they are relatively longer and finer in the female and infant, and are but little developed in mongoloid races (Branca, 1912). They are arranged in two or three irregularly placed rows. Those in the upper lid are larger, longer, more closely set, and more numerous than in the lower; in the former they number from 100 to 120 or even 150 (Donders, 1858), and are from 8 to 12 mm. long; in the lower lid they are from 50 to 75 in number, and 6 to 8 mm. long; they are larger in the middle region of either lid. In the

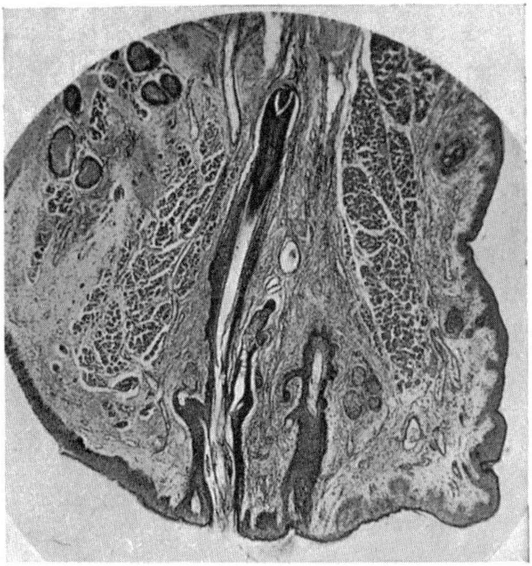

FIG. 84.—The Upper Palpebral Margin, showing **ciliary follicle** with glands of Zeis; sagittal section, enlarged; skin surface to right, conjunctiva on left. The *pars marginalis* of the *orbicularis oculi* muscle and its relation to the cilia, as described on p. 134, are well shown.

lower lid they are implanted more on its anterior face than on the actual margin; they curve upwards in the upper lid, downwards in the lower, and their convexities meet in occlusion of the lids; cylindrical in early life, they become a little flattened in the adult, a change in shape attributed to the compression exerted upon them by the marginal muscle fibres. The follicles are from 1·5 to 2·5 mm. deep, and are obliquely planted in a dense fibrous tissue derived from the tarsal plate, up to which they extend; they are surrounded by a rich plexus of sensory nerve fibres (the 'marginal plexus' of von Mises, 1882), and their relation to the ciliary muscle bundles is seen in Fig. 84. From the sides of the primary follicles secondary ones develop and give origin to new lashes, which replace the older ones after their life of about three months; certain follicles dating from foetal life may develop late and out of their proper order.

It has already been pointed out (p. 159) that the tarsal glands are to be regarded as greatly modified glands of primitive cilia which have been aborted; similarly it is found that the other skin glands at the lid margin differ from the normal.

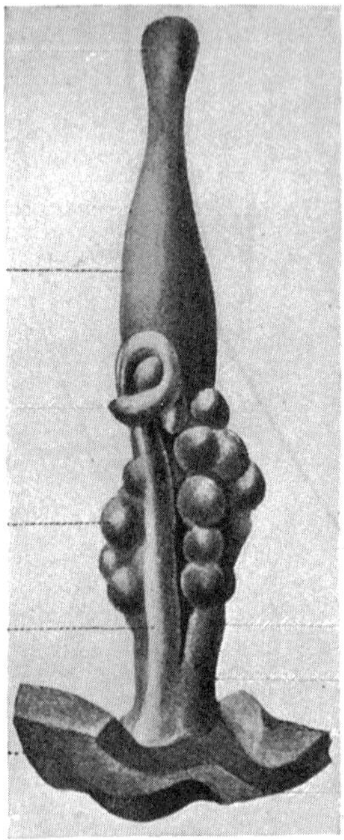

FIG. 85.—Reconstruction of the **Bulb of an Eyelash,** with its appendages. (Contino, 1907.)

The sebaceous glands are larger than those of the hairs elsewhere, and are known as the **glands of Zeis** (1835; generally spelt incorrectly as Zeiss); there are usually two for each follicle, developed by outgrowths of its epithelial wall, and they pour their secretion round the base of the hair. They consist of a short wide duct leading from a cluster of small rounded saccules (Fig. 85), which are lined with cells loaded with particles of fatty substance, and are supported by a basement membrane reinforced with a covering of connective tissue; they are better developed in the infant than in the adult.

The sweat glands are distinguished as the **glands of Moll** (1857) or 'ciliary glands' (a name which should be applied more appropriately to the glands of Zeis); they are ordinary sweat glands retarded in development, the secretory tubule forming a spiral instead of a glomerulus, but having a wider duct, and they open either into some of the ciliary follicles close to the surface or into the duct of a Zeis gland, or on to the surface of the lid margin between the eyelashes; usually there is one placed between each two cilia (Sättler, 1877). They extend deep into the lid, even as far as the tarsal plate (Fig. 86), and are better developed in the lower eyelid. The secretory part of the gland is lined by an epithelium formed by a single layer of cuboid columnar cells which increase to

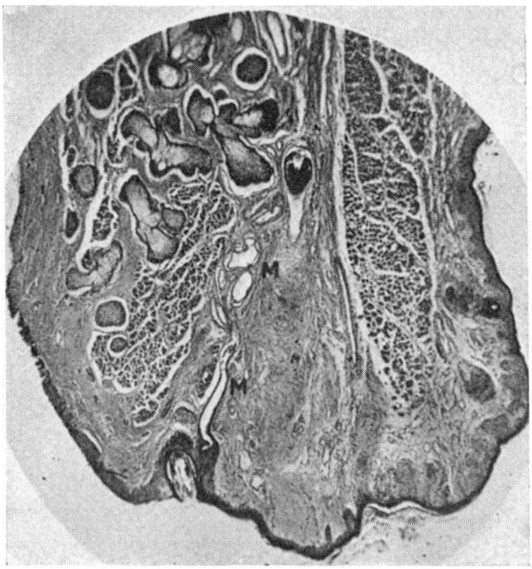

FIG. 86.

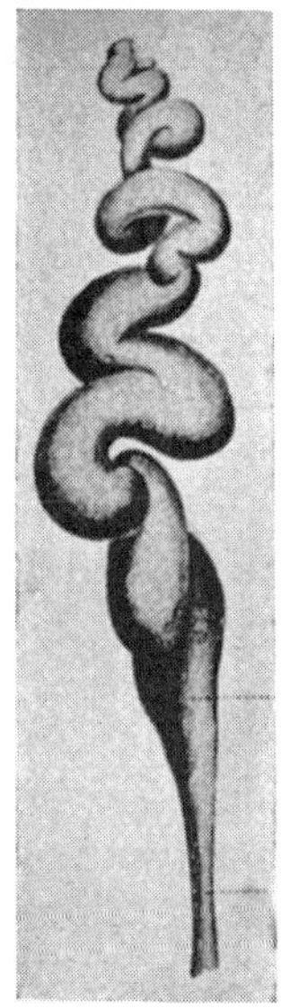

FIG. 87.

FIG. 86.—The Upper Palpebral Margin, sagittal section enlarged, showing a **Gland of Moll** (M., M.), and the various parts of the *pars marginalis* of the *orbicularis oculi* muscle (see p. 134 for description). Skin surface on right, conjunctiva and tarsal plate on left.

FIG. 87.—Reconstruction of **Gland of Moll** (Contino, 1907).

two or three layers in the duct, becoming keratinized and continuous with the epidermis at the orifice; it is supported by a basement membrane, and surrounded by smooth muscle fibres and a felting of elastic tissue (Carlini, 1907).

For the development of the lid margin see p. 189.

The **differences between the two eyelids** may here be summarized: the lower lid is thicker, shorter, and less mobile, having no special levator muscle; the size of its tarsal plate and glands and the length of its cilia are less; the orbicularis muscle is thicker over the lower lid and its marginal fasciculi are less differentiated; the inferior palpebral involuntary muscle is shorter and less developed than that of the upper lid. A section is shown on p. 198.

The **Lacrimal Caruncle** (*caruncula lacrimalis*, from the Latin diminutive of *caro*, flesh) is a small reddish oedematous-looking body about 5 mm. high by 3 mm. broad, lying in the lacus lacrimalis at the medial angle of the eye; from its mucoid appearance and position it looks as if it were a formation of the conjunctiva, but developmentally and structurally it is a detached and modified part of the lower lid margin (see p. 191). It is more prominent when the eyeball is directed laterally, being dragged forwards by the plica semilunaris of the conjunctiva with which it is connected on its under surface, and it is deeply recessed when the eye is turned medially; this recessed position is normal in infancy, but may follow as a disfigurement after operation on the medial rectus muscle, probably owing to the fact that fibres of the fascial sheath of the muscle enter its base, and so can draw it backwards (Fig. 156, p. 289). Its prominence serves to stem the flow of tears into the groove between it and the more laterally placed plica semilunaris, into which groove the lacrimal papillae are directed.

In structure the caruncle presents many traces of its developmental origin; it consists of a modified epithelium, hairs with sebaceous glands, an accessory lacrimal gland (ectodermal structures), and connective tissue and fat (mesodermal structures). The surface epithelium resembles that of the ciliary border, being of stratified pavement type but not keratinized, though Peters (1918) reports twenty-six cases of cornification of the caruncle by thickening of its epithelium. At the margins it shows a transition to the conjunctiva by a cylindrical formation and the presence of goblet cells; 'intra-epithelial glands' of simple mucoid alveolar type are described by Enslin (1905), but their existence under normal conditions was denied by Stieda (1890). The surface presents some fifteen colourless hairs which are directed towards the nose; they may be developed to such an extent as to become a nuisance and require removal, and may even resemble aberrant eyelashes in size (Berger, 1893); into the hair follicles there open glands like those of Zeis, and other sebaceous glands are seen, larger and resembling the acini of tarsal glands. These glands are the source of the white secretion that collects in this region of the eye, and are usually largely developed. Sweat glands were found by

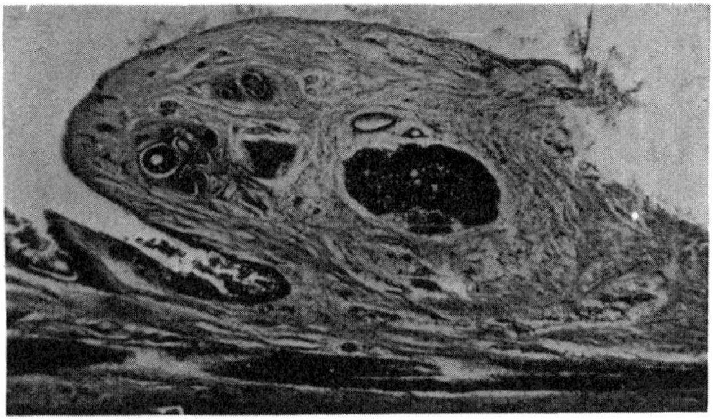

FIG. 88.—Section through the **Lacrimal Caruncle,** greatly enlarged, showing an accessory lacrimal gland in the centre, and sebaceous glands resembling the meibomian glands of the tarsal plate towards the left. The *plica semilunaris* just appears on the left side of the preparation.

Waldeyer, but have not been observed by Stieda and Terson, and could not be found in the writer's serial sections; they occur, however, in the embryo (Contino), and when persistent in the adult are perfectly formed (Virchow). The most conspicuous object in microscopical sections is the large centrally-placed 'acino-tubular gland'; structurally this resembles the accessory lacrimal glands of the conjunctiva described by Krause; it varies greatly in volume, and there may be from one to four present, though sometimes none can be found; in the writer's sections there was one large central gland with two smaller ones above it; it opens upon the surface by a large tortuous excretory canal. It is not a relic of Harder's gland, which is absent in man (*vide* plica semilunaris, p. 202), though both are present in the calf. A few smooth but no striated muscle fibres are present.

The blood-vessels are derived from the superior palpebral artery, and on account of the connective tissue through which they run they gape on section, and free haemorrhage occurs; the lymphatics drain into the submaxillary nodes. The sensory nerve supply is provided by the infra-trochlear branch of the ophthalmic division of the trigeminal nerve.

(3) **The Vessels of the Eyelids.**—The blood-vessels are derived from two sources, those of the face and those of the orbit, but there are free anastomoses between the two systems at the orbital margin, and since the facial vessels are extra-cranial whilst the orbital are intra-cranial in origin, these junctions, especially of the veins, are of considerable importance. The arteries of the facial group are all derived from the external carotid, the orbital (ophthalmic) from the internal carotid. The facial system of veins drains into the internal and external jugular veins, the ophthalmic veins into the cavernous sinus of the intra-cranial dural system (see Fig. 169, p. 317).

It will simplify description to retain here the older and more commonly known names of facial artery and facial vein, instead of the newer external maxillary artery and anterior facial vein, for these two collaterally placed vessels.

(1) The facial vessels are the facial and superficial temporal arteries and veins, and their branches supply most of the circum-orbital region and the superficial parts of the lids, as shown in the next figure.

The **facial artery** (*arteria maxillaris externa*) arises from the external carotid in the upper part of the neck, and reaches the face by passing over the ramus of the mandible just in front of the anterior border of the masseter muscle, where its pulsations can be felt in life, and where it may be compressed against the bone. It pursues a sinuous course obliquely upwards and medially towards the naso-labial angle, being crossed by the risorius and zygomaticus muscles, which separate it from the vein; it then passes vertically upwards, beneath or rarely over the levator muscles of the upper lip (quadratus labii superioris) and below the infra-orbital foramen and nerve, to the medial angle of the eye, where it is termed the 'angular artery'; in this part of its course it runs along the line of the naso-jugal skin-fold, and passes superficial to the medial palpebral ligament with the facial vein on its lateral side. Both vessels lie about 5 mm. anterior to the lacrimal fossa and are covered by skin, and also in most cases

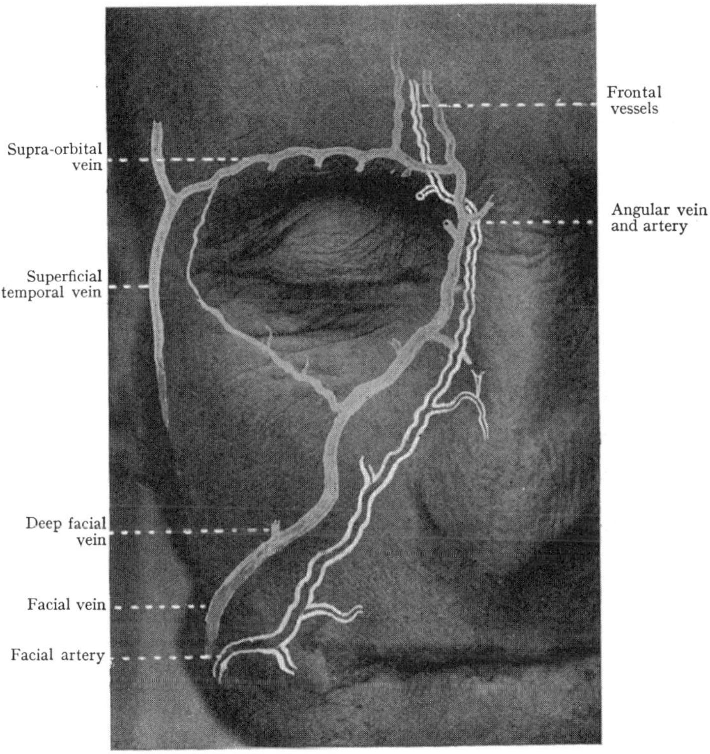

FIG. 89. Approximate positions of main **vessels of the face.**

by fibres of the orbicularis muscle. The facial artery anastomoses with the transverse facial and infra-orbital arteries and with its fellow of the opposite side, and after piercing the septum orbitale between the trochlea and the medial palpebral ligament it terminates by becoming directly continuous with the nasal branch of the ophthalmic artery. It supplies in the ocular region the orbicularis oculi muscle, the lower eyelid, and the lacrimal sac. In the upper part of its course it may be very small, the blood-supply to the lower eyelid being then derived from the vessels with which it anastomoses.

The **facial vein** (*vena facialis anterior*) is formed at the medial end of the eyebrow by the junction of the frontal and supra-orbital veins; it runs obliquely downwards and back-

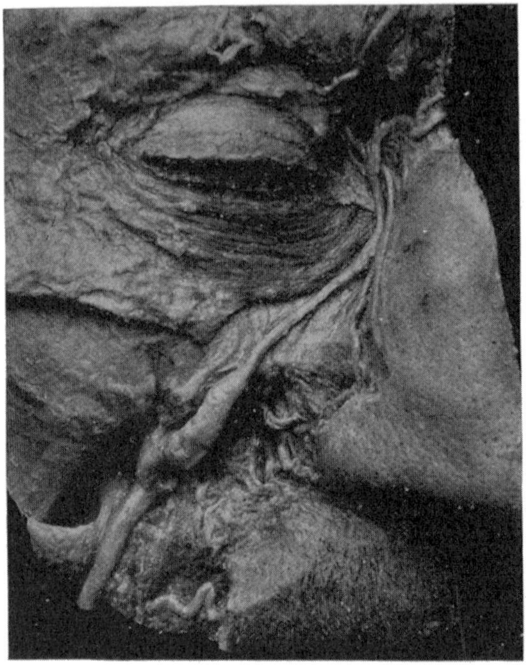

FIG. 90.—The **Facial Vein** (*vena facialis anterior*), showing varicosities. Its angular portion is seen crossing the medial palpebral ligament, with the corresponding angular artery lying to its nasal side; the overlying zygomaticus muscle has been removed.

wards across the face, along the anterior border of and then over the lower part of the masseter muscle, to cross the mandible and end in the internal jugular vein in the neck. In the first part of its course between the eye and the nose it is called the 'angular vein', and presents relations already described in connexion with its companion artery (see Fig. 59, p. 127); as it passes over the medial palpebral ligament it is distant about 8 mm. from the angle of the lids, and in this position often forms a prominence in the living; its position may also be indicated by the dark colour of the contained blood showing through the skin.

The facial vein lies lateral and posterior to the artery throughout, but takes a straighter line down the face and is more superficial, being covered only by the zygomaticus muscle. It is less flaccid than most of the superficial veins, and therefore tends to remain patent after section; it is a fairly large channel and may exhibit varicosities, as illustrated in the figure opposite. Such a condition would explain the swelling sometimes seen in the living along the line of the upper part of its course when the vessel is compressed and the blood-flow retarded by the contraction of the overlying muscle, as in laughing. Among its TRIBUTARIES the following are important:

(i) The supra-orbital vein does not accompany the artery of the same name down the forehead, but runs horizontally along the margin (hence the name marginal vein or arcade sometimes applied to it) from the temporal side, where an anastomosis is formed with the middle temporal branch of the superficial temporal vein; it lies beneath the orbicularis oculi muscle, which it pierces below the head of the eyebrow to join the frontal and so form the angular vein (Figs. 89, 91). In its course it sends a large branch through the supra-orbital notch and above the trochlea to join the superior ophthalmic vein within the orbit, and as this branch passes through the notch it receives the frontal diplöic vein, which drains the mucous membrane lining the frontal air-cell and may establish a communication with the superior longitudinal blood sinus.

(ii) The frontal vein (or fronto-nasal, represented as separate vessels by Merkel in Fig. 91) runs vertically down the forehead immediately beneath the skin, in company with the frontal or supra-orbital artery and nerve; it communicates with the superior longitudinal sinus through small foramina in the frontal bone (Piersol), and at the root of the nose is connected with its fellow of the opposite side by a short transverse 'nasal arch' (seen in Fig. 90).

(iii) The superior and inferior superficial palpebral veins

are small veins which drain the skin of the eyelids and surrounding parts and enter the angular vein; the lower one is sometimes seen to reach a considerable size.

(iv) The angular vein, like its tributary the supra-orbital, communicates freely with the superior ophthalmic vein by a branch passing into the orbit below the trochlea; it is chiefly by the junction within the orbit of these two communications from the angular and supra-orbital veins that the superior ophthalmic vein is formed.

(v) At about the middle of its course, where it passes beneath the zygomaticus muscle, the facial vein forms another important connexion by receiving the deep facial (anterior internal maxillary) vein, as seen in Fig. 90. This vessel passes backwards between the masseter and buccinator muscles into the infra-temporal fossa to join the pterygoid plexus, a venous network which communicates with the cavernous sinus and with the inferior ophthalmic vein by branches passing through the inferior orbital fissure, as illustrated on Fig. 169, p. 317.

Thus the facial vein has two chief connexions with the cavernous sinus of the intra-cranial venous system: by its free union with the superior ophthalmic vein, and by a pathway through the pterygoid plexus. It is said to be devoid of efficient valves, though Merkel, Festal, and others state that they exist at its junction with the ophthalmic vein, so disposed as to ensure a flow in a forward or facial direction (such valves could not be found in the specimen photographed in Fig. 90). Usually, however, there appears to be no hindrance to the passage of at least part of the blood-flow from the frontal region into the orbit, and this is no doubt the usual course, since when the ophthalmic system is congested during laughter or expiratory or straining efforts it is common to see the frontal veins become prominent, and in cases of orbital or intracranial tumour the ophthalmic, angular, and facial veins become congested and tortuous from pressure interference with the blood-current. Moreover, the angular vein is frequently found very small on both sides, and obviously insufficient to carry the whole flow from the frontal region into the facial vein As a consequence of the valvular deficiency, septic diseases such as malignant pustule, furuncle, carbuncle, or cancrum oris, involving the face or forehead, or even the upper lip, are exceptionally dangerous, since the infection may spread by way of the ophthalmic veins or pterygoid plexus to the cavernous sinus. See also p. 318.

The **superficial temporal vessels** call for but brief description. The artery (*arteria temporalis superficialis*) is one of the terminal branches of the external carotid, and runs up behind the angle of the jaw and in front of the ear; among other branches it gives off (i) a transverse facial branch, which runs forwards below the zygoma to terminate near the lateral margin of the lower eyelid by anastomosing with the infra-orbital and lacrimal arteries; (ii) a zygomatico-orbital branch, which passes along the upper border of the zygoma to supply the lateral parts of the lids; it also anastomoses with the lacrimal artery, and sends branches into the cavity of the orbit; (iii) a frontal branch, large and tortuous, which passes upwards and forwards across the temple to supply the frontalis and orbicularis oculi muscles, and anastomoses with the lacrimal and supra-orbital branches of the ophthalmic artery.

The vein (*vena temporalis superficialis*) communicates with the frontal and supra-orbital veins, draining the blood from the lateral region of the upper lid; it runs across the temple, then downwards in front of the ear, joins the internal maxillary vein to form the temporo-maxillary or posterior facial vein, and enters the external jugular vein; it receives a middle temporal vein, which communicates with the lacrimal and facial veins at the outer angle of the eye, and runs along the upper border of the zygoma beneath the temporal fascia.

(2) The **intra-orbital system of vessels** is comprised by the supra-orbital, frontal, nasal, and lacrimal branches of the ophthalmic artery, which supply the eyelids and the supra-orbital region in its medial half, and the general arrangement of which is shown in Fig. 164, p. 305. (i) The supra-orbital artery leaves the orbit by passing through the supra-orbital notch, or, like the nerve of the same name which it accompanies, may emerge in two branches, frontal and supra-

orbital proper, as in Fig. 91; it supplies the skin of the forehead and upper eyelid, anastomosing with the superficial temporal artery and with its fellow of the opposite side. (ii) The frontal artery has a similar but more medial distribution than the supra-orbital. (iii) The nasal artery (*arteria dorsalis nasi*) leaves the orbit by passing over the medial angle of the margin between the trochlea and medial palpebral liga-

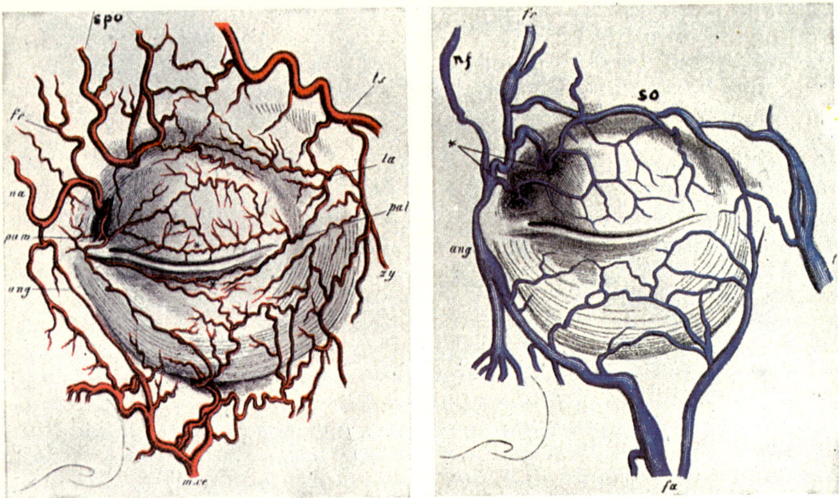

FIG. 91.—The **Arteries and Veins of the Eyelids** (Merkel, 1901).

In the first figure of the arteries: *ang.* = angular, *na.* = *dorsalis nasi*, *fr.* = frontal, *spo.* = supra-orbital, *ts.* = superficial temporal, *la.* = lacrimal, *zy.* = zygomatic, *mxe.* = facial or external maxillary with the infra-orbital just above, *pam.* and *pal.* = medial and lateral palpebral, and the asterisks mark the superior and inferior tarsal arcades.

In the second figure of the veins: *ang.* = angular, *fr.* = frontal, *t.* = temporal, *fa.* = facial, *nf.* = naso-frontal, *so.* = supra-orbital or marginal arcade, and the asterisk points to the anastomoses with the ophthalmic vein.

ment; it is the direct continuation of the main ophthalmic vessel, and becomes continuous with the angular artery on the face. (iv) The lacrimal artery, after supplying the lacrimal gland, gives off terminal twigs to the lateral ends of the eyelids. The ophthalmic veins have a similar distribution. The intra-orbital origin and course of the vessels are described later (p. 303); and an account of the infra-orbital artery, which, though extra-cranial in origin, is intra-orbital in position and supplies part of the orbital contents and the lower eyelid, will be found on p. 310.

(3) The skin of the lids and the orbicularis oculi muscles are supplied by freely-anastomosing branches of all the above-mentioned vessels; the deeper parts of the lids are supplied by the specially denoted **palpebral branches** of the nasal and lacrimal vessels, which are medial and lateral respectively.

The medial palpebral arteries (*arteriae palpebrales mediales*) are two vessels which arise, usually by separate origins, from the nasal branch of the ophthalmic just before it leaves the orbit below the trochlea; they descend behind the pars lacrimalis muscle and lacrimal sac, and one enters the medial end of each lid. Before doing so the superior vessel supplies the caruncle, lacrimal canaliculi, lacrimal sac, and pars lacrimalis muscle; the inferior is joined by an orbital branch from the infra-orbital artery, and supplies the naso-lacrimal duct as far as its entrance into the inferior meatus of the nose. On entering the lids each palpebral artery bifurcates into a larger marginal and a smaller peripheral branch, which run along in relation to the borders of the tarsal plates towards the lateral extremities of the lids, where they unite and anastomose with two corresponding lateral palpebral branches (*arteriae palpebrales laterales*) provided by the lacrimal artery, making connexion also with the zygomatico-orbitalis branch of the superficial temporal artery; in this way two arterial arcades (*arcus tarseus superior vel inferior*) are formed in each lid, an inferior or marginal and a superior or peripheral, a distinction much better marked in the upper and broader eyelid. The marginal tarsal arcade with its accompanying vein forms a conspicuous feature in sections of the eyelids (Fig. 79, p. 156); it lies about 3 mm. from the free border of the lid, at the base of the ciliary region, directly on the tarsal plate and beneath the orbicularis muscle; it supplies the ciliary follicles and glands. The peripheral tarsal arcade pursues a sinuous course along the convex border of the tarsal plate, and in the upper lid is found in the pre-tarsal space between the aponeurosis of the levator and the superior palpebral involuntary muscle. Most of its twigs pass as 'posterior conjunctival arteries' through the latter muscle to supply the conjunctiva of the fornix and bulbar region (Fig.

109, p. 204). The peripheral arcade of the lower lid is inconstant, being replaced by branches of the superficial vessels, and it does not supply the conjunctiva. Minute twigs pass from one arcade to the other, forming plexuses in front and behind the tarsal plates; the pre-tarsal plexus supplies all the structures in front of the plate and ramifies between the tarsal glands, whilst the post-tarsal plexus (retro-tarsal or sub-conjunctival) supplies chiefly the conjunctiva, as described on p. 203 (Fuchs, 1878; Langer, 1878). The circulation at the margins of the lids is terminal, so tending to stagnate and cause redness.

The veins of the lids are larger and more superficially placed than the arteries, of which they run independently; in the deeper parts of the lids they form less regular arcades and plexuses, which present similar relations to the tarsal plates. The arcades drain medially into the angular vein, and laterally into the superficial temporal and lacrimal veins. The pre-tarsal plexus drains blood from around the tarsal glands, ciliary follicles, and free border of the lid; the superficial venules of the skin pass backwards between the fasciculi of the orbicularis oculi muscle to enter this plexus, and consequently are occluded in severe blepharospasm, so causing transudation and subcutaneous oedema. The retro-tarsal or sub-conjunctival plexus in the upper lid drains into the superior arcade, and thence into the superior ophthalmic vein; it communicates with the veins of the muscles of the globe, but also with the cutaneous veins; in the lower lid the sub-conjunctival veins pass entirely into the inferior ophthalmic vein, and do not communicate with the cutaneous veins. These veins have been described in detail by Festal (1887).

The **lymphatic vessels** of the eyelids, like the blood-vessels, are divided into superficial and deep groups, the former draining the skin and orbicularis oculi, the latter forming plexuses in relation to the tarsal plates and the conjunctiva; but from a practical point of view they are better divided into medial and lateral groups, which collect into larger trunks and pass by separate paths to different lymph nodes.

The medial group of vessels (which are connected with those of the opposite side of the face) drains the superficial parts of the medial half of the lower lid, the medial com-

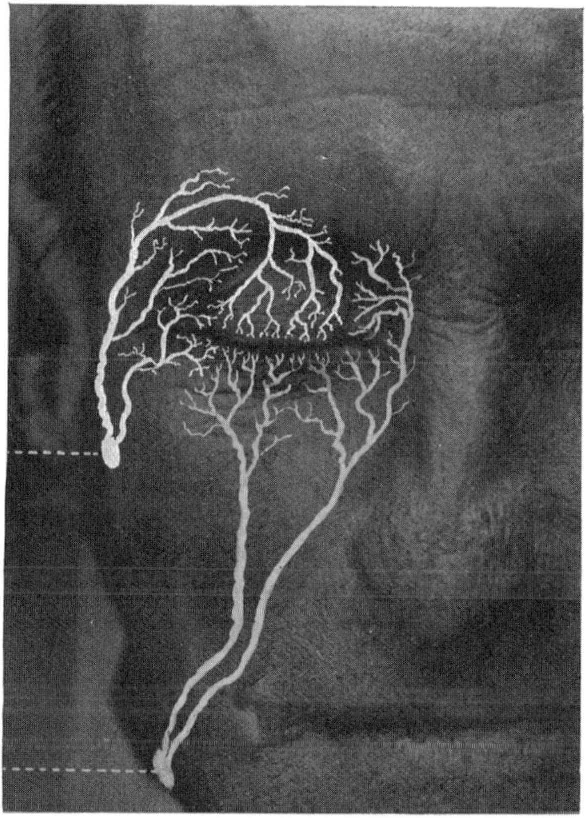

FIG. 92.—To illustrate the areas of the Skin and Conjunctiva of the Eyelids which drain by lateral and medial groups of **lymph vessels** into the parotid and submaxillary lymph glands respectively.

missure and medial end of the upper lid, and by deeper branches the conjunctiva of the medial two-thirds of the lower lid, and the lacrimal caruncle; from this field two superficial and two deep trunks follow the course of the facial vein to enter the submaxillary lymph nodes close beneath the lower border of the mandible. The superficial vessels enter a node lying in front of the vein, the deep ones into a node just behind it. Interruptions in the course of the vessels by either facial, infra-orbital, lateral nasal, or buccal nodes appear to be rare, according to most observers; the writer has not found any present in his series of dissections.

The lateral group of vessels drains a larger field; the superficial ones come from the skin of nearly the whole of the upper and the lateral half of the lower lid, the deeper ones from the conjunctiva of the whole of the upper and lateral third of the lower lid; they pass as three or four trunks across the upper part of the cheek to enter the upper superficial and deep parotid nodes in front of the ear, the first and chief regional gland being situated superficially in the parotid gland at the level of the external acoustic meatus. Both these and the submaxillary nodes are connected with the deep cervical glands near the junction of the common facial with the internal jugular vein (Most, 1905; Bartels, 1909; Dewey, 1920).

The deeper lymphatic vessels of the lids form pre-tarsal and retro-tarsal or sub-conjunctival plexuses which communicate with one another through the thickness of the tarsal plate (Grunert, 1901); there is also a rich plexus along the convex border of the plate, the presence of which explains why the conjunctiva in this region is more readily invaded by pathological processes. The lymph round the tarsal glands passes into the sub-conjunctival plexus, the vessels of which are provided with valves, whereas the pre-tarsal vessels are devoid of them, so that the deep lymph-flow is towards the conjunctiva and is in the opposite direction to the blood-flow (Fuchs, 1878). See also p. 205 (conjunctiva).

(4) **The Nerves of the Eyelids.**—The nerves are motor, sensory, and sympathetic; the motor nerve is the facial, and it innervates the orbicularis oculi, corrugator supercilii, and frontalis muscles in this region; the sensory nerves to the skin are provided by numerous branches of the trigeminal nerve; the sympathetic nervous system supplies the involuntary palpebral muscles, the skin glands, and the vessels, &c.

The **facial** or seventh nerve (*nervus facialis*) arises from a long group of cells in the lower part of the pons, forms a curved loop over the nucleus of the abducent or sixth nerve, and emerges on the surface of the brain at the lower border of the pons, as is indicated in Fig. 209, p. 407 (see also p. 187. The morphology of the nucleus is considered by Kappers, 1929). The nerve then passes into the opening of the internal acoustic meatus of the temporal bone in company with its sensory part, the *nervus intermedius*, and the acoustic nerve; it traverses the facial canal or aqueduct of Fallopius and leaves the bone by the stylo-mastoid foramen; it passes on to the face below the external ear and behind the ramus of the mandible, and divides in the substance of the parotid gland into numerous radiating terminal branches, which are distributed to all the muscles of the face and the platysma of the neck. These branches are grouped into two main divisions of the nerve, an upper temporo-facial and a lower cervico-facial division.

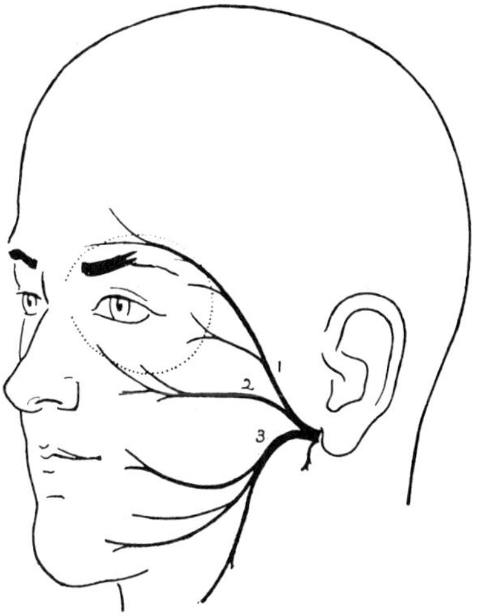

FIG. 93.—Diagram of the main branches of the **Facial Nerve** (from a dissection by G. Farmer, in the Oxford Anatomical Museum).

1 and 2 are the temporal and zygomatic branches respectively of the temporo-facial division of the nerve; 3 is the cervico-facial division with its three branches, buccal, mandibular, and cervical. The orbicularis oculi muscle is outlined by a dotted line.

Of the upper division, the temporal branches pass forwards and upwards over the zygoma in the deeper stratum of the superficial fascia, to be distributed to the upper part of the orbicularis oculi, the corrugator supercilii and the frontalis muscles, which they enter from the lateral side and upon their deep surfaces; the branch to the upper part of the orbicularis (superior palpebral branch) runs a little above the lateral angle of the eye and parallel to the supra-orbital margin as far as its mid-line. The zygomatic branches (malar and infra-orbital) pass across the cheek at a deeper level ($\frac{1}{4}$–$\frac{1}{2}$ inch) embedded in fat, to supply the lower part of the orbicularis oculi and lower facial muscles, the chief branch running along the lower border of the zygomatic bone (Friteau). The most important innervation of the muscles round the eye is therefore seen to be provided by the temporal branch of the facial nerve.

It may be noted that the levator palpebrae superioris is innervated by the upper division of the oculomotor or third nerve, in common with the superior rectus muscle of the eyeball.

All the motor branches of the facial nerve communicate with sensory branches of the trigeminal; the temporal branch with the auriculo-temporal, zygomatico-temporal and supra-orbital nerves, the zygomatic branch with the zygomatico-facial, infra-orbital, infra-trochlear and nasal nerves; by the union of the zygomatic branch and infra-orbital nerve an extensive sensori-motor plexus is formed below the infra-orbital foramen, which is the most important anastomosis between the facial and trigeminal nerves.

Davis (1923), and Ruskin (1928) adduce clinical and experimental evidence to show that the facial is a mixed nerve in man, containing certain afferent or sensory fibres whose cells of origin lie in the geniculate ganglion and transmit pressure and pain sensations and muscle-sense fibres (proprioceptive) from the face through the *nervus intermedius*.

The cortex of the brain is connected with the motor facial nucleus by fibres proceeding from cells situated in the lower third of the praecentral convolution (Fig. 204, p. 394); they traverse the genu of the internal capsule and the cerebral peduncle in reaching the pons. The facial nucleus also receives fibres from the spinal tract or descending root of the trigeminal nerve by means of which the orbicular reflex path is formed; the nucleus is connected with the auditory nerve, and also with the visual and auditory centres, probably by paths through the medial (posterior) longitudinal fasciculus, by means of which the facial nerve cells respond to impulses of sight and hearing, as shown by the automatic closure of the lids to a threatened blow or loud sound. See also p. 414

The **trigeminal** or fifth nerve (*nervus trigeminus*) is the great sensory nerve of the face and forehead, and is also the motor nerve to the muscles of mastication; its sensory part consists of three divisions—ophthalmic, maxillary, and mandibular (the last of which is joined by the motor part of the nerve). The ophthalmic nerve supplies the skin of the eyelids and surrounding region by its supra-orbital, supra-

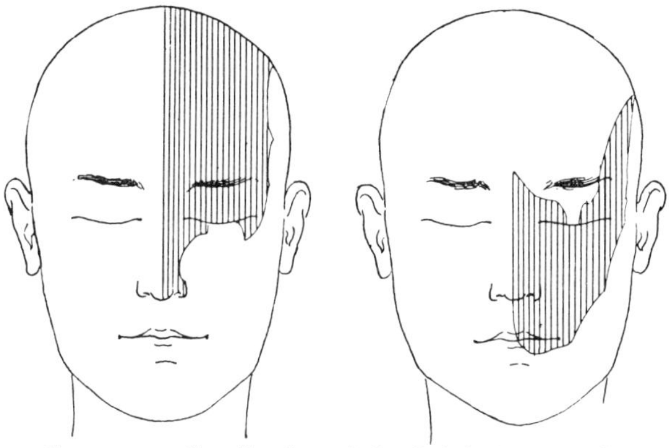

FIG. 94.—**Cutaneous distribution** of the Ophthalmic (left-hand figure) and Maxillary (right-hand figure) Divisions of the Trigeminal Nerve, after Zander (1897). The shaded areas show loss of touch and pain after section of the respective nerves. Note that there is considerable overlapping of the areas.

trochlear, infra-trochlear, lateral nasal and lacrimal branches; the maxillary nerve contributes infra-orbital, zygomatico-facial and zygomatico-temporal branches; the mandibular nerve contributes no branches to the palpebral region, and is not related to the orbit, being destined for the muscles of mastication, the teeth, and other parts in connexion with the lower jaw. A description of the origin and course of the trigeminal nerve and its divisions will be given in connexion with the nerves of the orbit on p. 347 and of its nuclei and intra-cerebral connexions on p. 413, but here it may be noted that all the above-mentioned terminal cutaneous branches of the ophthalmic and maxillary divisions have passed through the cavernous sinus and the orbital cavity before reaching the face; the first group (ophthalmic) enter the orbit through the superior orbital fissure, are mostly related to its roof, and pierce the septum orbitale at the orbital margin; the second group (maxillary) enter by the inferior orbital fissure, are related to the floor and lateral wall of the orbit, and emerge on to the face through bony canals.

The cutaneous areas supplied by the ophthalmic and maxillary divisions of the trigeminal nerve, as given by Zander (1897), are shown in Fig. 94, which demonstrates the areas of loss of touch and of pain after section of the nerves,

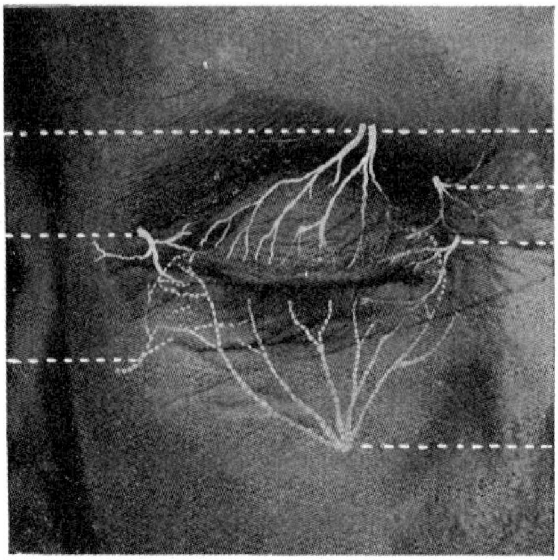

FIG. 95.—The **Palpebral Branches of the Ophthalmic Nerve** (solid lines) and of the Maxillary Nerve (dotted lines), right eye.

but he states that the cutaneous distribution is so inconstant that he never found two identical fields, even on the two sides of the same head; the areas supplied by the two nerves largely overlap on the nasal side.

The skin of the upper eyelid is mainly supplied by palpebral branches derived from the supra-orbital branch of the frontal (ophthalmic) nerve, the skin of the lower lid by the palpebral branches of the infra-orbital branch of the maxillary nerve; the medial extremities of both lids are supplied by the supra- and infra-trochlear nerves, and the lateral extremities by the lacrimal nerve, all three of which are branches of the ophthalmic (Fig. 189, p. 355). The lacrimal nerve is stated by Frohse not to supply the skin (Quain), but the writer has several times traced its cutaneous branches, as in Figs. 64, 71. The nerves run at first beneath the orbicularis, then traverse it to reach the skin; in each lid a plexus is formed on the face of the tarsal plate, from which branches pass backwards through it to supply the tarsal glands and conjunctiva (Bach, 1895). The nerve supply is especially abundant in the free margins of the lids, forming a 'marginal plexus' (von Mises, 1882), which is a special sensory apparatus connected with the cilia; terminal nerve corpuscles analogous to those of Meissner and Krause have been found present between the posterior edge of the lid margin and the orifices of the tarsal glands (Dogiel, 1895).

Irritation or excitation of the **sympathetic** system in the neck causes widening of the palpebral opening by its stimulatory action upon the involuntary palpebral muscles, and section of the nerve causes a drooping of the upper eyelid and narrowing of the fissure by relaxation of the muscles. The nerve fibres to these muscles may either accompany the arterial twigs or be incorporated with the nerves to the levator and inferior oblique muscles.

(5) The **Action of the Muscles** and movements of the eyelids.—The muscles which affect the movements of the eyelids are the orbicularis oculi, corrugator supercilii, frontalis, levator palpebrae superioris and the involuntary palpebral muscles. Their individual actions are as follows:

The orbicularis oculi is a sphincter muscle, and is seen to act as a whole in forced occlusion of the lids, in the presence of strong light or in protecting the eye from a blow: the palpebral margins are closely approximated, the skin is pursed up all round the fissure, the eyebrow is drawn downwards, and even the cheek is raised, the zygomaticus and levator muscles of the upper lip participating in the exaggerated movement. (It is also stated that in strong contraction of the orbicularis the eyeball is rolled upwards, and hence the superior rectus and inferior oblique muscles are also thrown into action.) The muscle also supports the orbital contents and prevents congestion of the globe in such forced actions as accompany vomiting, straining, etc., in which the intra- and extra-ocular pressure is raised. In partial closure of the lids, as occasioned by a glaring light, the lacrimal portions of the palpebral margins come in contact over the lacus lacrimalis, so that the whole palpebral opening is much reduced in horizontal width as well as in height, as represented by Fig. 96; the effect, however, is not seen in childhood, where the curve of the lacus lacrimalis is relatively wider.

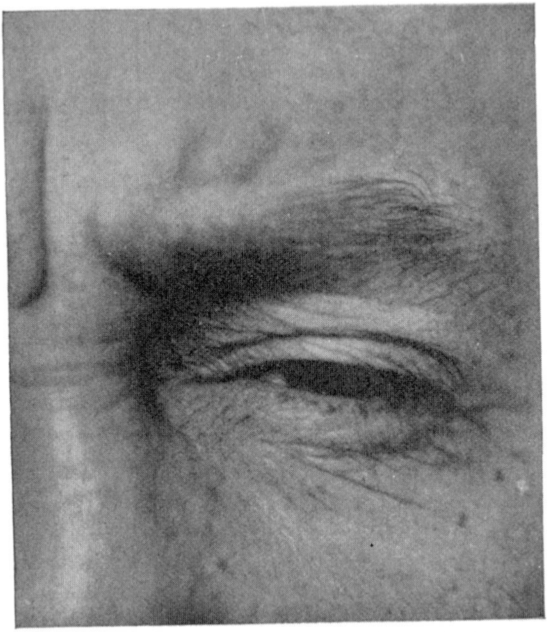

FIG. 96.—Photograph showing the **action of the Orbicularis Oculi** Muscle on exposing the eye to glaring light; the lacrimal portions of the lid margins are closely approximated by contraction of the pre-tarsal fibres, so that the palpebral fissure is narrowed in the horizontal direction as well as in the vertical. Note the wrinkling of the skin above the head of the eyebrow, caused by the *corrugator supercilii* muscle, and compare with Fig. 52, p. 113.

As regards the various subdivisions of the muscle, a distinction in their actions is to be noted. A difference between the orbital and palpebral parts of the muscle is illustrated by the act of holding an eyeglass: it is kept in place by the voluntary contraction of the pars orbitalis, whilst the pars palpebralis retains its independent function of moistening the surface of the globe by involuntary periodic contraction; moreover, paralysis of the one part may occur without the the other being affected, illustrating the fact that a voluntary muscle may be paralysed for a voluntary action and yet be perfectly free to act in an involuntary action. In gentle voluntary closure the movement can be confined to the palpebral part, and in the involuntary act of winking the tarsal subdivision alone contracts. The pars tarsalis keeps the tarsal plate in close contact with the globe (as explained on p. 155), whilst its lacrimal fibres (Horner's muscle) more especially direct the puncta of the lacrimal papillae backwards towards the lacus lacrimalis in which the tears collect, so that when this part of the muscle is inactive the lower lid tends to fall away from the surface of the globe, and the condition known as 'ectropion' is produced. The effect of action of this part of the muscle on the canaliculi is also referred to on p. 227. The fibres of the muscle nearest the palpebral border (pars ciliaris) have the important function of keeping the lid edges in close approximation to the globe, so contracting the mouth of the conjunctival sac and preventing the entrance of foreign bodies; these fibres, from their close relation to the terminal portion of the tarsal glands, may be concerned with the secretion of the latter, and those immediately overlying the bulbs of the cilia may, by their pressure in contraction, determine the curve of the outgrowing eyelashes. The orbicularis is antagonized in its palpebral part by the levator palpebrae superioris, in its orbital part by the frontalis in the upper eyelid, and in the lower lid by the inferior rectus and by gravity.

The corrugator supercilii muscle acts in harmony with the orbicularis, drawing the eyebrow downwards, as in shielding the eye from strong light; it also draws the brow towards the root of the nose and so contracts the skin of the forehead into vertical folds; it is the special muscle of frowning, an essentially human gesture to be noted even in the infant, though the muscle itself is not peculiar to man (Darwin, 1872). Sir Charles Bell, indeed, ranks it as the most remarkable muscle of the human face, since frowning irresistibly conveys the idea of mind and gives the countenance an aspect of intellectual energy.

The frontalis muscle raises the eyebrow and throws the skin of the forehead into horizontal folds, a movement that may become fixed in the aged; as regards facial expression, it is the muscle of attention and astonishment. It is antagonized locally and moderately by the orbicularis oculi when it takes its action from the galea aponeurotica as a fixed point; distally and more strongly it is counteracted by the occipitalis, the simultaneous contraction of which upon exaggerated raising of the eyebrows can be felt by placing the hand on the back of the head.

The levator palpebrae superioris muscle is the special elevator of the upper eyelid; by the contraction of its muscle belly the transversely-disposed aponeurosis is swung backwards over the globe like the vizor of a helmet, pulling with it the skin of the lid into which its terminal fibres are inserted, and so deepening the superior palpebral fold; at the same time the tarsal plate is raised both by the terminal fibres attached to the lower part of its face and by the conjoint action of the involuntary palpebral muscle inserted into its margin, and the loose conjunctiva of the superior fornix is pulled up by the facial attachment of the levator muscle sheath. The levator acts in harmony with the superior rectus muscle of the eyeball, with which it shares the same nerve supply, though it also acts independently of it, as in winking; it can further act in unison with the frontalis, as in expressing astonishment, and all three muscles act together in looking upwards in extreme excursion of the globe. The light degree of synergy which exists between the actions of these muscles is sometimes used to remedy cases of ptosis by connecting the eyebrow and eyelid more closely together, and so bringing the lid under direct influence of the frontalis. The levator is antagonized by the palpebral part of the orbicularis oculi muscle and by gravity. The action of the palpebral involuntary muscles is to widen the palpebral fissure in the vertical direction.

The **movements** of the eyelids.—The lids are closed by the action of the orbicularis oculi, the upper lid being also lowered by gravity and slightly by relaxation of the palpebral involuntary muscle; they are separated by contraction of the levator and relaxation of the orbicularis. Since the tone of the orbicularis is stronger than that of the levator, the lid is lowered in sleep. Neither lid can move quite independently of the other, since they are connected at the commissures. The movements of the globe are to some extent transmitted to the eyelids, the lower lid appearing to be pushed forwards and slightly raised in turning the eye upwards, though this movement may be caused by its commissural attachments to the upper lid. The association of these muscle movements is tabulated by Gowers (1879) as follows (with additions):

	Looking down.	Looking up.	Closing gently.	Closing forcibly.
Orbicularis . .	inaction	relaxation	contraction	contraction
Levator . .	relaxation	contraction	relaxation	relaxation
Superior rectus .	relaxation	contraction	inaction	contraction
Inferior rectus .	contraction	relaxation	inaction	relaxation

As regards types of movement there are three: voluntary (as in occlusion with effort), automatic (winking), and reflex (blinking). The reflex path for automatic involuntary winking is from the cornea (naso-ciliary nerve) and conjunctiva (infra-trochlear and lacrimal nerves) by afferent sensory branches of the trigeminal nerve to the cerebral centres, and thence by efferent motor branches of the facial nerve to the orbicularis (p. 414). There are two reflex arcs in spontaneous winking:

(i) Trigemino-facial, the reflex excitable stimulus being cooling and dryness of the cornea or touching of sensory parts.

(ii) Optico-facial, the excitable stimulus being strong light or proximity to danger (Lans, 1901).

Certain facts concerning the relation of these muscles to vision are of **clinical interest.** The frontalis is affected in certain forms of imbalance of the muscles of the globe, and a horizontal wrinkling of the forehead is a characteristic symptom in 'hyperphoria' (the name given to the condition when one pupil is higher than the other); moreover, the connexion of the muscle by means of the galea aponeurotica and occipital bellies with the deep fascia at the back of the head and neck (Fig. 51, p. 110) would explain the occurrence of an occipital headache, which in some cases is relieved by the use of suitable glasses, and the relief of severe headaches associated with efforts of accommodation of the lens which is said to follow complete severance of the muscle (see also p. 414). The corrugator contracts in the presence of too strong a light, in most myopic subjects in far vision, and in nearly all subjects in close work; indeed one of the symptoms of actual excessive accommodation of the lens is vertical wrinkling of the forehead. On such grounds the frontalis, orbicularis, and corrugator have been grouped together under the name of the **accessory muscles of accommodation,** and they become of clinical importance in connexion with certain forms of ocular headache (Howe, 1907). These muscles (which are supplied by the facial nerve) are called into action by the efforts of the ciliary muscle of the lens (supplied by the oculomotor nerve) which accompany excessive accommodation, and corroborative evidence of connexion between the two sets of muscles supplied by the two nerves is further shown by the occurrence of cases of total ophthalmoplegia (paralysis of the muscles of the globe, both extrinsic and intrinsic) accompanied by paralysis of the frontalis and orbicularis (Aldren Turner, 1890). Moreover, a differentiation between these muscles (the upper facial group) and the rest of the facial muscles (the lower group, comprising the muscles of the nose, cheek, and mouth) is indicated by the fact that the former commonly escape in facial nerve paralysis of cerebral origin, though the peripheral part of the nerve supplies both groups of muscles and so all are affected in paralyses of extra-cerebral origin. (See p. 187.)

Various explanations of the above conditions are given. An explanation of the headache is that the ophthalmic division of the trigeminal nerve, which supplies the skin of the forehead, represents the sensory somatic nerve of the same primitive segment of the neural canal from which the oculomotor nerve is derived, and frontal headache in disorders of accommodation is caused by reflex pain. An afferent path from the ciliary region of the eyeball to the brain is afforded either by the naso-ciliary nerve or by the communicating branch which the oculomotor nerve receives from the ophthalmic nerve in the cavernous sinus, or by both; in each case the brain may refer the pain not to the site of origin but to some other sensory field of the trigeminal (of which the naso-ciliary and ophthalmic are branches), just as it does in a neuralgia of dental cause, for example (see p. 350).

As regards the synergy of action of the upper group of facial muscles with the ciliary muscles, whilst no actual branch of the oculomotor nerve which supplies the latter has been traced to enter the facial group, it has been suggested that nerve-fibres destined for their innervation, although passing out of the brain in the substance of the facial nerve, either (i) actually arise from cells of the lower part of the oculomotor nucleus and then pass *via* the medial longitudinal bundle to emerge as part of the facial nerve itself (Mendel, 1887; Starling, 1920; an explanation denied by Bishop Harman on morphological grounds); or (ii) arise from a more dorsally placed group of facial nerve-cells which is distinguished as the 'superior facial nucleus' from three vertical sub-groups, and although specially supplying the upper facial muscles by its peripheral fibres, is connected by means of the medial longitudinal bundle with the oculomotor nucleus, and so can act harmoniously with it (Bruce and Pirie, 1908; Von Gehuchten; Villiger; Bailey, 1923). As regards the escape of the upper facial muscles in a paralysis, apart from the possible exemption of such segregated cells in a nuclear lesion, it is generally considered that the superior facial nucleus receives a bilateral innervation from the motor centres of both cerebral hemispheres, and the muscles it supplies will therefore be unaffected by a unilateral cortical lesion.

The phenomenon of 'jaw-winking', when there are associated movements of the jaw and eyelids, is considered on p. 416.

(6) **The Development** of the eyelids.—The face begins to be developed towards the end of the third week of foetal

life; about this time five processes arise from the base of the primitive cerebral capsule, and by the end of the second month they have completely united to form the facial part of the head and surround the stomodaeum or primitive mouth. Of these five processes one, the fronto-nasal, is median in position, and is composed of symmetrical right and left halves, the others are placed laterally, two on each side,

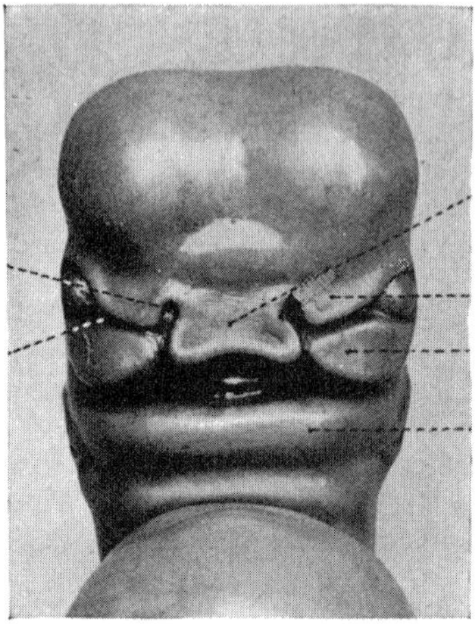

FIG. 97.

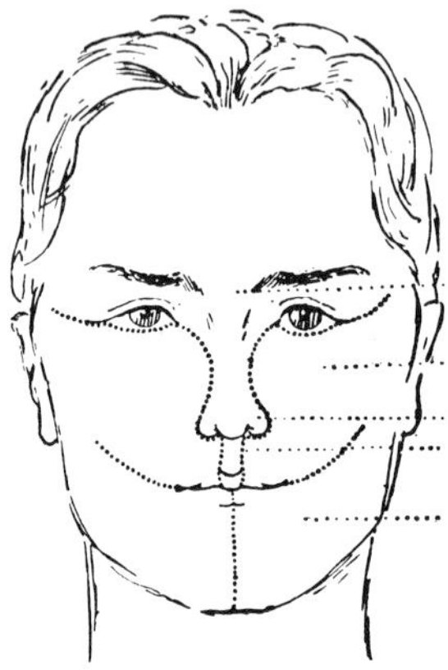

FIG. 98.

FIG. 97.—Model of the **Face of an Embryo** (reconstruction by His) to show the processes which form it. The two mandibular processes have fused at the stage taken.

FIG. 98.—Showing the **corresponding parts of the Adult Face.**

and are called the maxillary and mandibular processes (Fig. 97). The fronto-nasal process becomes subdivided by the appearance of two olfactory pits into a median and a lateral nasal process on each side. The cleft between each lateral nasal and the adjacent maxillary process leads upwards to the already developed eyelids, and is called the naso-optic groove; in its lower part the naso-lacrimal duct is developed, and its upper part is continuous with the future orbit; it is obliterated by fusion of the adjacent processes in the second month of foetal life. The parts of the adult face derived from each process are shown in Fig. 98.

The eyelids are formed in the earlier weeks of the third

month of foetal life by folds of epiblast, which project above and below the front part of the eyeball and enclose mesoblastic tissue in their thickness. The lower eyelid is derived from the primitive maxillary process; the upper eyelid is developed from the fronto-nasal process in two parts, medial and lateral, and occasionally in the adult a notch on the margin may mark their point of union. The epiblast on the outer surface of the folds develops into the skin of the lids, whilst that on the ocular surface becomes modified into the conjunctival membrane, which comes into contact with the anterior surface of the globe and is thus continuous with the epiblast of the cornea. The edges of the folds become the palpebral margins, and ingrowths from the epiblast in this region into the substance of the lids form the ciliary follicles and the tarsal and other glands, which are all therefore ectodermal in origin. From the mesodermal tissue enclosed between the skin-folds there are developed the orbicularis oculi muscle, tarsal plates, vessels, and connective, elastic and fatty tissues.

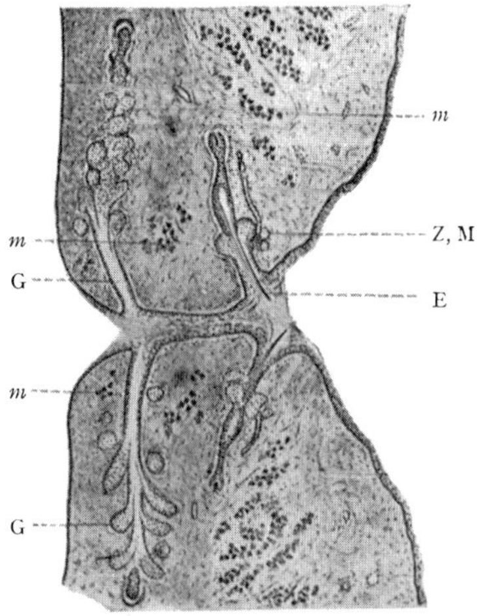

FIG. 99.—**Developing lid margins** at the time of separation in a foetus of 225 mm. A tarsal gland (G.) and an eyelash (E.) with glands of Zeis (Z.) and Moll (M.) are seen in each lid, and the subdivisions of the *pars marginalis* of the *orbicularis oculi* muscle (*m.*) are shown. (Contino, 1907.)

About the beginning or middle of the third month (that is, as observed in an embryo of 31 mm. by Ask, 1908) the edges of the lids meet, adhere, and do not separate until the end of the sixth month; the fusion begins at the two extremities of the palpebral fissure, and was seen quite complete in an embryo of 37 mm. (Contino, 1907); it is after the fusion of the lids that the epiblast on their deep surfaces becomes modified into the conjunctiva. The separation of the eyelids begins at the end of the fifth month on the nasal side, and is effected by a process of cornification of the fused edges, aided by the pressure of the growing eyelashes and secretion of the glands (Fig. 99); the actual date of separation presents individual variations, and Oppenheimer records a case where it was not complete at term, a condition which is normal in kittens, rabbits, and mice. The details of the development of the lid margin will be found in articles by the authors quoted above, and the comparative anatomy of the lids has been specially studied by Eggeling (1904).

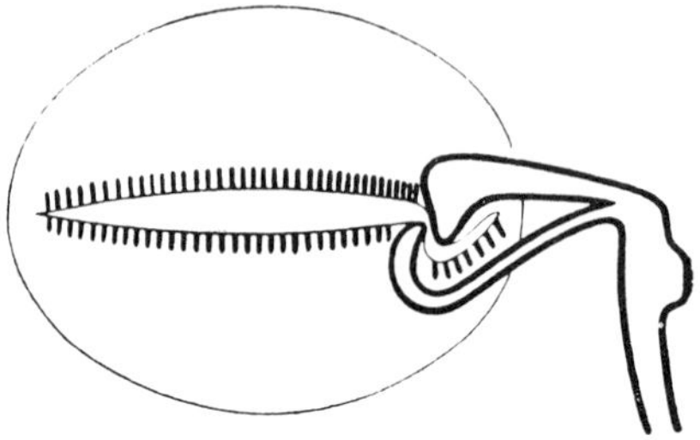

FIG. 100.—Development of the **Lacrimal Caruncle.** Scheme of the modification produced in the medial part of the Eyelids by the growth of the Lacrimal Canaliculi; the lower one cuts off a part of the lid margin which becomes the lacrimal caruncle. The lids are really fused at this stage (embryo of 170 mm.) but are shown separated for the sake of more clearly representing the meibomian glands (the vertical lines). (From Ask, 1908.)

The tarsal or meibomian glands begin to develop in the embryo at about the twelfth week, arising as inward buddings from the epithelium of the posterior edges of the fused margins of the lids; those of the upper lid exceed the lower ones in length only after the middle of embryonic life. The glands of Zeis and Moll arise as pouchings out of the basal cells of the *anlage* of the cilia at about three months. The tarsal plates appear as a condensation of mesoderm at about ten weeks, but develop more clearly when the secretory activity of the tarsal glands is established. The septum orbitale and aponeurosis of the levator palpebrae superioris begin to be differentiated from mesoblastic tissue at about six months. The orbicularis appears at ten weeks, and its ciliary bundle is separated off by the ingrowing bulbs of the eyelashes at twelve weeks. The development of the vessels of the lids has been studied by Engel, of whose work a résumé is given by Picou (1912).

The lacrimal caruncle appears immediately after the fusion

of the lids, and is formed from the medial end of the lower lid margin by a portion being cut off in the development of the inferior lacrimal canaliculus (Fig. 100). Hence the caruncle is not developed in animals, such as the pinnipedes, which have no lacrimal passages (Broman and Ask, 1910). It is definitely isolated in about five days, when its ectodermal covering becomes modified, as does its mesodermal body. In many animals, such as, for example, the calf, it retains a cutaneous formation and is still connected with the lid margin by a bridge of skin, a primitive condition that has been seen to persist in man (Peschel, 1903). Its cutaneous origin is also shown by the occasional presence of small papillomata. Reitsch (1929), in a consideration of this region, shows the gradual recession of the caruncle and plica semilunaris from their more prominent position in the new-born to that in the adult.

(7) The **congenital abnormalities** of the eyelids are explained by reference to the above short account of the development. Only chief examples are here noted, and for further details the reader should consult the articles by Lang and Treacher Collins (1900), and Hardy (1914)

(i) The eyelids may be absent, when the conjunctiva fails to develop and the front of the globe is covered by skin; all the lid appendages, including the lacrimal gland and passages, are absent in the condition known as cryptophthalmia; part of either lid may be absent, but the deformity usually affects the medial part of the upper one.

(ii) The margin of either lid may be notched or fissured, sometimes up to the orbital margin, and more commonly on the nasal side of the upper lid; the deformity may be caused by imperfect descent or union of the embryonic processes, and is called a coloboma of the eyelid. Associated with it there is frequently found a small nodule of skin on the surface of the globe, forming a congenital mole or dermoid of the bulbar conjunctiva; this nodule does not represent a detached part of the skin which should have filled the gap in the eyelid, but is an islet of the embryonic skin-fold over the eyeball which has not become metamorphosed into conjunctiva owing to its exposed position (Fisher).

(iii) The palpebral fissure may be wider than usual (lagophthalmos), a condition that has been found to be a congenital malformation in four generations by Peters; on the other hand, the fissure has been seen narrowed by incomplete separation of the margins (Hoeve), or the lids may even be completely united by vascularized tissue, the pre-natal separation not having taken place, causing the condition known as 'ankyloblepharon'.

(iv) Dermoid cysts may occur at the inner angle of the eye or even within the orbit, but more commonly develop near the tail of the eyebrow (one of the favourite sites chosen, curiously enough, for the affixation of a 'beauty spot'); they lie beneath the orbicularis oculi muscle, and may hollow out the frontal bone, or even be connected with the dura mater lining of the skull by an opening through it; they are caused by an inclusion of epiblast between the growing bones; occasionally they are present in the middle of the upper eyelid, where the two parts from which it is formed fuse, but here they have no deep connexions.

(v) The muscles of the lids may fail to develop, and congenital ptosis result. The superior rectus is not uncommonly defective as well as the levator. The condition may be hereditary, and Briggs (1919) reports sixty-four such cases. Refer also to Natale (1929).

(vi) A reduplication of the conjunctival membrane lining the upper tarsal plate, known as the 'epitarsus' or 'double tarsus', is described on p. 207.

Abnormalities of the orbicularis oculi muscle are given on p. 135, and of the levator palpebrae superioris on p. 149.

A reduplication of the caruncle has been recorded by Stephenson (1896); a case of a 'supernumerary eyelid', formed by an outgrowth of the caruncle and containing both palpebral and caruncular elements, is reported by Shoemaker (1914).

3. THE CONJUNCTIVA

The conjunctiva (*tunica conjunctiva*) is the mucous membrane lining the posterior surface of the eyelids and reflected from them on to the anterior face of the eyeball, thus forming a union between the lids and globe. In its entirety the membrane forms the 'conjunctival sac', the mouth of which is bounded by the free margins of the lids and so corresponds to the palpebral fissure, but the walls are in the closest apposition throughout, and the sac is only a potential one. At the lid margin it is continuous with the skin through a transitional zone common to both layers. It lines the ocular surfaces of the tarsal plates and extends some short distance beyond their peripheral limits into the supra- and infra-tarsal regions as well as laterally before being reflected on to the globe; the fold of reflection forms a circular cul-de-sac round the globe known as the conjunctival *fornix*. The membrane becomes smoother as it passes over the front of the eyeball, where it is structurally continuous with and represented by the epithelium of the cornea. On the medial side of the globe it forms a fold called the *plica semilunaris conjunctivae*. The ducts of the lacrimal glands open into the upper lateral region of the conjunctival sac, and on the medial side the puncta lacrimalia lead from it into the lacrimal passages, through which the tears are drained into the nose.

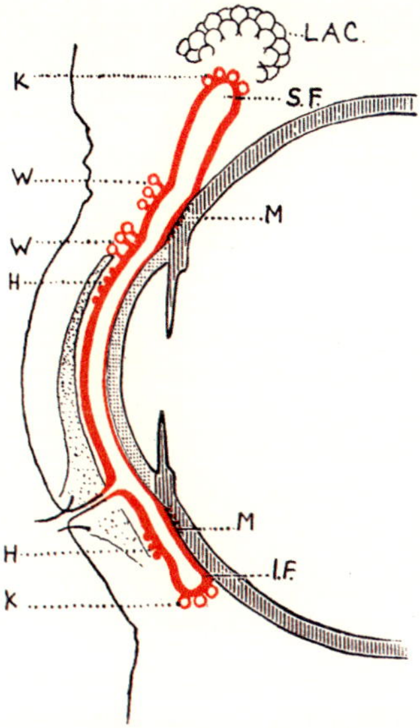

FIG. 101.—Scheme of a sagittal section through the Eyelids and Eyeball to show the **Conjunctival Sac** and the position of its Glands (after Dubreuil, 1908).

S.F., I.F. are the superior and inferior fornices respectively; LAC. = the lacrimal gland proper; K. = the accessory lacrimal glands of Krause, and W. = those of Wolfring; H. = the crypts of Henle; M. = the 'glands' of Manz.

The conjunctiva is a continuous membrane throughout its whole extent, but since it presents special characters and relations in different regions, it is divided topographically into:

(i) The palpebral conjunctiva, lining the eyelids;
(ii) The conjunctiva of the fornix, in its fold of reflection;
(iii) The bulbar conjunctiva, covering the front of the eyeball;
(iv) The plica semilunaris.

The palpebral and bulbar portions of the membrane are fixed, but where it forms the plica and in the region of the fornix it is mobile.

(i) The palpebral conjunctiva (*tunica conjunctiva palpebrarum*) is more vascular than the rest of the membrane, and its condition is referred to as a sign in anaemia; it is smooth and intimately adherent to the posterior faces of the tarsal plates, the meibomian glands of which can be seen through it on everting the lids; a slight groove, 'the subtarsal sulcus', in which foreign bodies commonly lodge, may exist a short distance from the lid margin, and minute folds can be seen on this surface by the aid of a lens. At the puncta lacrimalia the membrane is directly continuous with the lining of the lacrimal passages, and through them with the nasal mucous membrane, an important continuity of structure in view of the spread of infection from one region to the other. Near the convex border of the tarsal plate the surface presents a network of grooves limiting minute islets, the zone of plateaux and grooves described by Virchow (1905); the same author states that the surface of the membrane over the plates is covered by papillae, but their presence is denied by others. Beyond the borders of the tarsal plates the palpebral conjunctiva of the supra- and infra-tarsal regions is looser, being thrown into horizontal folds, which are folds of movement since they only appear after birth and become effaced in occlusion of the lids; the membrane is here related on its deep surface to the involuntary palpebral muscles. In the lateral region in the upper lid there rests upon it the palpebral lobe of the lacrimal gland, the ducts of which pierce it.

(ii) The conjunctival fornix (Latin, *fornix*, a vault or arch) is the circular cul-de-sac formed by the fold of reflection of the membrane from the eyelids on to the globe; its exact limits are somewhat indefinite, but in the different regions it is referred to as the superior, inferior, lateral, or medial fornix. The distances from the orbital margin, corneal limbus and lid edges are given in Fig. 102, though great individual differences occur.

The superior conjunctival fornix corresponds nearly with the superior palpebral skin-fold, but may lie at a higher level; the centre of the inferior fornix is about 6 mm. above the bend of the orbital margin; on the medial side the plica semilunaris and caruncle lie in the fornix between the palpebral and bulbar conjunctivae. The membrane in the region of the fornix is thicker and more loosely disposed than elsewhere, being thrown into folds when the eye is opened, and allowing the globe to move freely independent of the lids; its looseness also allows it to be easily raised by subjacent infiltrations. It is surrounded by connective tissue or 'retinacula', formed by expansions of the fascial sheaths of the recti muscles of the eyeball and containing smooth muscle fibres (Charpy, 1912), so that its folds are drawn back and its recess is deepened by the action of these muscles in turning the globe. If an instrument be thrust through the summit of the superior fornix it will lie in the plane between the levator and superior rectus muscles, but will impinge on the mass of connective tissue by means of which their fused fascial sheaths are attached to it, as can be seen in Figs. 67 or 73. Through the inferior fornix the instrument would strike the inferior oblique muscle covered by a fascial expansion prolonged from the inferior rectus muscle. In the diagonal regions of the fornix, that is, in the intervals between the muscle sheath expansions, the connective tissue behind the membrane abuts on the orbital fat, and it is through these areas of less obstruction that, according to Charpy, artificial injections within the orbital cavity reach the fornix, and can then extend beneath the bulbar conjunctiva nearly up to the corneal margin; it is along these paths that the large sub-conjunctival ecchymoses due to haemorrhage into the orbit following a fractured skull travel (see also p. 302).

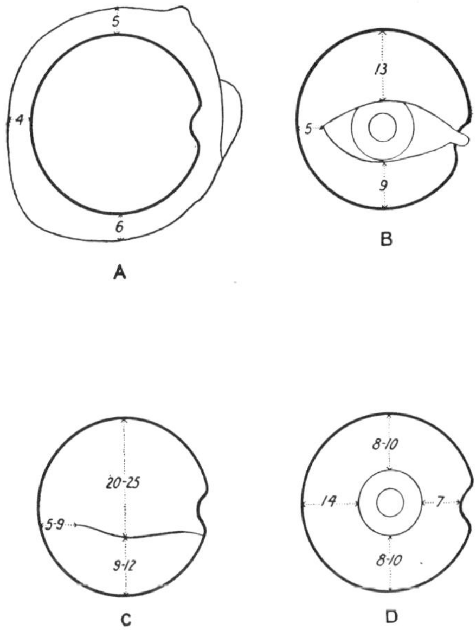

FIG. 102.—Diagrams to show the distance in millimetres of the **Conjunctival Fornix:**—
A. From the orbital margin (figures according to Gerlach).
B. From the palpebral margins with the eye open (figures as given by Baker).
C. From the palpebral fissure (eye closed) (after Henle, Dwight, Charpy, Richet).
D. From the corneal margin (as given by Testut and Merkel).
The conjunctiva is indicated in each case by the heavier outline.

(iii) The bulbar conjunctiva (*tunica conjunctiva bulbi*) is that hinder part of the membrane which is reflected onto the anterior face of the globe, of which it covers approximately one-third; it consequently presents a scleral and a corneal portion. The conjunctiva covering the sclera is thinner and less vascular than in either the palpebral or fornical regions, so thin, indeed, that the white of the eye shows through it, and sub-conjunctival haemorrhages in this region retain their red colour, the oxygen in the air enabling the blood to maintain its arterial character. It is separated from the globe by the anterior part of the fascia bulbi (capsule of Tenon), with which it blends as it approaches the corneal margin, the junction of the two membranes taking place about 3 mm. away, and being sometimes marked by a slight ridge, which is more apparent in certain ophthalmic catarrhs. In consequence of their union these two layers, bulbar conjunctiva and anterior part of the fascia bulbi, are raised together in exposing the scleral surface of the globe and the recti muscle tendons, and the capsule, filled with delicate episcleral tissue, is opened into (Fig. 148, p. 269, and Fig. 103, below). At the sclero-corneal junction the fascia bulbi ceases, but the conjunctiva, now reduced to its epithelial layer only, passes on to become directly continuous with the epithelium covering the face of the cornea, which thus forms the corneal portion of the bulbar conjunctiva. This direct continuity of the epithelium of the bulbar conjunctiva with that of the cornea explains the spread of disease from one part to the other.

(iv) The ***plica semilunaris conjunctivae*** (plica lacrimalis or semilunar fold) is a narrow fold of the bulbar conjunctiva, situated in the medial angle of the eye on the lateral side of and deep to the caruncle, together with which it intervenes between the bulbar and palpebral conjunctivae, and forms the floor of the lacus lacrimalis. It is vertically elongated and crescentic in form, with the concave and free border directed towards and projecting a little over the globe; its lower horn reaches the middle of the inferior fornix, the upper is not quite so long; its extent and connexion with the bulbar conjunctiva is made manifest by turning the eye laterally. It first appears in the upper lid in embryos of 18 mm. long (Popoff, 1912), is relatively large in the foetus, and is well developed in infancy; usually it is about 2 mm.

broad, but in one Malay tribe it is found 5 mm. broad (Wiedersheim). In structure it resembles the rest of the conjunctiva, but in its base or adherent border is embedded a fatty lobule, and some smooth muscle fibres are present, vestiges of the lateral palpebral muscle detached from the medial rectus; these fibres are striated in marine mammals (Groyer). In ruminants, anthropoids, and most mammals it contains a plate of hyaline cartilage, and a small nodule has been found in Negroes (in 12 out of 16 by Giacomini), in Japanese (25 per cent. by Adachi), in Caucasians (3 out of 548 by Giacomini), and very rarely in Europeans (5 out of 1,096 by Testut). The vestiges of glands which may be found in the plica semilunaris are described on p. 202.

The plica semilunaris has generally been held to represent the third eyelid of lower vertebrates and the nictitating membrane of birds and reptiles, but Stibbe (1927–8) states that in man and mammals it differs entirely from them in structure and function. He considers that it is not so vestigial as is suggested, and by bridging across the medial conjunctival fornix it diverts foreign bodies across the caruncular region, and hence might better be termed the *plica intercipiens*. It is of interest to add that Treacher Collins (1901) says that Lindsey Johnson saw in a youth a case of an obvious membrana nictitans which extended nearly as far as the cornea and was capable of slight movement.

The **Structure** of the conjunctiva:—Like other mucous membranes, it consists of two main layers, a surface epithelium, with a connective tissue basis or tunica propria; its special features are the scantiness of the mucous secretion and the presence of large serous lacrimal glands. The epithelium is composed of several layers of cells limited by a basement membrane; the cells are stratified squamous in character at the lid borders, like the epidermis with which they are continuous; in the palpebral region and fornix they are columnar or cylindrical in two or three layers, with islands of cuboid cells among them over the tarsal plates; in the bulbar region they gradually assume a squamous formation as the cornea is approached, and they then form the stratified pavement epithelium of its surface, some 5 cells deep. Clear oval mucous secretory or goblet cells are normally present (Leedham Green, 1894), especially in the bulbar region and fornix, where they may be aggregated to resemble glandular crypts, but they are rarer in the palpebral conjunctiva and absent towards the lid margins. Pigment

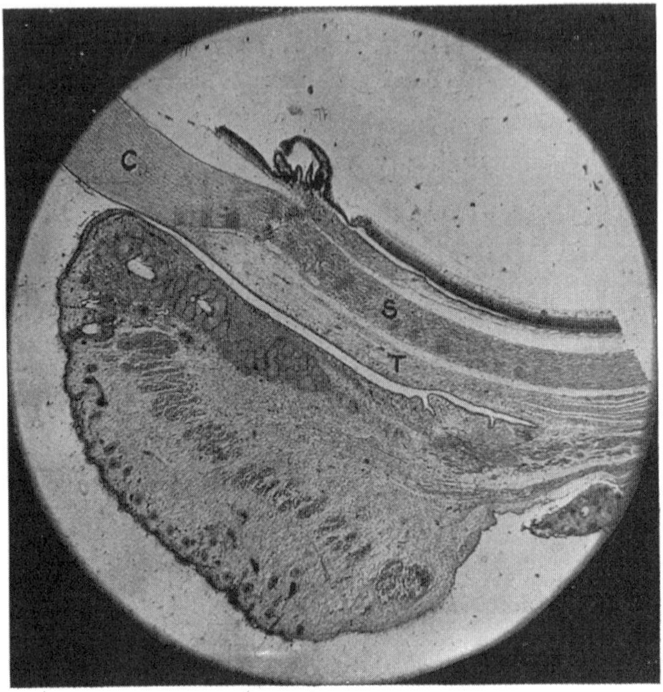

FIG. 103.—Section through the **Lower Eyelid** and part of Globe. In the inferior conjunctival fornix there are seen the folds or crypts described as 'glands' by Henle (see Fig. 106 for enlarged view).

C. = cornea; S. = sclera; T. = the anterior part of the *fascia bulbi* or capsule of Tenon.

granules occur in the deeper cell layers of the bulbar conjunctiva, especially near the corneal margin in the triangular space of the bulbar surface exposed to light by the opening of the palpebral fissure; here they may even form pigment spots (Pergens, 1898; Coppez, 1905; Fischer, 1905; Kusel, 1907), as found in Negroes, Indians, Chinese, and Japanese; they are also present in the palpebral conjunctiva of trachomatous subjects (Steiner, 1906). Hiwatari (1921) finds pigment also in the corneal margin itself. The tunic apropria corresponds to the dermis or chorion of all the integuments; it is reduced over the tarsal plates, well developed in the fornix, and absent over the cornea, and consists of a thin adenoid layer beneath the epithelium and a thick fibrous deeper layer with an abundance of elastic tissue. The adenoid layer is loose in texture, and is invaded throughout by white corpuscles, though less freely in the bulbar part; this lym-

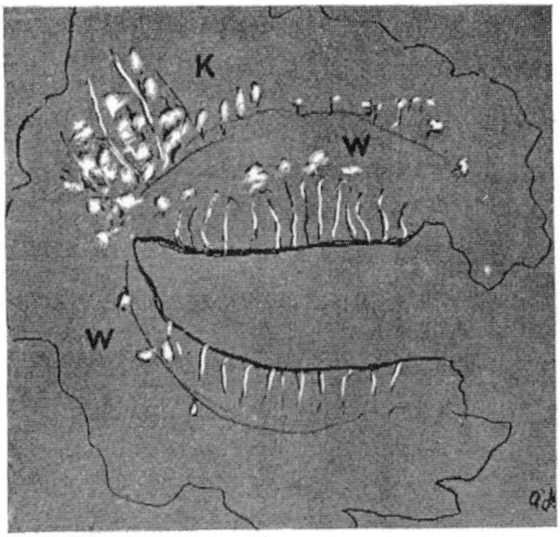

FIG. 104.—The **Accessory Lacrimal Glands** of Krause (K.) and of Wolfring (W.) as figured by Terson (1903).

phatic infiltration is especially abundant in the region between the convex border of the tarsal plates and the fornix, and principally towards the commissures of the lids, where it is concentrated to form lymph nodules (*noduli lymphatici conjunctivales*), termed 'trachoma glands' by Henle. Papillae of the dermis occur in a circum-corneal zone of about 2 mm. broad (Villard, 1896; Nakagawa, 1903; Virchow, 1905; but denied by Hiwatari, 1921), and near the margins of the lids, but the appearance of papillae described elsewhere, as by Virchow in the tarsal region, may only be due to circumscribed islets of mucous membrane. The deeper fibrous layer of the dermis is not a true constituent of the mucous membrane, being really formed by sub-conjunctival connective tissue, reinforced in the palpebral region and fornix by fascial expansions of the sheaths of the levator and recti muscles, and in the bulbar region by the anterior part of the fascia bulbi; it does not exist in the tarsal region. Near the corneal margin the adenoid layer of the bulbar conjunctiva is absent, that membrane being represented by its epithelium only, supported by the fascia bulbi. The comparative histology of the conjunctiva is given by Zietzschmann (1904).

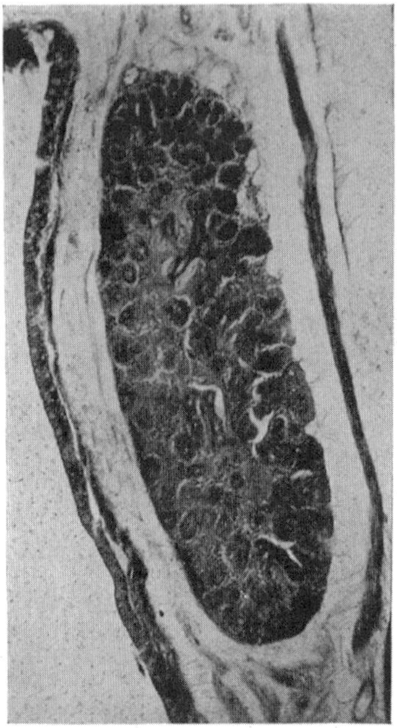

FIG. 105.—An **Accessory Lacrimal Gland** (gland of Krause), from the conjunctival fornix of the upper eyelid. × 60.

The **glands** of the conjunctiva are either lacrimal glands of various degrees of development or simple epithelial formations; the true glands comprise:

(i) The lacrimal gland proper (*glandula lacrimalis*), the large mass in relation to the upper lateral region of the fornix, which will be described in accordance with custom as part of the lacrimal apparatus (p. 208).

(ii) The accessory lacrimal glands (*glandulae lacrimales accessoriae*), which are smaller and are found in the fornix elsewhere; they are sometimes referred to as the glands of Krause, Wolfring, or Ciaccio.

The glands of Krause (1854) are microscopic acino-tubular glands or single acini scattered along the conjunctival fornix, especially laterally, where they are more numerous in the upper eyelid (8 to 20, or even 42, according to Parsons) than in the lower (2 to 8). These are doubtless what is meant by the *glandulae mucosae Krausei* of the B.N.A. terminology, though they are not mucous glands. Those described by Wolfring (1872) and Ciaccio (1874) are similar acino-tubular glands situated along the middle of the convex border of the tarsal plates close to, and even between, the upper extremities of the tarsal glands; they are larger, but fewer in number, than those of Krause, being from 2 to 5 in the upper lid, with rarely a single one in the lower (Terson). All these accessory lacrimal glands are structurally identical with the lacrimal

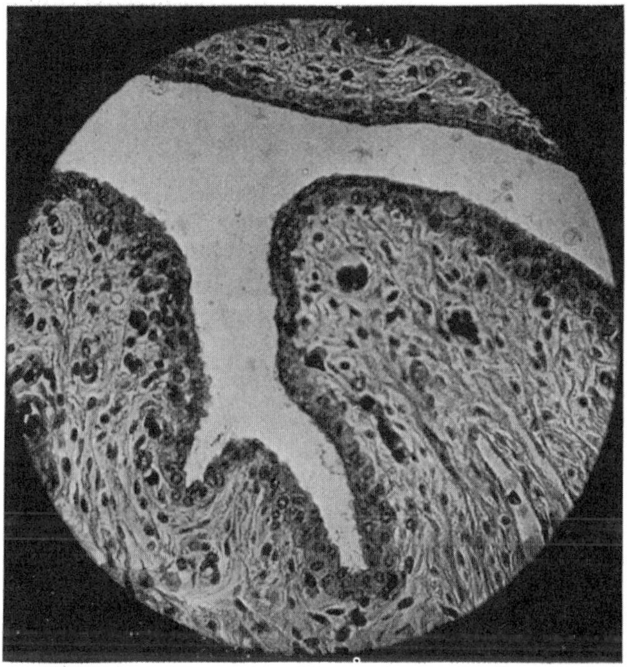

FIG. 106.—**Conjunctival Crypt**, sometimes described as a 'gland of Henle', or 'mucous gland of Reich'. Magnified from Fig. 103.

gland proper, and differ only in possessing large sinuous excretory canals. Hocevar (1900) described 'sterile lacrimal glands' without ducts in the same region.

The 'glands' described by Henle (1853) in the palpebral conjunctiva of both lids between the edges of the tarsal plates and the fornix appear to be simple epithelial invaginations; Zaluskowski (1887), Ciaccio, and Reich (1875) consider them 'mucous glands' formed of goblet cells, but many authors regard them not as true glands but as folds of the conjunctiva seen in transverse section (Stieda, Waldeyer, Krause, Sättler,

Terson), and they are well named 'conjunctival crypts' by Dubreuil (1908).

Similarly the 'glands of Manz' (1859) are utricular formations in the circum-corneal region of the bulbar conjunctiva, described by him in the pig, and since found in other animals; they are not hollow diverticula, but resemble crypts filled with epithelial cells ('epithelial nests') in the dermis; they are said to exist in man by Stromeyer, Henle, and Ciaccio, and more recently by Bartels (1908), and they have been thought to be connected with trachoma, but their presence and function is uncertain.

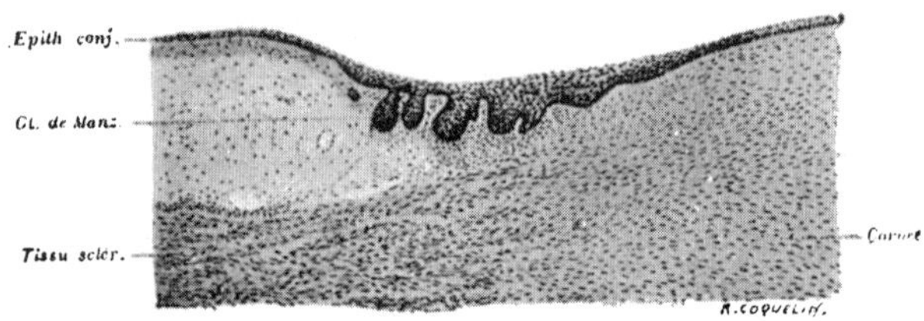

FIG. 107.—**Epithelial Nests** at the Sclero-Corneal Junction, sometimes described as the 'glands of Manz', as figured by Druault (1912), from a seven months' foetus. × about 55.

The position of these various glands in the conjunctiva is shown in Fig. 101, as given by Dubreuil (1907–8), who has made a special study of the lacrimal glands in man and animals.

(iv) In addition there have been described in the plica semilunaris of the human foetus, and as persisting in extremely rare cases in the adult, vestiges of glands found normally in lower animals. In all cases where the plica is well developed, as in birds, ruminants large and small, pigs, carnivores, and rabbits (Law, 1905), there is found a special acino-tubular gland which secretes a fatty substance; this gland was first discovered by Harder in 1694, and has recently been investigated by Löwenthal (1892); a rudiment of it has been found in a case of malformation in man (Fleischer, 1907), and also in a Bushman (Giacomini), and both Ask and Contino consider that a glandular organ which appears in the plica semilunaris of the human embryo towards the eleventh week, but usually disappears about the eighth month, is a transitory form of Harder's gland, and that in rare cases it persists in the adult.

Finally, the so-called 'infra-orbital gland' of lower animals, such as the sheep and the rabbit, which is an acino-tubular gland of the lacrimal type but phylogenetically more ancient, is said to be represented in the inferolateral fornix of the human foetus of 17 cm. (Löwenthal, 1892, 1894, 1910). Probably this 'infra-orbital gland' is only one of the outlying accessory lacrimal glands of Krause, as, indeed, Popoff (1912) regards any rudiments of glands found in the human plica, and with respect to their various positions the diagram shown in Fig. 108 of the phylogenetic migration of the lacrimal gland proper, as given by Wiedersheim, is of interest (see also p. 222).

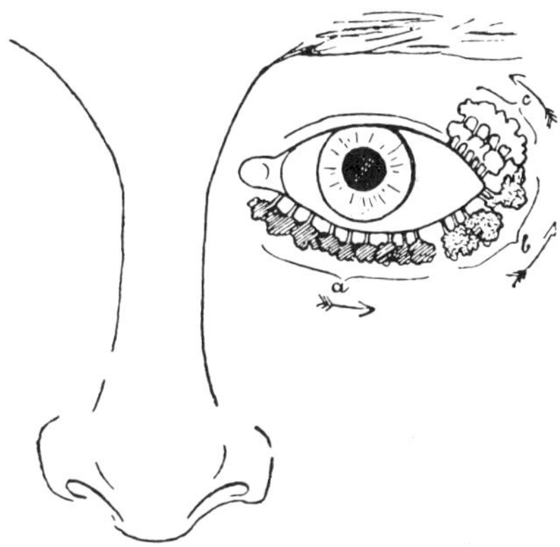

FIG. 108.—Diagram to illustrate the shifting of the **Lacrimal Gland** in the course of Phylogeny. (From Wiedersheim, 1895.)

a. = position in amphibians; *b.* = position in reptiles and birds; *c.* = the normal position in man.

The **vessels:** The arterial supply for the greater portion of the conjunctiva is derived from the two tarsal arcades formed by the palpebral branches of the nasal and lacrimal arteries in each eyelid; the marginal arcade and the posterior tarsal plexus afford a rich supply to the conjunctiva lining the tarsal plates, hence the red colour in this region; the peripheral arcade by its perforating 'posterior conjunctival' branches supplies both the fornix and the bulbar portion, and consequently is the more important; its branches become finer, and therefore the membrane paler towards and over the globe, where they are usually only visible in hyperaemic conditions. The details of the vascular supply in the region of the sclero-corneal junction are of considerable clinical interest. About 3 mm. from the edge of the cornea the posterior conjunctival branches anastomose with 'anterior conjunctival' ones derived from the anterior ciliary arteries, which latter come from the ophthalmic artery and accompany the recti muscles in their course from the back of the orbit. The two sets of vessels are in close relation, but the posterior conjunc-

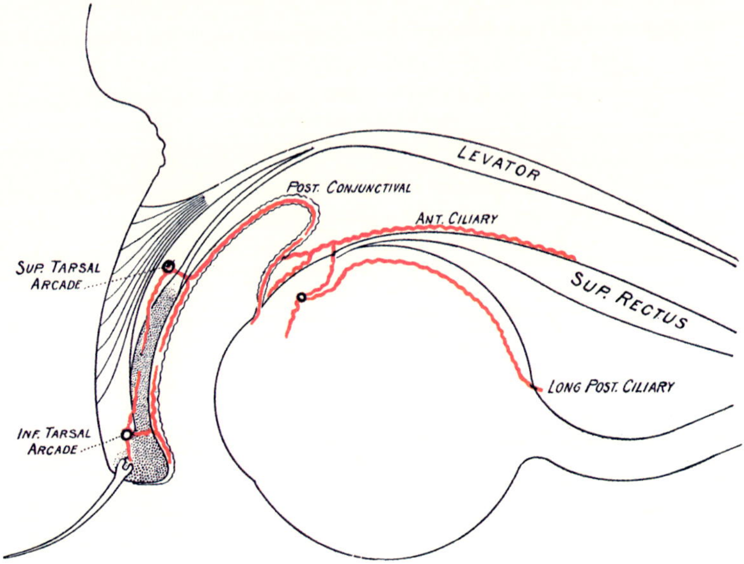

FIG. 109.—Diagram to show the deep **Arteries of the Upper Eyelid** and Conjunctiva and their anastomoses with those of the forepart of the Eyeball.

tival are more superficial and more movable than the more deeply placed immovable anterior conjunctival branches. The latter, after giving off their anastomotic twigs, form a 'peri-corneal plexus' about 4 mm. broad, which is disposed in two planes, a superficial or conjunctival and a deep or episcleral, and sends loops of minute vessels into the edge of the cornea. It is further to be noted that the anterior ciliary arteries, after giving off their conjunctival branches, turn abruptly to pierce the sclera just in front of the muscle insertions (the point of flexion being marked by the presence of pigment granules) and enter the base of the iris to form the *circulus arteriosus major iridis* (indicated by the unmarked circle in Fig. 109; see also Fig. 165, p. 307). The above arrangement of the vessels explains the significance of inflammation in this region:

(i) In inflammation of the conjunctiva the posterior conjunctival vessels, normally scarcely visible, appear as a close network over the bulb, fading away towards the corneal margin; they move freely with the conjunctiva and do not disappear under pressure.

(ii) In superficial inflammation of the cornea the corresponding plane of the peri-corneal plexus formed by the anterior conjunctival vessels is engorged, and spreads in a freely branching manner into the corneal margin.

(iii) In affections of the iris and deeper parts of the cornea the episcleral plane of the peri-corneal plexus, normally invisible, appears as a pink band of fine straight vessels round the cornea; the vessels do not move with the conjunctiva and disappear on pressure.

The veins of the conjunctiva generally accompany the arteries, one or two to each. The greater number drain into the palpebral vessels through the retro-tarsal venous plexus; near the corneal margin, where they form an episcleral network apparent in hyperaemia, the venules accompany the anterior conjunctival arteries, draining backwards by the veins of the recti muscles into the ophthalmic vein and cavernous sinus.

The lymphatic vessels form two communicating networks in the dermis of the conjunctiva; a smaller, superficial, and a larger, more deeply placed in the fibrous layer (Leber, 1876). Most (1905) found the lymphatic vessels of the lower eyelid easier to inject than those of the upper; they are minute in the tarsal region, and at both margins of the plates they anastomose with those of the skin of the lids. Near the corneal margin an extremely minute plexus of vessels was stated by Recklinghausen to communicate directly with the lymph spaces of the cornea, but Koeppe (1918), who has made clinical studies of the conjunctival vessels with the corneal microscope, denies this; he finds, however, the blood-vessels of the bulbar conjunctiva and sclera near the limbus to be surrounded by lymph spaces. All the vessels pass into the deeper layer of the conjunctiva and drain towards the palpebral commissures, where they join those of the eyelids, as described above on p. 175.

The **nerve supply** of the conjunctiva is provided medially by the infra-trochlear branch of the naso-ciliary nerve, and laterally by the lacrimal nerve, which supplies a larger area of the conjunctiva than of the skin of the lids; the epithelium covering the cornea and a narrow zone of the bulbar conjunctiva round it are supplied by the ciliary nerves, which have followed an intra-ocular path before piercing the sclera to gain this region (p. 366). The details of the distribution to the cornea are given by Boucheron (1890). The nerves terminate in the conjunctiva either by free ends, which are probably pain receptors (Herrick), or in tactile corpuscles of Krause. In the former case, after losing their nerve sheaths, they form a 'sub-epithelial plexus' (Dogiel, 1895), especially well marked at the free margins of the lids and near the convex borders of the tarsal plates (Bach, 1895), and containing formations resembling ganglion cells. The corpuscles of W. Krause (1867) (which occur in the skin and mucous membrane elsewhere) are round or oval bodies of a granular substance, with a nucleated membrane of connective tissue; one or more nerve fibres approach the body, lose their sheaths, and end as delicate fibrils with enlarged heads. They lie close beneath the epidermis, especially in the supero-lateral region of the conjunctival sac supplied by the lacrimal nerve. They are also found in the corneal limbus on the ends of the ciliary nerves, and at the margins of the eyelids; they have again been investigated by Ciaccio (1874) and Poncet (1875), and the latter points out that the presence of the tactile corpuscles in the lateral part of the conjunctiva of the upper lid can be realized by touching the skin over it with a blunt point such as that of a pencil; in this area the sensation is more poignant than elsewhere, and it feels as though sand were beneath the lid. Other types of sensory nerve terminations, such as tufts and ribbons, have been found by Crevatin (1903).

The **development** of the conjunctiva has been referred to in connexion with that of the eyelids on p. 189; the ectoderm lining the primitive palpebral folds begins to differentiate into a mucous membrane after the eyelids have met and fused at about the third month of foetal life. All the glands of the conjunctiva, including the lacrimal, are formed by invaginations of the ectoderm; the mucous cells appear after the seventh month; the glands of Henle first appear as simple depressions of the surface, and Ask found them in the bulbar as well as the palpebral regions in an embryo of 39 cm. Dubreuil saw glands of Manz clearly defined in a seven months old foetus. The plica semilunaris has been seen in an embryo of ten weeks by Contino, earlier than the caruncle and quite independent of it. The development of the lacrimal gland follows on p. 220.

Abnormalities of the conjunctiva.—The so-called 'epitarsus' or 'double tarsus' is a congenital malformation in the upper lid formed by a reduplication of the lining conjunctival membrane. First described by Schapringer (1905), numerous instances of its occurrence have since been recorded by Radinski ('double tarsus'), Mathewson ('accessory eyelid'), Robertson ('third eyelid'), Lloyd, &c.; it usually takes the form of a triangular fold arising by its base from the superior fornix and attached at its apex to the palpebral margin, and is regarded either as a developmental malformation or as the result of an amniotic adhesion or as a cicatricial result of an ophthalmia neonatorum (Fuchs).

Congenital dermoids (see p. 191), naevi and fatty growths may be found.

On the bulbar conjunctiva, along the line between the medial, and less often the lateral, side of the cornea and the corresponding angle of the lids, there may be formed a horizontally elongated yellow mass known as a 'pinguecula' (Latin, *pinguis*, fat); such bodies occur most commonly in adults subject to conditions which irritate the conjunctiva, the site being in that triangular part of the sclera exposed by the opening of the palpebral fissure; they are pigmented in tropical races. The mass, according to Parsons, is not of a fatty nature, but is composed of vascular conjunctiva thickened by fibrous and yellow elastic tissue; Fuchs states that besides an increase in the number and size of elastic fibres there is an additional formation of numerous concretions of a yellowish hyaline substance. It may develop into a 'pterygium', which is a larger wing-like growth extending over the cornea towards the pupil. Pingueculae are present on both sides of the cornea in the subject photographed in Fig. 60, p. 128, where they appear as dark areas in the position described above.

4. THE LACRIMAL APPARATUS

The lacrimal apparatus consists of a **gland,** which secretes the tears into the conjunctival sac, and the **lacrimal passages,** a system of canals, through which they are drained away into the nasal cavity.

I. The **lacrimal gland** (*glandula lacrimalis*) is, as previously noted, the chief mass of conjunctival glands, there being also present 'accessory lacrimal glands' which, though microscopic in size, are of similar nature. It forms a small flattened and lobulated mass, yellowish-pink in colour and a little darker than the orbital fat, from which it is not easily distinguishable at first sight. There are to be described its situation and relations, form, attachments, structure, vessels and nerves, development and variations.

It lies deep beneath the lateral part of the upper eyelid, covered in front by the septum orbitale, orbicularis oculi muscle and skin, and is placed just within the orbital margin, wedged in between the lateral angular process of the frontal bone and the globe (Fig. 60, p. 128); the bone is slightly hollowed out for its reception, so forming the *fossa glandulae lacrimalis,* in the lower and forepart of which it mainly lies; posteriorly the gland is adherent to a mass of orbital fat which fills the hinder part of the fossa, termed the *loge accessoire* by Rochon Duvigneaud (1903), and sometimes a slight undulation can be felt across the surface of the bone demarking the regions occupied by the two masses. The gland extends downwards as low as the zygomatico-frontal suture (which, as illustrated in Fig. 11, p. 25, may be raised in a low ridge limiting the fossa below), and it rests upon, and is moulded by, the eyeball.

The gland is crossed in front and deeply grooved by the lateral horn of the aponeurosis of the levator palpebrae superioris muscle, round which it is folded; so deep is the groove that the gland is almost divided into two parts or lobes, an upper and a lower (Fig. 110). They are distinguished as the *glandula lacrimalis superior* or orbital lobe, and the *glandula lacrimalis inferior* or secondary or palpebral lobe, but the writer has met with no instance of entire separation of the two parts of the gland, and the term 'lobe' is preferable to gland, since they are not even as completely separated as are the lobes of the lungs for example; they differ in size, position, and relations.

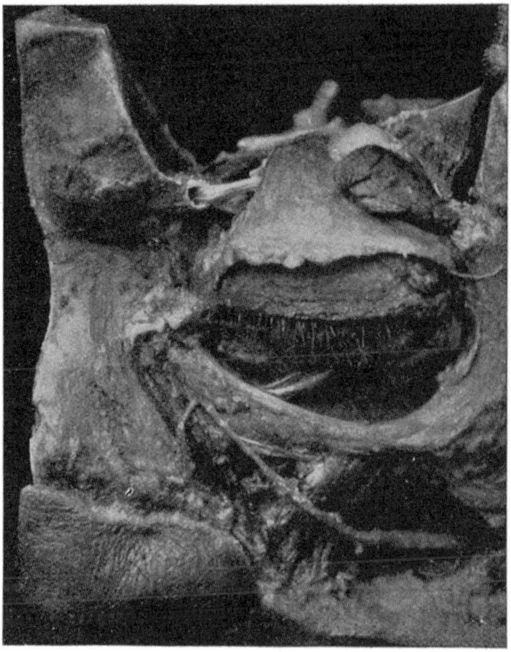

FIG. 110.—The relations of the **Lacrimal Gland**. Dissection of the Left Orbit from above and also in front (the same preparation as Fig. 69, viewed from in front) to show the aponeurosis of the *levator palpebrae superioris* muscle and the lacrimal gland. The pulley of the superior oblique and its tendon are also seen. Natural size.

The superior or orbital lobe (*glandula lacrimalis superior*) is the larger, and its lobules, owing to its confined position between the globe and the bone, are more closely pressed together; it forms an oval mass rather like a haricot bean, about 20 mm. long by 12 mm. wide and 5 mm. thick, but the size and shape, like the weight, which is given as on an average 0·78 gr. by Göz (1908), are very variable (Fig. 112). Its orbital or lateral surface is smooth and convex, in harmony with the contour of the bony fossa in which it is closely lodged and from which it is separated only by the periorbita; the forepart of the ocular or under surface lies on the aponeurosis of the levator muscle, but the hinder part, which is confluent with the palpebral lobe, lies on the fascial expansion which passes from the sheath of the superior rectus to that of the lateral rectus muscle, whereby it is separated

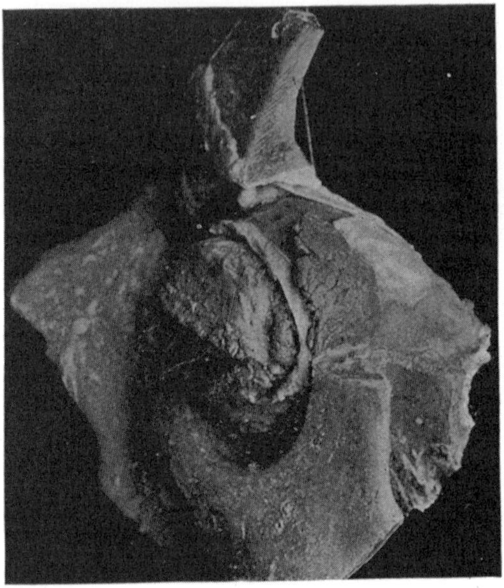

FIG. 111.—Dissection of Left Orbit viewed from the side to show the **lacrimal gland** subdivided by the aponeurosis of the levator, the moulding of its upper lobe by the frontal bone (removed), and its close connexion behind with the orbital fat. The absence of any capsule allows its lobulations to be seen clearly. Natural size.

from the fascia bulbi covering the eyeball. The long axis of this upper lobe is curved, and reaches from the lateral margin of the muscular part of the levator above to the zygomatico-frontal suture below; its upper extremity or pole is usually free, but the deep surface is often connected to the fascial sheath of the levator; the lower pole is fixed by its stroma to the periorbita in the angle between the lateral expansion (check ligament) of the sheath of the lateral rectus muscle and the zygomatico-frontal suture, a fixation often strengthened by the entrance at this point of the anastomotic branch from the zygomatic (temporo-malar) nerve to the lacrimal (Fig. 77, p. 154). The anterior edge of the lobe is thin and free, and runs parallel to the orbital margin, just behind which it normally lies, and is separated by a slender roll of fat from the septum orbitale; the posterior edge, thick and ill-defined, extends to a variable distance backwards into the orbit, and is not free but is connected with the mass of orbital fat lying between the superior and lateral recti muscles, as is seen in Fig. 111, and at about its middle there enter the lacrimal nerve and blood-vessels (Fig. 69, p. 141, and Fig. 77, p. 154).

The inferior, palpebral, or accessory lobe (*glandula lacrimalis inferior*) is from one-half to one-third the size of the superior lobe; it forms a thin plate with diffuse limits, but is usually about 15 mm. wide, and its lobules are more loosely connected; they number from 15 to 40 (Sappey), and the writer has never found only one or two present, as Fuchs states is the normal number. Behind, it is confluent with the superior lobe, and presents the same relations; in front, it extends well beyond the orbital margin, beneath the aponeurosis of the levator and onto the lateral part of the upper eyelid, where it lies partly on the superior palpebral muscle, but chiefly on the palpebral conjunctiva, to which it is firmly adherent and through which its ducts open; these can be seen by raising its fore edge, which extends a few millimetres over the convex border of the tarsal plate. It is not uncommon to find a few lobules of this part of the gland extending downwards between the lateral raphe of the orbicularis muscle and the lateral palpebral ligament.

The gland (presumably its orbital lobe) is described as being enclosed in a delicate but definite **capsule** (Picou, 1912), which separates it almost completely from the surrounding parts and most definitely behind from the orbital fat, but the writer has never found anything comparable to such a structure: the lobules of the gland appear bound together by a connective tissue stroma, which is slightly more abundant on the surface and is certainly continuous behind with the tissue between the adipose lobules of the orbital fatty mass, to separate the gland from which a careful dissection is needed.

Similarly, various **ligaments** have been described, none of which, however, present any special degree of development, and to which the application of the term 'ligament' is pretentious and misleading. These are as follows:

(i) A suspensory 'ligament of Soemmering', fixing the upper surface of the gland to the orbital wall. This consists merely of loose, delicate strands of connective tissue passing from the stroma of the gland to the periorbita, and they can easily be destroyed with a blunt instrument.

(ii) An inferior 'ligament of Schwalbe', fixing the lower part of the gland in the angle between the expansion of the sheath of the superior rectus muscle and the lateral orbital wall at the zygomatico-frontal suture. This is the most definite attachment of the gland.

(iii) A 'posterior ligament' passing to the periorbita at the point of entrance of the lacrimal nerve into the gland (Holmes).

(iv) An 'internal ligament' accompanying the excretory ducts of the gland. This, like the last, is neither a true ligament nor a means of fixation of the gland.

In dissecting out the whole gland, as in the preparations made for Fig. 112, the first structure to be severed is the lateral horn of the aponeurosis of the levator palpebrae superioris, which is often fixed to the gland where the latter is folded round its hinder edge; then the upper part of the gland requires freeing from the fascial expansion of the sheath of the same muscle, and farther back from the orbital fat, where its nerve and blood-vessels must be cut; the gland can then be pulled forwards and freed at its lower angle, the most fixed point, from the so-called inferior ligament; the deep surface of the palpebral lobe of the gland is strongly adherent to the conjunctiva, upon which it rests, and is separated only with difficulty from it.

The gland is kept in position between the globe and frontal bone not by any such ligaments, but by being folded round the lateral horn of the aponeurosis of the levator which lies in front of it, and by the support afforded to it by the mass of orbital fat behind; as just noted, it is most firmly fixed at its lower pole, but elsewhere a certain amount of movement is possible and probably does take place when the aponeurosis is swung back by contraction of the levator in raising the lid. In the specimen described on p. 150, where muscle offshoots passed from the edge of the levator to the gland, traction on the muscle pulled the gland backwards, even though the preparation had been hardened in spirit. The gland regains its more forward position upon relaxation of the muscle owing to the resiliency of the orbital fat against which it has been withdrawn. The movement is slight, but may be a means of stimulating or aiding the expulsion of the tear fluid.

FIG. 112.—Four **Lacrimal Glands** dissected out entirely; the two left-hand ones are from right orbits and are viewed from the lateral aspect; the upper right-hand gland is from a left orbit; the lower right-hand gland is viewed from its ocular or medial aspect. All show clearly the subdivision into superior and inferior lobes. Natural size.

Structure and Ducts.—The lacrimal gland consists of a mass of lobules, each about the size of a pin's head; they are more loosely aggregated in the palpebral portion, but present the same detailed structure in each lobe. It is a tubular-racemose gland—that is, each lobule consists of a small mass of ramifying tubes, and each tube branches further into tubules or acini of secreting cells, as illustrated in Fig. 113.

The acinus has a basement membrane supporting two kinds of cells arranged round a central canal (Fig. 115); one kind of cell is flat, basal, and contractile (myoepithelial,) the other is cylindrical and secretory, and the protoplasm appears under three different aspects according to the stage of secretory activity; thus the cell in charging contains refringent granules and is vacuolated, and in resting shows obscure details (Noll, 1901; Fleischer, 1904; Dubreuil, 1908). Fat granules have also been observed in the secretory cells of normal glands (Axenfeld, 1898).

By contraction of the protoplasmic network of the cell, the secretion is expressed into the central canal of the acinus, and thence passes into the excretory canals. These resemble the acini in structure, but soon become lined by two rows of epithelial cells, the deeper of which are contractile. At first they are intra-lobular in position (Fig. 114), but join with others to form larger extra-lobular or collecting canals, which eventually become the ducts of the gland (*ductuli excretorii glandulae lacrimalis*). These are constructed by an adventitia outside a basement membrane which encloses an epithelium of two rows of flat cells lined by cylindrical cells. The chief characteristic of all the ducts is their relatively large lumen; they are white, delicate-walled vessels, not easily distinguishable at sight from the blood-vessels or nerves. In number they seldom exceed a total of 12, of which 2 (Gosselin, 1843) or from 3 to 5 (Sappey, 1867) drain the orbital lobe, and from 6 to 8 (Gosselin) or 2 to 5 (Sappey) the palpebral portion. Some of the upper may join those of the lower group, but nearly all open separately into the supero-external region of the conjunctival sac along a line just in front of the superior fornix and about 4 to 5 mm. above the convex border of the tarsal plate. One duct, the largest, opens behind the lateral commissure, and one or two from the palpebral lobe may

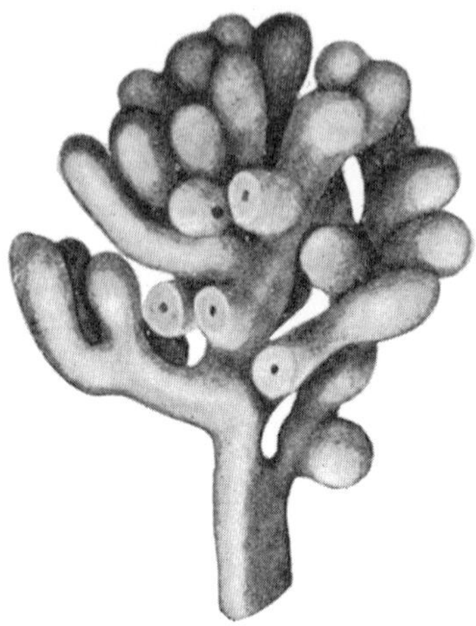

FIG. 113.—Reconstruction of a Lobule of the human **Lacrimal Gland** (Maziarski, 1902). × 170.

open as low down as the corner of the inferior fornix below the level of the commissure. From a practical point of view, it is important to note that the ducts of the upper or orbital lobe traverse or pass in close contact with the lower or palpebral lobe before they reach the fornix, so that removal of the palpebral lobe through the conjunctiva of the everted eyelid severs the ducts of both lobes, and is equivalent to extirpation of the whole gland as far as its secretory function is con-

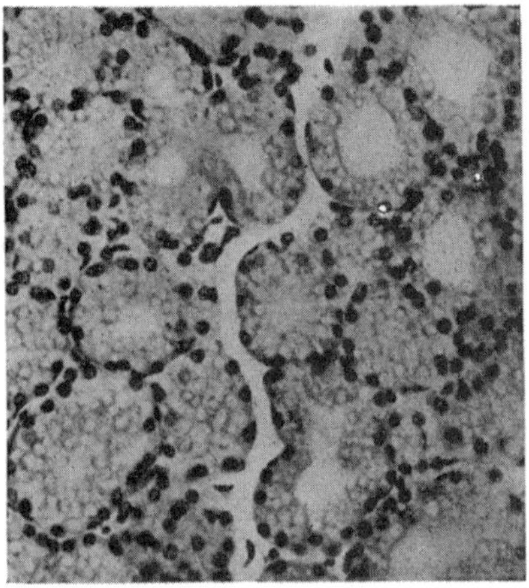

FIG. 114.—Section of human **Lacrimal Gland,** showing Acini; running down the middle of the section is an intralobular secretory canal. × 200.

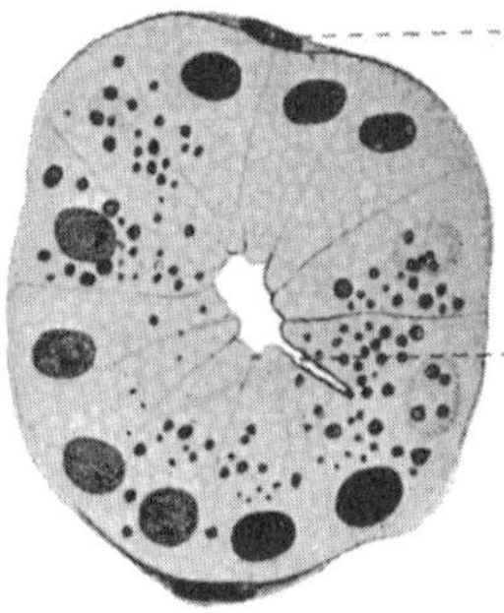

FIG. 115.—Microscopical section of an Acinus of the human **Lacrimal Gland** (Schirmer, 1904).
B = basal cell; S = secretory capillary.

cerned (Schirmer, 1904); it will be realized, however, that such an operation does not entail dryness of the conjunctiva, owing to the presence of the accessory lacrimal glands and its own mucous cells, though the condition may temporarily occur.

The stroma or supporting tissue between the lobules (less developed than in the salivary glands, with which the lacrimal gland is often compared, but from which it differs in the shape of its secretory cells and by the absence of crescent cells of Gianuzzi) is mesodermal tissue derived developmentally from the dermis of the conjunctiva, and like it consists of loose connective, elastic, and lymphoid tissues; the lymphoid tissue is absent in the new-born, sparsely present at puberty, but well developed in the adult, especially in the female; in old age it may hypertrophy; Stanculeanu (1900) found it present around the ducts and vessels, and chiefly concentrated towards the lower pole of the gland. Plasma cells have been found present by Hannes (1911).

The **lacrimal secretion** is a clear fluid, salty, and slightly alkaline; the exact composition varies according to the condition of the epithelium which secretes it, and, it is worthy of note, whether the fluid analysed is the secretion of the gland itself collected from its ducts (as in Magaard's analysis which follows), or the 'tears', which are the product of all the secretory cells including the mucous cells of the conjunctival sac. The details of the chemical composition are given by Schirmer (1904), from whose tables the following are quoted:

In 100 parts of Tear Fluid.	(Frerichs.)	(Magaard.)
Water	99·06	98·1200
Epithelium	0·14	1·4638 (Epithelium, Albumen, Mucus and fat)
Albumen	0·08	
Mucus and fat	0·3	
NaCl	0·42 (NaCl, Phosphates, &c.)	0·4160
Phosphates, &c.		0·

The normal quantity excreted in twenty-four hours is 6·4 gr. (Magaard), of a specific gravity of 1·0086 at 20° C. (Arlt).

The possession by the fluid of any bactericidal action is uncertain. It remains sterile indefinitely if collected under aseptic conditions, and is a bad culture medium; some authors consider it to have a bactericidal action lost on boiling (Kalt, Starling), others deny this property (De Bono and Frisco, 1901; Nedden, 1907). Schirmer considers it to have an insignificant antiseptic action against the majority of bacteria, but none on the tubercle bacillus or bacillus coli. Lindahl (1907) studied its effect on various bacteria, and found a potency against some owing to the presence of an albuminoid substance of the nature of an enzyme, which is secreted by the lacrimal gland; the action is feeble on the staphylococci, and is lost on heating. Ridley (1928) found an anti-bacterial body present, the enzyme 'Lysozyme'. The conjunctival sac is never free from organisms, save in the new-born, the most common being the bacillus Xerosis, which was found present by Lawson (1899) in 118 out of 159 cases examined, and in over one-half of these unaccompanied by other bacteria, and designated by him 'bacillus conjunctivae communis' in consequence; a staphylococcus albus of low virulence is also found (Parsons). Lucic (1927) criticizes the methods of bacterial culture used in such observations.

Vessels.—The arterial supply is derived from the lacrimal (p. 308), a branch of the ophthalmic artery, which enters the gland at its posterior margin and passes superficially on to the eyelids; an additional supply is often afforded by the infra-orbital branch of the internal maxillary artery. The venules form a lacrimal vein (p. 318), running backwards as a tributary of the ophthalmic vein to enter the cavernous sinus. The lymphatics are probably limited to the surface of the gland, and enter the conjunctival and palpebral systems, whence they will drain into the pre-auricular nodes. In malignant tumours of the gland the facial and pre-auricular lymph nodes have been found enlarged.

Nerves.—The gland is supplied by the lacrimal nerve, the facial nerve, and the sympathetic nervous system.

The lacrimal nerve is one of the three branches of the ophthalmic division of the trigeminal, and its course and relations within the orbit are described on p. 352; it anastomoses with the zygomatic branch of the maxillary division of the trigeminal nerve, and enters the gland with the blood-vessels; part of its fibres end in the gland substance as fine non-medullated ramifications between the secretory cells of the acini and round the ducts and vessels (Dogiel, 1893), and on some of the fibrils small ganglion cells are present (Puglisi-Allegra, 1903); the bulk of the fibres traverse the gland and supply the conjunctiva and skin of the eyelids (pp. 206, 180).

The facial nerve supplies parasympathetic fibres through a long and complicated path; they arise from cells of the geniculate ganglion of that nerve in the facial canal (aqueductus Fallopii), but may have a central connexion with the glosso-pharyngeal nucleus in the brain (Parsons, 1902); under the name of the greater superficial petrosal nerve (parasympathetic), they join the deep petrosal nerve from the carotid plexus (sympathetic) to form the nerve of the pterygoid canal or Vidian nerve, and so enter the nasal (sphenopalatine or Meckel's) ganglion; thence they are generally considered to pass into the zygomatic (temporo-malar) branch of the maxillary nerve, and reach the gland through anastomosis between the zygomatico-temporal division of this branch and the lacrimal nerve. That this anastomosis is of importance is indicated by the fact that in cases where the lacrimal nerve is absent a branch from the zygomatic nerve takes its place (see p. 352); the whole path is displayed in Fig. 116. Further, it is stated that the minute orbital rami passing directly from the ganglion into the orbit supply the lacrimal gland (Cunningham, 1931).

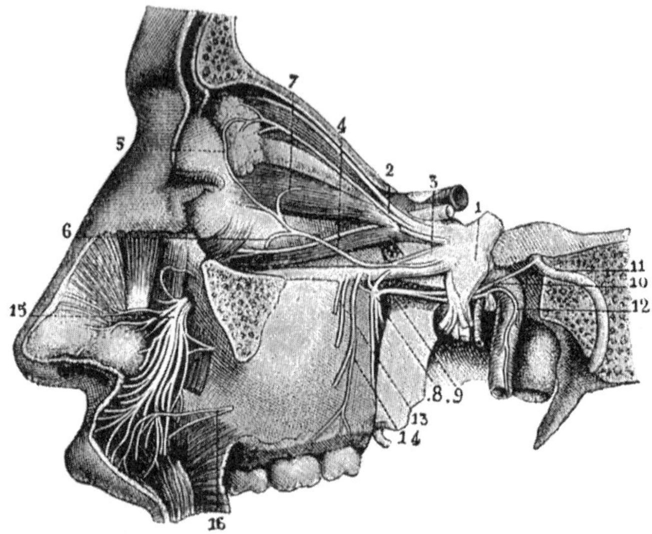

FIG. 116.—The **Maxillary Division** of the Trigeminal Nerve, its branches and their connexions. (From Sappey, after Hirschfeld and Leveillé.)

1 = semilunar or Gasserian ganglion; 3 = the maxillary nerve; 4 = the zygomatic branch with its anastomosis (5) with the lacrimal nerve (2); 11 = the facial nerve; 10 is the greater and 12 the deep petrosal nerve which join to form the Vidian nerve (9) and enter the spheno-palatine ganglion (8). The other branches can be followed from the text.

The sympathetic nerve supply is derived from the superior cervical ganglion through the carotid plexus and its extension along the lacrimal artery as well as by the great deep petrosal nerve noted above; other such fibres which pass from the cavernous plexus to the ophthalmic division of the trigeminal probably supply the gland incorporated with the lacrimal nerve.

As regards the unsettled problem of the actual source and path of the fibres conveying secretory impulses to the gland, it is first to be noted that the secretion is provoked:

(i) as a normal physiological act, in the slight amount necessary to prevent drying of the conjunctiva;

(ii) as a reflex act following stimulation of the terminal nerve-endings of branches of the trigeminal nerve to the conjunctiva or nasal mucous membrane, as in irritation by a cold wind or a foreign body;

(iii) by action of a supposed cortical tear secretory 'centre', whereby psychical weeping, which is a human peculiarity, occurs;

(iv) by impulses passing through the retina and optic nerve, as when the eye is exposed to bright light.

It may be added that the tears which are forced in yawning, coughing, laughing, etc., more probably result from stimulation by co-innervation than, as is sometimes stated, from mechanical pressure of the orbicularis oculi muscle on the gland.

Further facts bearing upon the question are, first, that there is a secretion of tears on stimulation of either the lacrimal nerve itself, or of the facial nerve, or of the cervical sympathetic nerves (Starling, 1920). Next, that the trigeminal nerve represents the reflex centripetal path, since there is a secretion of tears upon stimulation of its trunk or of almost any of its branches (Czermak quoted by Tepliachine, 1894); indeed the secretion following stimulation of the lacrimal nerve may be but a reflex action, since irritation of the area of conjunctiva supplied by this nerve produces the same effect. Then it is to be noticed that the muscles innervated by the facial nerve are usually thrown into action during weeping; and finally, that a connexion with the sympathetic system is apparent since its stimulation produces tears even after section of the lacrimal nerve (Kalt, 1903), and the lids become swollen and the face red in crying. Merkel, following Campos (1897), supposed that the excitation travels by all three nerves, fifth, seventh, and sympathetic. Wilbrand and Saenger (1921) suggest a double path: for reflex secretion of tears by the fifth, and for emotional or psychic weeping by the seventh nerve. Most authors consider the facial nerve as the source of the parasympathetic secretory fibres, the path being through the greater superficial petrosal nerve and spheno-palatine ganglion as described above (Schirmer, Lagrange, Gaskell, Quain, Dubreuil, Blum, Parsons, Müller and Dahl, Müller, 1924), and according to Kuntz (1929) this function has been demonstrated both clinically and experimentally; Ruskin (1928, 1930) cites five cases in support of this view.

Kalt (1903) objects to the complicated facial route on the grounds that (i) the zygomatic nerve leaves the maxillary nerve just after the latter has emerged from the foramen rotundum and before it is connected with the spheno-palatine ganglion—that is, before the Vidian fibres enter the main nerve; (ii) the anastomosis between the zygomatic nerve and the lacrimal is not constant; (iii) total paralysis of the facial nerve does cause loss of the action of weeping but not of all lacrimal secretion, as, indeed, Ruskin found blocking of the spheno-palatine ganglion did, and Campos found that section of the great superficial petrosal nerve in monkeys did not cause loss of lacrimal secretion on the same side.

The most probable conclusion is that the lacrimal nerve merely contains centripetal fibres to the brain, as do other sensory branches of the trigeminal, conveying impulses called forth by the movement of the gland in the involuntary act of winking or by irritation of the conjunctiva; that the sympathetic system provides fibres responsible for normal secretion; and that the facial nerve supplies parasympathetic fibres which occasion the superabundant secretion of weeping. Finally, certain drugs may affect the gland through the blood-stream, such as pilocarpine, histamine (Beattie).

The site of the central ganglion cells that control the tear reflex has not been definitely localized, but is supposed to lie near the seventh nerve nucleus in the brain stem (Müller). A vaso-motor automatism of central origin which rhythmically affects the lacrimal ducts is described in the rabbit by Benjamins and Rochat (1916, 1917).

It may be of interest to add that an article has been written by Rutherford (1913) on the behaviour of the lacrimal gland during the various stages of surgical anaesthesia, wherein the amount of the secretion is shown to bear a definite relation to the depth of the narcosis.

Development.—The lacrimal gland is formed by the

ramification and agglomeration of simple tubular buddings from the epithelium lining the conjunctival fornix in its supero-lateral region; they push their way along the path of least resistance between the globe and the orbital wall. The outgrowths appear in two successive tides; the earliest were seen as knobs in embryos of 22 to 26 mm. long by Keibel and Mall (1912), and in embryos of 32 mm. (ten weeks) by Speciale-Cirincione (1908), and by increase in length they ultimately form the more deeply placed orbital lobe of the gland; the secondary buds appear a week later (in embryos of 40 to 60 mm.), are shorter, and form the palpebral lobe; the tubules become hollow in embryos of 50 to 55 mm. long. In the mesodermal tissue between the two masses of outgrowths, the aponeurosis of the levator muscle begins to be developed as an interglandular septum in embryos of 38 mm., and was found complete in one of 60 mm. The accessory lacrimal glands appear later, and have been first seen in a foetus of 17 cm. (Ask, 1910). In the new-born the gland is only a quarter to two-thirds the adult size, and differs also in the appearance of its cells; the early growth is rapid, but the full structural development is not complete until the age of three to four years (Kirschstein, 1894). After the tenth year there is a gradual change in the histological structure, the glandular epithelium becoming decreased in height, the lumen of the

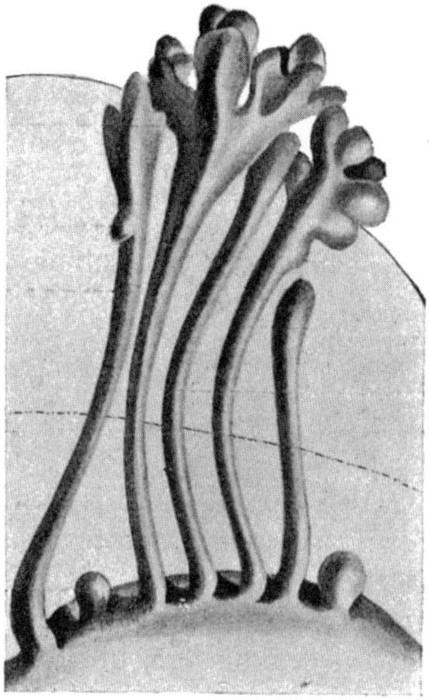

FIG. 117.—A reconstruction of the **developing Lacrimal Gland** in an Embryo of 40 mm. The transverse curved line indicates the position of the aponeurosis of the levator, developed between the two successive tides of outgrowths from the conjunctival fornix; the upper larger ones form the orbital lobe, the lower knobs form the palpebral lobe of the gland. (Speciale-Cirincione, 1908.)

acini widening, and the interstitial and adenoid tissue gradually increasing, though less rapidly, after the age of forty (Göz, 1908); in old age the glandular substance atrophies and the interstitial tissue increases.

Considerable VARIATION is found in the size and form of the gland, as is to be expected from the mode of development. Bock (1896) found it as usually described only in 9 out of 20 cases; the orbital lobe extended to a very variable distance backwards into the orbit, and varied from 11 to 22 mm. in length and from 5 to 12 mm. in breadth; the palpebral lobe was undeveloped in seven cases. The gland is relatively larger in the child, as in the female where the upper eyelid, of which the gland is but an annex, is also relatively larger and more mobile than in the male. Terlinck and Gallemaerts describe cases of congenital lacrimal fistula in which the normal secretion was discharged by an opening onto the face of the upper eyelid.

The lacrimal gland is found first in the amphibia, and in air-breathing vertebrates it appears first in the lower lid, as in the triton, and the higher the series examined the farther the gland is found to have migrated towards and round the lateral commissure, until the outer part reaches the upper eyelid and there predominates (Fig. 108, p. 203). The phylogeny of the gland in mammals has been studied by Broman and Ask (1910).

II. **The Membranous Lacrimal Passages.**—The tears excreted by the lacrimal gland into the supero-lateral region of the conjunctival sac are spread over the surface of the globe by the movement of the eyelids aided by gravity, and collect in the lacus lacrimalis at the medial angle of the eye; they then pass by capillarity through the lacrimal puncta, the minute openings on the summit of the lacrimal papillae, into the lacrimal canaliculi, which run horizontally through the thickness of the lacrimal portions of the lid margins and conduct the fluid into the lacrimal sac lying in the lacrimal fossa; the membranous passage formed by the sac is continued directly downwards as the naso-lacrimal duct, which traverses the bony naso-lacrimal canal to open by its lower extremity into the inferior meatus of the nasal cavity (Fig. 118). The lacrimal passages thus lead from the conjunctival

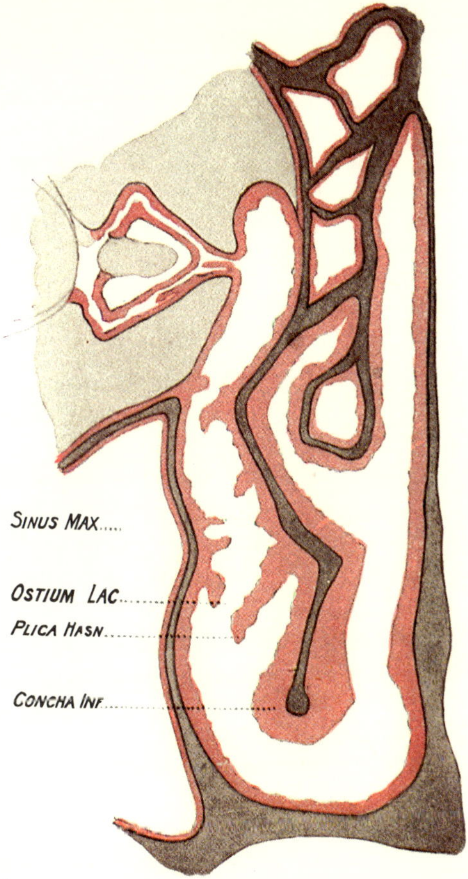

FIG. 118. Sketch from a frontal section through the Right Nasal Cavity viewed from in front, to show the relation of the **lacrimal passages** to the ethmoidal cells, the inferior concha, and the maxillary sinus. The lining mucous membrane is coloured to emphasize its continuity. The lacrimal sac is represented as usual in an abnormal dilated condition. (After Gérard, 1907.) × about 2.

sac to the nasal cavity, and may conveniently be described in three parts:

(i) **The lacrimal canaliculi.**
(ii) **The lacrimal sac.**
(iii) **The naso-lacrimal duct.**

(i) The lacrimal **papillae** (*papillae lacrimales*) are placed one on each lid margin at the junction of its ciliary and lacrimal portions, and on the posterior edge. They are pale, being but little vascularized; their prominence is increased in the aged owing to atrophy of the muscle fibres which encircle them at their base. The lacrimal point or punctum (*punctum lacrimale*) is the minute circular or oval opening of the cana-

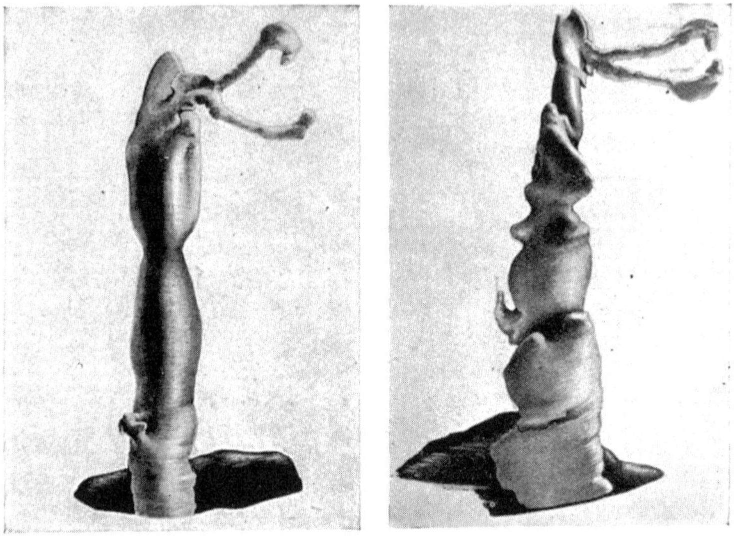

FIG. 119.—Reconstructions of two types of the **Naso-Lacrimal Passages.** The left one is from an adult aged 60 years, and shows a regular type. The right one is from a man aged 65 years, and shows the marked irregularities these passages may present. (From Schaeffer, 1912.) × 2.

liculus which appears on the summit of the papilla, and since the papillae are directed backwards the lids must be slightly everted to see the openings. They are well displayed in Fig. 82, p. 158, and Fig. 126; they lie in the same plane as the orifices of the tarsal glands, the nearest of which is only ½ to 1 mm. distant. The inferior punctum lies 6·5 mm. and the superior 6 mm. lateral to the medial canthus, so that when the lids are shut the points are not superimposed but the inferior lies just to the lateral side of the superior; the reverse arrangement has, however, been seen to occur (Dieulafé, 1905). They are plunged into the lacus lacrimalis, the inferior orifice, like the papilla on which it is placed, being directed backwards into the groove between the globe and the plica semilunaris, whilst the superior papilla with its punctum is directed downwards as well as backwards into the groove between the plica and the caruncle; both puncta touch the cornea when the eye is turned strongly nasalwards. The orifices are encircled by a ring of dense connective tissue, so that they always remain patent and capillarity is ensured.

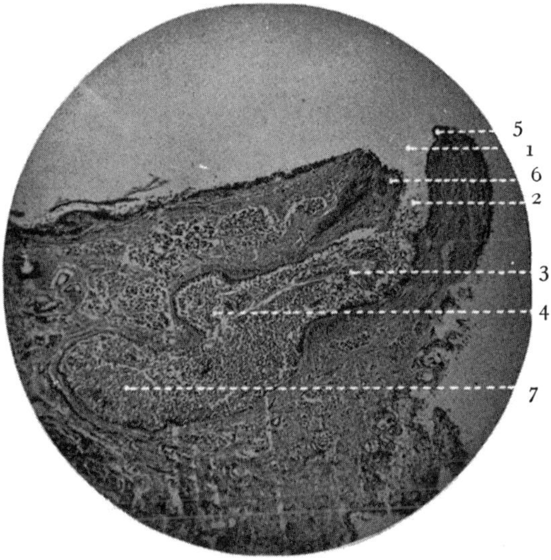

FIG. 120.—Magnified section of the **Punctum** and vertical portion of the Lacrimal Canaliculus of the Lower Eyelid.

1 = *infundibulum*; 2 = *angustia*; 3 = piriform dilatation, and 4 = sacculiform dilatation, the two forming the *ampulla*; 5 = 'valve' of Bochdalek; 6 = 'valve' of Foltz; 7 = commencement of horizontal part of canaliculus.

The lacrimal **canaliculi** or ducts (*ductus lacrimales*), into which the puncta lead, are two minute canals, each running in the thickness of the lacrimal part of the lid margin behind the medial commissure to enter the lacrimal sac. They therefore constitute the palpebral part of the lacrimal passages. The total length is about 10 mm. (Gerlach, 1880), the lower one being a little longer than the upper, in harmony with the more lateral position of its punctum. Each is bent nearly at right angles into two unequal portions, a shorter vertical part lying in the papilla, united by a widened angle to a longer horizontal part which runs along the lid margin.

The vertical part (*pars verticalis*) begins with an inverted funnel-shaped lumen, the *infundibulum* of Foltz, of which the base corresponds to the punctum and the neck is called the *angustia* of Gerlach; it leads into the dilatation of the angle or *ampulla*, which is subdivided into two secondary piriform and sacciform recesses, and then narrows slightly as it turns into the horizontal part; there may be further recesses of the ampulla directed in the opposite direction towards the ciliary part of the lid margin, as found in a model reconstructed by the writer. The total height of the vertical part is from 1·5 mm. to 2 mm., of which 0·5 mm. is formed by the infundibulum (Heinlein, 1875); the diameter of the punctum is from 0·2 to 0·3 mm., that of the ampulla is 1 mm. (Merkel).

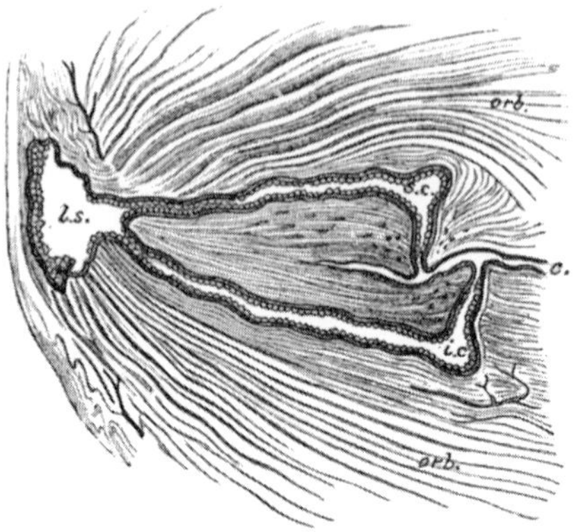

FIG. 121.—Section of the medial part of the Eyelids passing along the **Lacrimal Canaliculi.** The figure, taken from Gerlach (1880), is drawn from a sixth-month foetus, a fact that is not mentioned in the numerous reproductions of this illustration. × 13.

c = medial end of palpebral fissure; *orb.* = the *orbicularis oculi* muscle; *s.c.*, *i.c.* = the two canaliculi; *l.s.* = the lacrimal sac.

The horizontal portion (*pars horizontalis*) runs from the base of the vertical portion medially along the lid margin to the lacrimal sac; this part is not strictly horizontal, but inclines towards its fellow of the opposite lid as it courses medially; the lower one is also slightly curved. The length is from 6 to 7 mm. (Gerlach) or from 7 to 9 mm. (Henle), and the diameter is from 0·3 to 0·6, but can be dilated to 1·5 mm., the inferior being 0·5 mm. longer and a little wider than the superior; though the lumen is small the elasticity of the walls allows of a dilatation to three times the normal calibre, and for the same reason a probe will easily straighten out the angle between the vertical and horizontal portions, but the sclerous nature of the punctum does not favour its regaining the normal size after passage of an instrument, and the capillarity of the opening may thereby be interfered with.

The base of the papilla lies buried in the fibres of the orbicularis oculi muscle (pars lacrimalis), which cross one another and are disposed in a more or less sphincter-like manner (Fig. 132, p. 245), forming the *musculus quadrangularis constrictor puncti lacrimalis* (Halben, 1903; Gerlach, Merkel), which is considered to have a definite sphincter action, but owing to the density of the walls the lumen is probably not reduced in normal winking, though it may be in blepharospasm. The horizontal portion of the canaliculus is likewise surrounded on all sides by the same muscle fibres (as seen in Fig. 127), which some authors describe as being spirally arranged round the tube and even attached to its walls by elastic fibres (Klodt, 1893; Krehbiel, 1878; Heinlein, 1875), though this arrangement was not found by Halben and could not be confirmed by the present writer, who found the fibres running parallel with the canaliculus on all sides, though less abundant posteriorly; contraction of these fibres draws the papillae medially and directs them into the lacus lacrimalis, a fact that can be verified by everting the medial parts of the lids so as to expose the points, and then instructing the patient to try and blink; the action may also increase the calibre of the canaliculi (cf. p. 251).

The canaliculi usually pierce the lacrimal fascia separately, as shown in Fig. 124, and then unite to form a common canal of union about 0·8 mm. long, which narrows slightly before it enters the lacrimal sac; a conjoint termination of the canaliculi into the sac is described as the normal by some writers (Sappey, Foltz, Hyrtl), and has been found to occur in 109 cases out of 112 by Lesshaft (1868), who has made a special study of the question; on the other hand, Huschka (quoted by both Merkel and Schirmer) found a common opening only in 14 per cent. The different views may be accounted for by a distinction not having been made between the lacrimal fascia and the wall of the sac itself. The canal of union is really a diverticulum of the sac, as shown by its structure, and it is often enlarged to form a so-called 'sinus of Maier', as found in 14 per cent. by Huschka; the position is shown in Fig. 131 and an example in Fig. 119. The point of entrance into the sac is situated a little above the middle of its lateral wall about 2½ mm. from the upper end; it lies nearly opposite the mid-point of the medial palpebral ligament, but 2–3 mm. deep to its surface; the ligament covers the medial third of the canaliculi and their common canal of union, and behind this part of their course lies the pars lacrimalis or Horner's muscle (Fig. 127, p. 234).

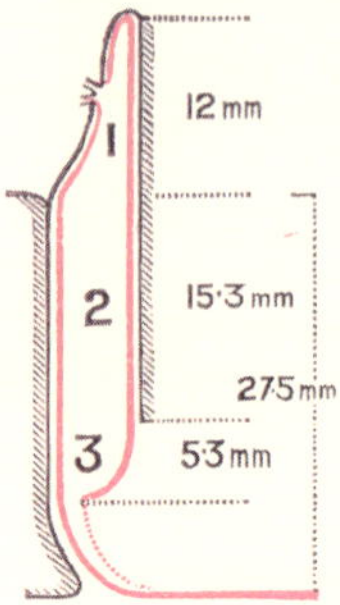

FIG. 122.—Diagram to show the **Orbito-Nasal Tube** subdivided into three parts according to its external relations, and seen in profile. The figure is drawn to scale, natural size, after the measurements of Power and Aubaret.

1 = the lacrimal sac, covered by the lacrimal fascia.
2 = the inter-osseous part of the naso-lacrimal duct.
3 = the meatal part of the duct; the rare condition of a prolongation of the passage below the opening is indicated by a dotted line.

The lacrimal sac and the naso-lacrimal duct form one continuous membranous tube, and a distinction between them is hardly justified either on developmental, structural, or physiological grounds, though one must be drawn for practical purposes on their anatomical relations. It is really an 'orbito-nasal tube', of which the upper part lies beneath the periorbita or lacrimal fascia in a groove (lacrimal fossa) on the medial margin of the orbit, and the middle part is contained in an osseous channel (naso-lacrimal canal) in the maxilla; moreover, beyond the lower end of this part the tube often runs downwards some distance beneath the lining mucous membrane of the lateral wall of the inferior meatus of the nose before it opens into that space. Thus there may be demarked three parts of the tube, orbital, maxillary and meatal, which are respectively sub-periorbital, inter-osseous and submucous as regards their positions; the first part is called the lacrimal sac, the second and third together form the naso-lacrimal duct. The whole tube can be dissected out from the surrounding parts, when it will be realized that the differentiation is largely extrinsic and relational (Figs. 122, 123).

In the new-born infant the whole channel is regularly cylindrical in shape (Rochon-Duvigneaud, 1900; Aubaret, 1910; Lepage), but marked secondary modifications of the form, both internal and external, are usually found in the adult.

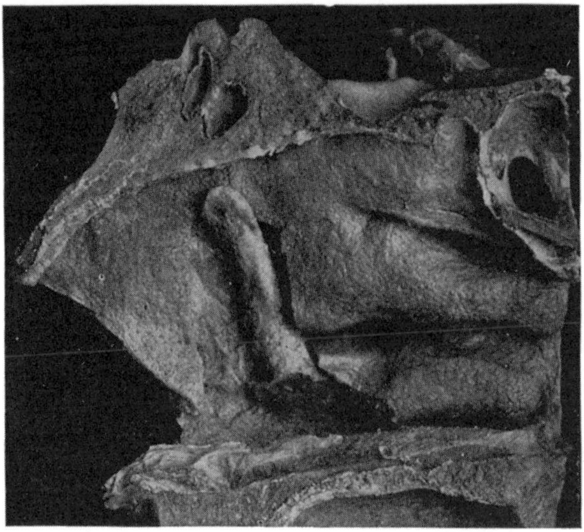

FIG. 123.—Right Lateral Nasal Wall viewed from inside, showing the lacrimal sac and **naso-lacrimal duct** lifted out of their osseous channels. Natural size.

At this point of full development two main types of tube are found:

(*a*) Regularly cylindrical in outward form, with no differentiation into sac and duct. This retention of the primitive shape was found by Aubaret to occur in 32 out of 116 specimens. The external diameter was about 5 mm.

(*b*) Irregular types; in some (28 out of 116) the upper end of the tube is enlarged, so forming a sac as usually described, though, it is to be noted, the increase in size is in an antero-posterior direction only; in others (26) the lower part of the tube presents a second dilatation, giving the classical hour-glass shape, but the constriction between the two parts does not coincide with the bony junction of fossa and canal, being situated below this point; in some (6) the whole tube is uniformly enlarged. In these irregular types the external diameter varies from 2 mm. in the narrowest to 9 mm. in the widest part.

It will be realized that the tube necessarily conforms in shape to the bony fossa and canal in which it is lodged, and the casts of these bony passages taken by Zabel (1900), like the reconstructions of Schaeffer, therefore serve to illustrate the outward form of the soft contents (Fig. 39, p. 78, and Fig. 119, p. 224). The internal configuration of the passages is described on p. 241.

Since the sac and duct into which the tube is conveniently divided present marked differences in their anatomical relations, they must now be described separately, the internal configuration and structure being subsequently considered as a whole.

(ii) The **lacrimal sac** (*saccus lacrimalis*) is lodged in the fossa sacci lacrimalis on the medial margin of the orbit. On its lateral or ocular aspect it is covered over by the periorbita, which splits at the posterior lacrimal crest, one layer following the bone and lining the fossa, the other bridging straight across it to reach the anterior lacrimal crest and being attached below to the incisura lacrimalis or upper edge of the bony naso-lacrimal canal. The lacrimal sac is roofed over completely by the periorbita, and so smoothly does this membrane maintain the contour of the medial orbital wall that after removal of the eyelids and the entire contents of the orbit the only indication of the position of the lacrimal sac is afforded by the orifices of the several canaliculi (Fig. 124). This layer of periorbita which roofs the lacrimal fossa is termed the 'lacrimal fascia' (*fascia lacrimalis*), and although it forms a strong definite layer of fibrous tissue readily demonstrable by dissection (Figs. 124, 126) and conspicuous in cross-sections (Fig. 127), it is not figured and but rarely mentioned in accounts of the anatomy of this region, though sometimes it is referred to by such misleading terms as 'deep' or even 'palpebral fascia'. Failure to realize its existence in passing a probe through an opening in the supposed wall of the lacrimal sac down the duct leads the student to force the instrument between the sac itself and the periosteal lining of the naso-lacrimal canal, and naturally a difficulty is experienced in reaching the nasal cavity. Usually there exists a narrow space between the lacrimal fascia and the sac, filled by a minute venous plexus continuous with that around the duct and draining into the angular or supra-orbital vein, and a separation between them is easily effected; occasionally,

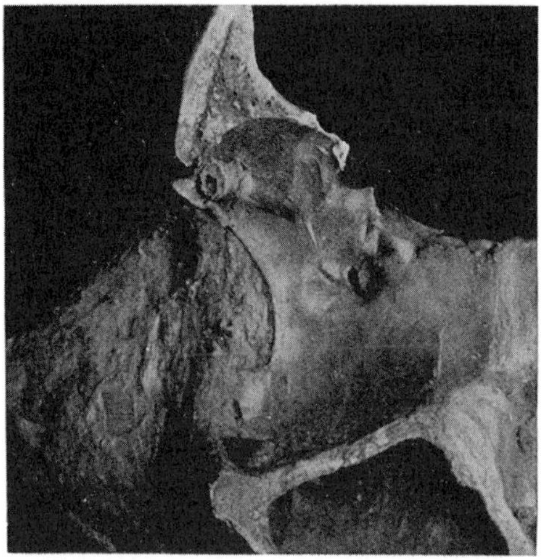

FIG. 124.—Dissection to show the **lacrimal fascia** (*fascia lacrimalis*). Medial Wall of Left Orbit viewed from inside. The lining periorbita is left *in situ*. The cut orifices of the canaliculi are seen in the roughened part of the periorbita which covers the lacrimal fossa and so forms the lacrimal fascia; the position of the lower part of the lacrimal sac is indicated by a slit in the lacrimal fascia, immediately below which is the cut end of the inferior oblique muscle. There is an **ethmoidocoele** in this wall. The *trochlea* is seen above. The *ostium maxillare* can just be seen. Natural size.

however, they are adherent, as indeed the sac generally is to the periosteum lining the medial wall of the fossa and always at its summit (fundus). The fascia is pierced separately by the two canaliculi, at which point it sends reflected sheaths along them nearly as far as the lacrimal papillae; it is also pierced in its lower part by a twig passing from the infra-orbital artery to the naso-lacrimal duct, and by filaments of the infra-trochlear nerve.

Thus confined by the tense lacrimal fascia, as is well shown in Fig. 127, the lacrimal sac naturally conforms to the fossa within which it lies; it is flattened from side to side in the shallow upper part, becoming more rounded below where it joins the cylindrical naso-lacrimal duct. It therefore does not represent the upper dilated end of the duct, as commonly described (e.g. in Figs. 118, 131), but the two together may be likened to the end of a quill pen seen sideways, as is well illustrated in the next figure and by Fig. 39, p. 78. Just as deep or shallow forms of lacrimal fossae are found in the skull, so variations in the size of the sac may be expected. The average dimensions are given on p. 240.

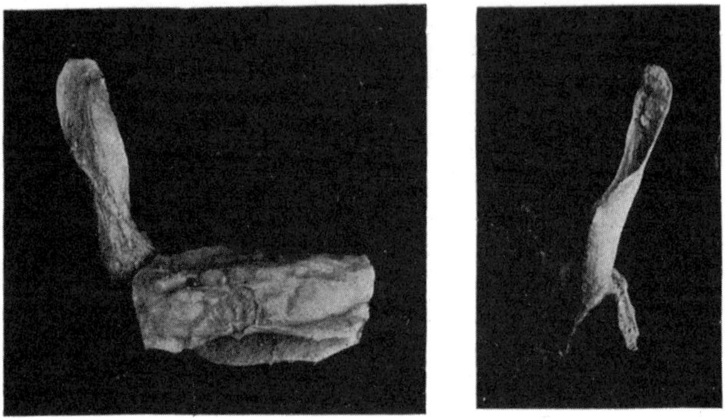

FIG. 125.—The Right Lacrimal Sac, the **Naso-Lacrimal Duct,** and the Mucosa of the Inferior Meatus of the Nose, dissected out entire and viewed from the medial side (left figure) and from in front (right figure). The mucous membrane is atrophied, the specimen having been taken from an aged subject, but the normal flattened shape of the lacrimal sac and the inclination of the whole tube are well illustrated. Natural size.

The flattened upper part of the sac above the entrance of the canaliculi is termed the fundus (*fornix sacci lacrimalis*); its summit lies from 3 to 5 mm. above the line of the medial canthus and from 10 to 15 mm. below the trochlea of the superior oblique muscle, in the interval there being situated the anastomoses between the angular and ophthalmic vessels and the branches of the infra-trochlear nerve.

The relations of the lacrimal sac on its medial or nasal aspect are those of the corresponding wall of the bony fossa, which have been described above on p. 70. On its lateral or ocular side it is first covered over entirely by lacrimal fascia, a primary relation to be accepted throughout the whole of the following description. Outside this the sac presents different relations in approximately its upper and lower halves, and both in front and behind. In front of the upper half lies the anterior limb of the medial palpebral ligament (*tendo oculi internus*); this does not form a narrow band crossing the sac about its middle and allowing it to bulge out above as is

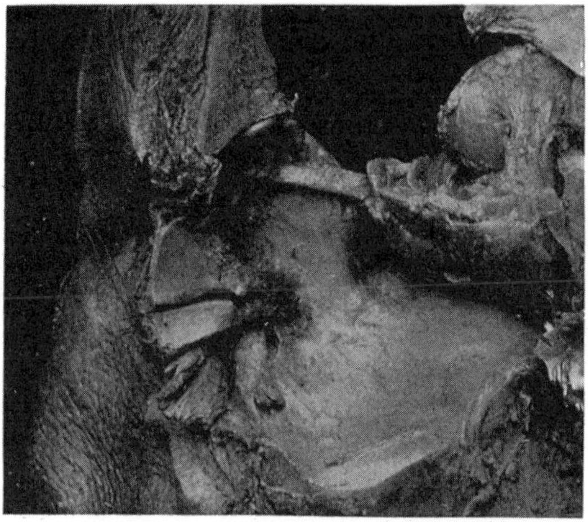

FIG. 126.—Dissection of the Medial Wall of the Left Orbit to show the relation of the **pars lacrimalis** of the *orbicularis oculi* or Horner's muscle to the upper half of the lacrimal sac; a window has been cut through the lacrimal fascia to show the position of the lower half of the sac. The relative positions of the two lacrimal puncta are clearly shown. Natural size.

sometimes depicted, but covers over the whole of this half of the sac, and is first attached to the upper part of the anterior lacrimal crest and then spreads out to blend with the periosteum of the bone beyond, as is well displayed in Figs. 60, 76, 148. Its lower border is thick and free, and is separated from the sac by fibres of origin of the orbicularis, as noted on p. 153; its upper part becomes thinner, and blends with the lacrimal fascia covering the fundus of the sac. This intimate connexion is noteworthy as affording a possible explanation of cases where the lids swell up on blowing the nose after a comparatively light blow on the eye, as in boxing: a sudden strain is thrown upon the ligament and the sac is torn (Fisher, 1904). Superficial to the ligament lie fibres of origin of the orbicularis oculi muscle, the angular vessels, and the skin.

Behind the upper half of the lacrimal sac there passes the posterior or reflected limb of the medial palpebral ligament, a thin sheet of fibres which is difficult to identify and must be of little importance. Deep to it lies the pars lacrimalis (Horner's muscle), of which, indeed, the reflected part of the

FIG. 127.—Relations of the **lacrimal sac.** Horizontal section through the Medial End of the Lower Eyelid, showing—

1, the lacrimal caruncle; 2, the lacrimal canaliculus; 3, the *pars lacrimalis* muscle attached to the posterior lacrimal crest (4); 5, the lacrimal fascia covering the lacrimal sac (6); and 7, the anterior lacrimal muscle fibres. Magnified.

ligament seems to form but a fascial covering. The muscle arises from the upper part of the posterior lacrimal crest (Fig. 65, p. 136, and Fig. 127), frequently encroaching forwards upon the lacrimal fascia, and passes forwards upon and laterally across the upper part of the sac from behind, as seen in Fig. 126, to divide and run along each lid margin as the pars tarsalis of the orbicularis oculi muscle. Deeper still behind the muscle organ lies the septum orbitale, separating it from the fascial expansion or check ligament of the medial rectus muscle of the eyeball and the orbital fat, as is shown diagrammatically in Fig. 156, p. 289. The chief structures, then, in relation to the upper half of the lacrimal sac are the medial palpebral ligament in front, and the pars lacrimalis muscle behind. As previously noted, when traced from the lateral side the muscle fibres in the lids lie in front of the tarsal plates, and hence are distinguished as the pars tarsalis, but at the medial angle of the eye they pass deep behind the sac and form the pars lacrimalis muscle, whilst the plates are continued on as the medial palpebral ligament which passes in front of the sac, as shown in Fig. 78, p. 155. The more posterior attachment of the muscle ensures the plates being kept in close contact with the curved anterior surface of the globe, as noted on that page, and it will be understood why, when the ligament is severed in operations giving access to the lacrimal sac, an ectropion or falling away of the lower lid margin from the globe will not result if the pars lacrimalis muscle lying behind the sac be uninjured (Whitnall, 1913).

The lower half of the sac presents simpler relations, since it lies altogether below both medial palpebral ligament and pars lacrimalis muscle, as seen in Fig. 76, p. 151, and Fig. 126 above. In front lie the skin, orbicularis muscle, septum orbitale, and lacrimal fascia; and access to the sac can readily be obtained from this aspect by exposing and separating (spanning) the muscle fibres along its line, and, if need be, by upraising the ligament. Pathological swellings or diverticula of the sac (Fig. 135, p. 250) are most commonly present in this region, that is, below the medial palpebral ligament. On the lateral side the inferior oblique muscle may arise partly from the fascia covering the sac, and in this region there enter the latter branches from the inferior palpebral and infra-orbital arteries; otherwise on this side and posteriorly it is in contact with the orbital fat.

The guides to the position of the lacrimal sac are the medial palpebral ligament, made conspicuous by drawing the lids laterally, and the anterior lacrimal crest, which can be felt with the finger-tip along the line of the upper part of the naso-jugal fold; there is sometimes a tubercle present at the base of the crest (Fig. 33, p. 69). It is often easy to identify a prominent posterior lacrimal crest as well as an anterior by placing the pulp of the finger flat against the side of the nose in this region, or the finger nail can be pressed into the hollow of the fossa. It is important to realize the parallelism of the antero-posterior axis of the fossa with the medial orbital wall. The surgical application of the above principles is the subject of a special paper by Bailey (1923).

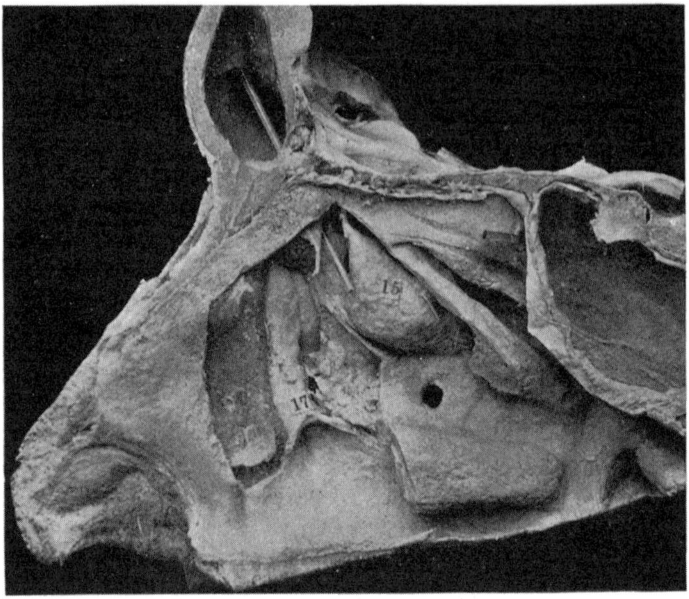

FIG. 128.—Dissection of Right **Lateral Nasal Wall,** showing the naso-lacrimal duct (17) exposed *in situ*; its slit-like opening into the inferior meatus can be seen. Part of the inferior and all of the middle conchae have been cut away. Compare with the earlier stage of a similar dissection as shown in Fig. 29, p. 60. Natural size.

16 = an anterior ethmoidal cell in relation to upper part of lacrimal sac.
15 = the *bulla ethmoidalis*.
Rods have been passed through the openings of the frontal and sphenoidal air-sinuses, and the *ostium maxillare* is conspicuous. The hole on the right of number 17 is artificial.

(iii) The **naso-lacrimal duct** (*ductus naso-lacrimalis*) is the direct continuation downwards of the lacrimal sac. It is contained in the naso-lacrimal canal, and leads likewise into the inferior meatus of the nose. Whereas, however, the bony canal always opens into the apex of the space, the duct may run for a variable distance lower down beyond it and beneath the mucous membrane lining the lateral wall of the meatus before opening into the nose (Fig. 122). The whole duct, therefore, is to be considered in its upper inter-osseous portion and in its lower meatal portion. The upper part necessarily conforms in outward shape and relations to the bony naso-lacrimal canal in which it is lodged, and these characters have been described on p. 77. The form is also illustrated by Figs. 36, 39, 123, 125, 128. The direction taken by the duct in its downward passage is described on p. 82.

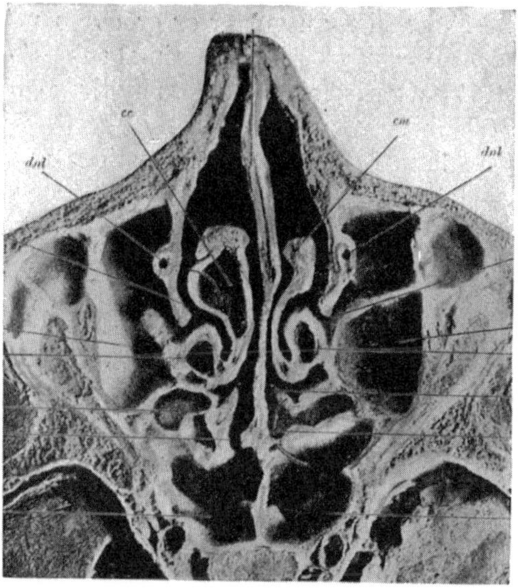

FIG. 129.—Horizontal section through the **Nasal Cavities** of an Adult to show the bulging (lacrimal protuberance) of the naso-lacrimal canal (*dnl.*) into the maxillary air sinus. (From Onodi, 1913.)

ec. = the *bulla ethmoidalis*; *cm.* = the middle concha.

The lower or meatal part of the duct lies buried in the mucous membrane of the lateral wall of the inferior meatus, and is therefore related on the one side to the maxillary sinus, into which it may bulge, and on the other to the meatus itself. It is the extent of this part that chiefly determines the variation in length of the whole tube.

The inferior opening, meatal aperture, or *ostium lacrimale* of the naso-lacrimal duct is most variable in shape and position, and is often difficult to discover by inspection. In the foetus up to eight months, often up to birth, and sometimes even in the new-born child, the duct is closed below by a thin partition formed by the approximation of the lining membrane of the nose to that of the duct, the ostium being subsequently formed by its dehiscence. The resulting opening, which is rarely as large as the lumen of the duct, may be round or linear (vertical or transverse), punctiform, or even duplicate. Holmes found it nearly always slit-like and very fine; it may be guarded by a valve, a flap, or a diaphragm.

It may be followed by a gutter or groove continuing the line of the duct downwards (the 'lacrimal sulcus' of Verga); sometimes this gutter is bridged across by threads of mucous membrane (Henle), or even completely closed over, so that the duct is continued down below the ostium, and the blind end may reach as far as the incisive foramen on the nasal floor (Bochdalek); in such cases a probe might be passed along this lower continuation of the duct without opening up the ostium (Fig. 122).

Aubaret (1910) emphasizes the importance of the degree of permeability of the opening and its influence upon both the shape and the lining membrane of the duct and sac; out of 139 specimens examined by him, 80 presented visibly open orifices and 59 appeared closed; but he concludes that whatever be the shape, dimension, or degree of opening, the ostium is permeable to the air of the nose, and if valves are present they are ineffectual. A widely patent opening would allow of a raising of the air pressure within the duct and sac each time the nose is blown, with a consequent expansion in the direction of least resistance during development and possibly a resulting evolution of the lining mucosa into an atrophic type; the greater the degree of permeability, the more will the air pressure affect the calibre of the duct. He found the regularly cylindrical types of tube correlated with a small inferior meatal orifice, whilst the irregular types all had permeable openings, and in those cases where the whole duct was uniformly enlarged there was a wide ostium. The writer's cases corroborate these observations, though a fine slit-like opening was more commonly found. It would therefore appear possible to ascertain to some extent the probable condition and size of the duct by inspection of the opening.

The position of the ostium is variable. Swerschewsky (1910) described three positions:

(i) In the roof of the inferior meatus of the nose, coinciding with the aperture of the bony naso-lacrimal canal, and found in 45 per cent. There was a fold guarding the opening in 27 per cent. of these, a diaphragm in 12 per cent. and a circular orifice in 6 per cent. Rochon-Duvigneaud, however, found this position to occur only in 5 per cent. of cases.

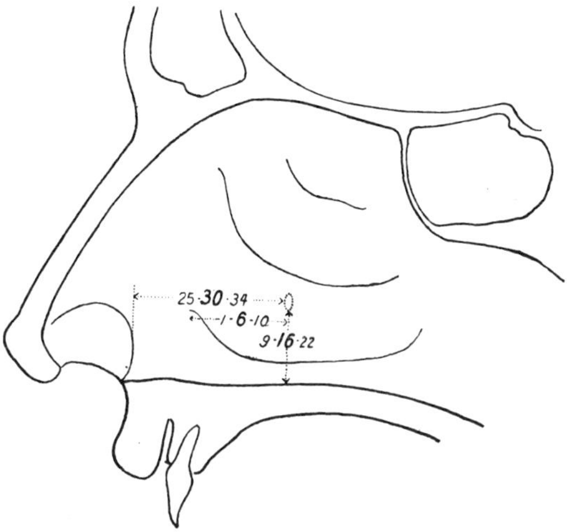

FIG. 130.—Diagram to illustrate the position of the opening of the Naso-Lacrimal Duct (**ostium lacrimale**) on the lateral wall of the inferior meatus as if seen through the overhanging inferior concha. The figures (taken from the measurements of Holmes, 1919) show in millimetres the average of 50 specimens in heavy type, with the extremes on either side in lighter type.

(ii) Opening below this point on the side wall of the inferior meatus in 49 per cent., the generally described position. The aperture was a wide one in 18 per cent., a half-open canal in 16 per cent., and a small groove in 15 per cent. of these. This last was the condition considered by him to be most favourable to disease.

(iii) Opening far in front or behind as a small canal in 3 per cent.

The generally described position is at about the middle of the side wall of the meatus, 30 mm. or $1\frac{1}{4}$ inches behind the lateral margin of the anterior nares, and opposite the junction of the first and second quarters of the attached border of the inferior concha which overshadows it. The detailed measurements of its position from various points are shown in the above diagram made from the figures given by Holmes (1919) after an examination of fifty specimens.

It would appear that Winslow's location of the position by intersecting lines, one drawn horizontally from the inferior border of the ala of the nose, the other vertically and passing through the second molar tooth or between the second and third molars, can give only approximate results.

The **dimensions** of the lacrimal sac and naso-lacrimal duct correspond with those of the bony fossa and canal given on p. 77; the dimensions of the sac are a little less than those of the bony fossa, being on an average 12 mm. ($\frac{1}{2}$ inch) long and from 4 to 8 mm. ($\frac{1}{4}$ inch) broad; the lateral width differs according to the part of the sac measured, but is about 2 or 3 mm. in the mid-region. The intra-osseous part of the duct measured in length on an average 12·4 mm. ($\frac{1}{2}$ inch), the meatal portion 5·32 mm. ($\frac{1}{4}$ inch), with extremes of from 1 to 12 mm. in sixty observations made by Aubaret. Onodi gives a total length of the tube as about 30 mm. (1$\frac{1}{4}$ inches), comprised by the sac of 12 to 14 mm., intra-osseous part of duct 10 mm., meatal portion 5 mm., as illustrated in Fig. 122, p. 228. A further measurement of practical value is the distance that the upper or orbital opening of the duct lies above the nasal floor, since the figure includes the extreme extent of any meatal part of the duct. Power (1886) found an average distance of 27·4 mm. in 292 European skulls; Gérard (1907) an average of 23·3 mm., with extremes of 25 mm. and 40 mm. in 69 skulls, and he emphasizes the importance of retaining this highest extreme of 40 mm. (1$\frac{1}{2}$ inches) as the least length that should be given to that part of the catheter destined to traverse completely the whole duct.

The external calibre of the duct will be in harmony with, and vary according to, that of the canal, as described on p. 78, but the lumen presents greater variations, since not only is the lining mucous membrane commonly thrown into folds and ridges (second type in Fig. 119), but, like the nasal mucosa, may vary from the hypertrophic condition found in chronic catarrh to the atrophic form of old age (see also p. 244). Normally the lumen of both sac and duct is a mere cleft, but will take a probe of 3·5 mm. diameter (Treves). The normal capacity of the sac is given by Hyrtl as 20 c.mm., but Arlt found that it could be injected until it measured 4 mm. in transverse width, with a capacity of 120 c.mm. Aubaret, again insisting upon the effects of a permeable lower opening, describes narrow ducts of from 1 to 3 mm. in internal diameter and dilated ones of from 3 to 7 mm., 41 out of 60 specimens being of the latter class; in the former the mucous membrane is thick (as in Fig. 133), in the latter thin (as in Fig. 127). His fusible metal casts of the sac and duct illustrate well the variable morphology of the passages, but a slightly artificial dilatation of the sac is to be expected by this means. Theobald obtained an average transverse width of from 4 to 4·7 mm. in ten ducts measured on the cadaver, slightly greater than the canals in the skull, and explained by the elasticity and yielding of the bones in the fresh condition. Onodi (1913) gives the following diameters (presumably internal) of the nasal duct in children: in the new-born infant between 1·5 and 2 mm.; in one of 2 months, 2 mm.; at 5 months, between 2 and 3·5 mm.; at 8 months between 3 and 8 mm.; in a child of 11 months it was 2 mm., in one of 12 months, 3 mm.; in one of 14 months, 1·5 mm.; in one of 3 years, 2·5 mm. The average diameter, therefore, in young children may be expected to be about 2 mm.

Internal Configuration of the lacrimal passages.—In the new-born the lining mucous membrane of the passages is utricular, irregularities being left in the walls after disintegration and absorption of the solid epithelial cord from which they are developed. In the adult most of the crypts and folds disappear, but some persist, and if correlated with a patent lower meatal orifice may develop under the influence of a raised air-pressure into so-called valves and sinuses; on the other hand, they may be occasioned by a gradual distension due to accumulation of tear fluid and mucus, with an insufficiently clear lower passage. The folds are of no special structure or anatomical regularity, but occasionally some are sufficiently well developed to be termed 'valves', though their role as such is doubtful.

They have been specially studied by Aubaret (1908), and the positions they occupy and the names by which they have been dignified are shown in Fig. 131; they appear, however,

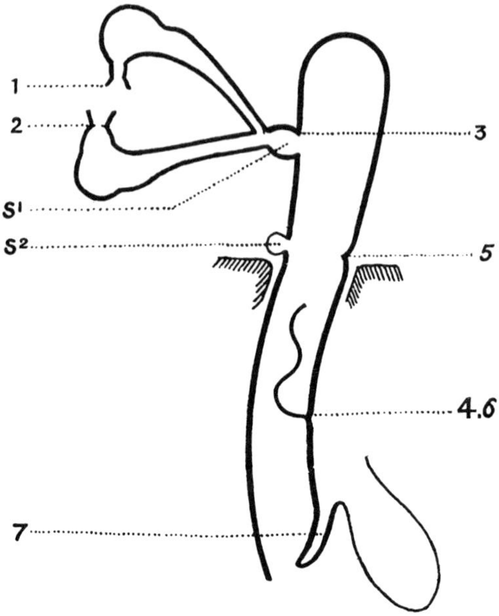

FIG. 131.—Diagram to show the position of the so-called **Valves** of the Naso-Lacrimal Passages (after Aubaret). The valves are numbered as follows, with the names by which they are known:

1 = Bochdalek; 2 = Foltz; 3 = Rosenmuller or Huschka; 4 = spiral valve of Hyrtl ending below in 6; 5, the valve of Krause or Béraud; 6 = Taillefer; 7 = Horner, Hasner, Bianchi, or Cruveilhier, the *plica lacrimalis*; S^1 = the sinus of Maier; S^2 = the sinus of Arlt.

The lacrimal sac is shown in the conventional dilated condition.

most inconstant, as for example the valve in the middle of the naso-lacrimal duct (Taillefer), which is said to occur only in 6 per cent. of cases. The most important and most frequently developed fold is one situated at the meatal opening of the naso-lacrimal duct, the *plica lacrimalis* or valve of Hasner, though previously described by others; it is the remains of the foetal membrane closing the lower end of the duct, and a blocking and distension of the tube may be caused by accumulation of epithelial debris above it; when present in the adult it is rarely efficient as a valve, since incompetency was found experimentally by Aubaret in 90 per cent. of cases. On the other hand, Bert (quoted by Aubaret) found that coloured fluids injected into the nose escaped from the lacrimal puncta only three times in eighteen experiments, whereas direct injections into the duct invariably appeared at these points, showing that the valves above Hasner's were always incompetent. Passages of the enlarged atrophied type associated with widely open ostia would explain the instances recorded of smokers who could cause fumes to escape from the puncta. From a clinical point of view, a paper by Tartuferi (1902) on catarrhal conditions of the passages, well illustrated by microscopical sections showing various folds, is here worthy of reference. Campbell (1922) has studied internal configuration of the passages by means of the röntgen ray after injection with bismuth.

The folds are considered to be the usual cause of obstruction to the course of the tears or the passage of a sound; and in connexion with this, it may be of interest to summarize the possible natural obstacles to the passage of a probe along the lacrimal passages (Gérard, 1907):

(*a*) Osseous.—Accentuation of the angle between the long axes of the lacrimal fossa and naso-lacrimal canal, such as occurs when the anterior nares is narrow; the development of a spur at the base of either the anterior or posterior lacrimal crest may narrow the upper aperture of the canal, or the presence of a well developed hamular process raise its level; the canal may be constricted about its middle (see p. 80); a prominent supra-orbital ridge would throw the upper end of the probe too far forwards.

(*b*) Membranous.—Mucous folds in the canaliculi, dilatations or sinuses of the sac, or valves in the duct, may be sources of difficulty; the ostium may be guarded by a diaphragm; the point of a probe could pass beyond the meatal opening and lie in a blind continuation of the duct.

The **Structure** of the Lacrimal Passages.—They are essentially formed throughout by a mucous membrane continuous with the conjunctiva at the lacrimal puncta, and with the nasal mucosa at the inferior orifice of the naso-lacrimal duct (Fig. 118, p. 223). This continuity of the membrane is of considerable clinical importance, since in Kuhnt's experience disease of the lacrimal passages has a nasal origin in 93·7 per cent. of cases, though micro-organisms of conjunctival origin may cause inflammation of the lacrimal sac (Onodi, 1913). This lining membrane is reinforced by elastic, connective, and muscle fibres in the canaliculi, by fibro-elastic tissue in the sac, and by cavernous erectile tissue in the duct. It differs from the conjunctiva in that its mucous tissue is rich in elastic fibres, is thicker, and more vascular; it gradually assumes the characters of the nasal mucosa as it approaches the nasal cavity, and it shares the degree of atrophy or hypertrophy of that lining membrane (cf. Figs. 125, 133).

The canaliculi are lined by an epithelium which is thicker here than elsewhere in the passages, stratified six to twelve layers deep on the surface, but cylindrical below; the presence of a basal membrane is disputed. Outside the epithelium of the canaliculi is a fibrous layer derived from the lacrimal fascia, which is reflected along them from the points where they pierce it; it is characterized by a rich development of elastic fibres, which spread out amongst the muscle fibres of the pars lacrimalis in which these channels are buried (Halben, 1903; Fig. 132). Round the vertical part of the canaliculus and papilla the elastic tissue is denser and the connective tissue of these portions is continuous with that of the tarsal plates; round the puncta the tissue is almost sclerous. The muscle fibres surround the base of the papilla in sphincter-like form, as described on p. 227, and as figured below.

The lacrimal sac and naso-lacrimal duct are lined by two or more layers of columnar epithelium resting upon a basement membrane; cilia are said to be present by some authors (Schäfer, Quain, Tourneux, Santos-Fernandez), but their presence is denied by others (Werncke, Halben, Rochon-Duvigneaud); goblet cells are present. There is a delicate dermis or tunica propria, infiltrated in the adult, though not in the new-born, by lymphocytes, which are often, though possibly abnormally (Parsons, Werncke), aggregated to form follicles. Krehbiel has several times seen a small lymph nodule in the wall of the sac immediately behind the point of entrance of the canaliculi. The outermost coat is of fibro-elastic tissue, containing a venous plexus especially well developed towards the meatal end of the tube, where it resembles the erectile cavernous tissue characteristic of the nasal mucosa with which it becomes continuous. On the

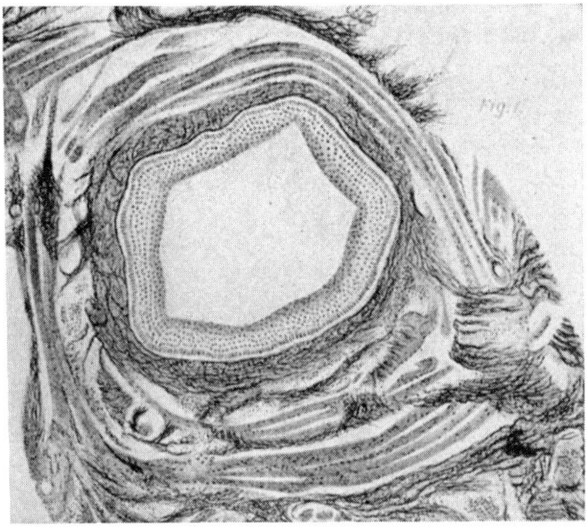

FIG. 132.—Cross-section of base of vertical part of **Lacrimal Canaliculus** of three year-old child, showing the sphincter-like arrangement of muscle fibres constituting the *musculus quadrangularis constrictor puncti lacrimalis,* and also the large amount of fine elastic tissue surrounding its wall. (From Halben, 1903.) Magnified.

whole, the elastic tissue is found to decrease from the puncta downwards, whereas the venous plexus and thickness of the walls increase. The presence of glands in the walls of the sac and duct is disputed; they have not been found either by Stanculeanu in the embryo or by Rochon-Duvigneaud in the adult, and such as have been described may be pathological formations or simple diverticula of the mucous membrane. On the other hand, definite mucous glands have been noted in normal sacs by Kuhnt, Arlt, Halben; and Werncke (1905), in a special investigation, found both composite and single tubular glands present in 8 out of 14 lacrimal sacs, whereas in 17 pathological sacs none were found; they were situated in the antero-medial wall near the summit, and contained mucous cells and crescents of Giannuzzi. Small serous glands, resembling those of Krause of the conjunctiva, have been found by Joerss, in 8 per cent. of cases in the submucosa of the fundus of the sac. The writer has found one instance of such in the lacrimal sac. Whilst the duct wall at its upper end can be separated from the lining periosteum of the canal in which it lies, as it passes downwards the two become more intimately blended to form a muco-periosteum, a close association which explains the extension of chronic inflammatory diseases of the duct to the surrounding bone.

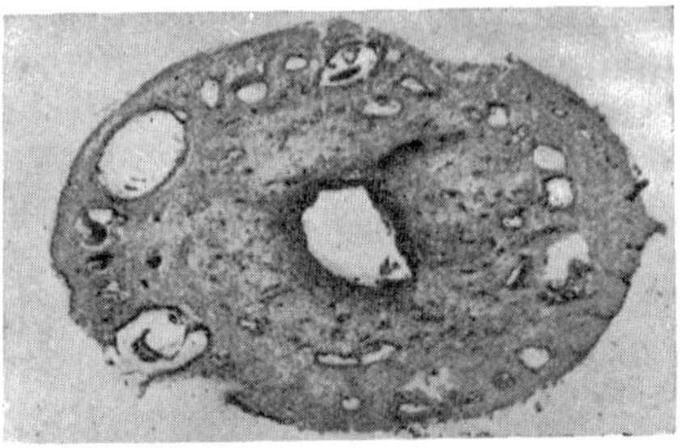

FIG. 133.—Transverse section through the middle of the **Naso-Lacrimal Duct** of an adult; hypertrophied type of mucous membrane. The actual diameters of this specimen were: outside transverse, 4 mm.; antero-posterior, 6 mm.; inside transverse, 1·1 mm. × about 9.

The **Vessels and Nerves** of the naso-lacrimal passages: The arteries are derived (i) from the ophthalmic, the superior palpebral branch supplying the sac, the inferior palpebral the duct; (ii) from the angular artery of the facial, which supplies both parts; (iii) from the infra-orbital artery, which in the forepart of the infra-orbital sulcus gives off a branch which sends twigs to the lower eyelid and inferior oblique muscle, then grooves or pierces the lateral margin of the upper orifice of the naso-lacrimal canal and supplies the sac and duct. The lower part of the duct receives a supply from the nasal branch of the spheno-palatine artery, which is a derivative, like the infra-orbital, of the internal maxillary artery.

The rich venous plexus which surrounds the duct, and to a lesser extent the sac, is connected above with the angular and inferior orbital veins and below with those of the nasal cavity, the latter draining through the spheno-palatine veins into the pterygoid plexus and internal maxillary vein; venules pass from the plexus round the lacrimal sac through the lacrimal bone or its sutural lines into the mucous membrane of the anterior ethmoidal cells.

The lymphatic vessels from the sac accompany the facial vein in their course and drain into the submaxillary nodes (Fig. 92, p. 175); those of the lower part of the duct join the lymphatic vessels of the inferior nasal meatus, which drain both anteriorly towards the skin of the anterior nares and thence to the submaxillary nodes, and posteriorly to the retro-pharyngeal and deep cervical nodes (Most, 1905).

The nerves of the sac are derived from the infra-trochlear branch of the ophthalmic division of the fifth nerve; the lower part of the duct receives a twig from the anterior superior alveolar branch of the maxillary division of the same nerve. There would appear to be some physiological relation, probably through the branches of the nerve, between the innervation of the lacrimal gland and that of the lacrimal sac, explanatory of the known clinical fact that destruction of the sac leads to diminution of the tear secretion; the epiphora of dacryocystitis is most likely caused by reflex irritation from the diseased sac.

Development and Abnormalities.—The lacrimal passages are developed along the line of the cleft between the lateral nasal and maxillary processes of the embryonic face (p. 188); they are formed, not by the cleft being primarily converted into a tube by the approximation and fusion of its boundaries, as was supposed until the observations of Born in 1876, but by canalization of a solid rod of ectoderm cells formed beneath the surface along this line. In later years the development has been an object of special study by many (Stanculeanu, 1900; Rochon-Duvigneaud, 1900; Monesi, 1904; Matys, 1905; Fleischer, 1906; Cosmettatos, 1906; Contino, 1907; Lang, 1911; Schaeffer, 1912; notably by Iwata, 1927, and Speciale-Cirincione, 1930). It may be described in two stages: a first, consisting in the formation of the solid epithelial rod; a second, the hollowing out of the rod to form the lacrimal passages.

The rod appears, after the cleft has become obliterated by the fusion of the processes which bound it, as a thickening of the deeper layers of the epithelium along the line of junction, the first appearance being at a point corresponding to the medial part of the conjunctiva of the lower lid; the lower end elongates and sinks beneath the surface, the whole rod following and becoming surrounded by mesenchymal tissue, but retaining its contact with the skin longest at the point of origin. The first appearance has been seen in an embryo of 9·5 mm. in length, and the rod was found completely separated from the surface in one of 15 mm. The canaliculi are formed secondarily by buddings from the upper end of the rod, and reach the surface of the lid margins about the same time that the lower end of the rod reaches the nasal cavity, as has been seen in embryos of from 18 to 24 mm. The inferior canaliculus appears to be the direct continuation of the main stem, the upper one being a secondary outgrowth. The inferior canaliculus in its growth upwards cuts off a part of the lid margin which is destined to form the lacrimal caruncle (Fig. 100, p. 190).

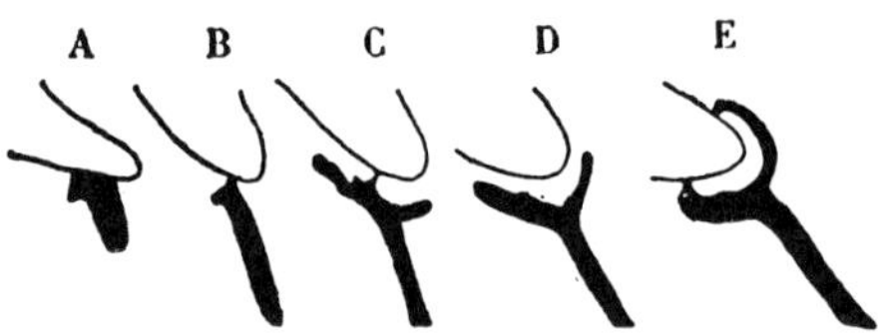

Fig. 134.—Schematic representation of the successive stages of development of the **Lacrimal Passages** to form the canaliculi. (From Matys, 1905.)

The process of canalization begins by disintegration of the central cells, first in the middle of the lower canaliculus, as has been seen in an embryo of 35 mm. long or at three months, then in the upper, and then in the lacrimal sac. In embryos of 6 cm. in length a lumen is established throughout the whole system, but is closed at the puncta and the inferior orifice of the duct; the former open in embryos of 13 cm. just before the eyelids separate towards the seventh month; the latter remains closed until the end of the eighth month, or even after birth. According to Monesi, after the sixth month the lower end of the duct is swollen by an accumulation of the debris of disintegrated cells, which eventually causes the perforation of the septum between it and the nasal fossa, and the remains of the septum form the plica lacrimalis.

The **abnormalities** are readily explained by irregularities in this disintegrative mode of development. The puncta may be supernumerary, two, three, or even four having been noted. Duplication of either canaliculus, but more often of the lower, has been observed, but such developments must be rare, since Wicherkiewicz found only one case of double punctum and canaliculus in 60,000 patients examined, and according to Schoute, not more than 22 cases had been recorded in the literature up to 1901. On the other hand, cases are recorded of congenital imperforation of the puncta, of non-development of the canaliculi, and of atresia of the whole duct or part of it. The lacrimal sac may present a congenital fistula.

Diverticula or sinuses of the passages are also found. The usual site is on the lateral wall of the lacrimal sac, where the resistance to expansion is least. A small dilatation at the point of entrance of the canaliculi is common (Fig. 131, p. 242), and is known as the 'sinus of Maier'. Another, called the sinus or 'recessus of Arlt', bulges outwards from the sac just beneath the medial tarsal ligament; Aubaret (1909) states that he found the latter present in 24 out of 50 specimens (a proportion quite at variance with the writer's experience, though it is not clear in Aubaret's paper whether or no he includes dilated sacs in this category); an example is photographed in Fig. 135. Such diverticula or folds, if largely developed, may cause the appearance of a bilocular or double lacrimal sac. The naso-lacrimal duct itself has been seen doubled throughout (Monesi), possibly due secondarily to amniotic bands forming a facial cleft (Ask and Van der Hoeve, 1921), or it may present two inferior apertures. The various folds or valves of the passages have been described above (p. 240).

For accounts of the lacrimal apparatus in domestic animals, papers by Lichal (1915) and Rochat (1915) may be referred to, and an admirable study of the histological structure of these organs in the ox will be found in a paper by Sundwall (1916).

The **Mechanism** of the drainage of the tears from the conjunctival sac through the lacrimal passages has been the subject of much discussion. Normally the lacrimal gland secretes fluid in quantity just sufficient to replace that lost by evaporation; the fluid is prevented from running over the edge of the lower eyelid by the greasy lubrication of the tarsal glands, and is supposed to be induced to flow towards the medial canthus and collect in the lacus lacrimalis by a slight 'spiral movement' of the lids in closure; more probably the direction of the flow is determined by the general contraction of the orbicularis muscle from its mobile lateral to its fixed medial side. For the tears to enter the lacrimal sac two conditions are essential: first, the puncta must be directed towards the floor of the lacus lacrimalis, that is, there must be no paresis of the marginal fibres of the orbicularis, or epiphora will result; second, the puncta must retain their capillary action. The

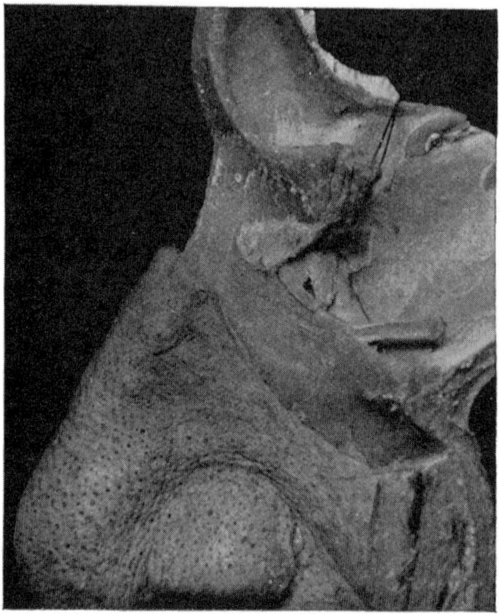

FIG. 135.—Dissection of forepart of Medial Wall of Left Orbit, to show relation of medial palpebral ligament (in front) and *pars lacrimalis* muscle (behind), both of which have been lifted upwards, to the lacrimal sac, which shows a well-marked anterior dilatation or **sinus of Arlt** (opened). Natural size.

canaliculi are elastic-walled tubes, and in the act of winking they are shortened and widened by the pull of such fibres of the pars lacrimalis muscle as are connected to the walls of their horizontal portions (Halben, 1903), and consequently the fluid can be sucked into them. The vertical portion of the canaliculus is too dense to be compressed by the action of the quadratus muscle surrounding its base in light closure of the lids, and the lumen always remains open in normal physiological conditions, but in blepharospasm or forced closure the muscle does act as a sphincter; the lumen is then effaced, and the tears are seen to accumulate in the lacus lacrimalis (see also p. 227). Frieberg (1918), West (1918), Rosengren (1929), and Ploman (1928, 30; who has studied the mechanism by the röntgen ray and use of canulae on the living subject), all consider that the canaliculi play the leading part, the tears being drawn in by capillarity and then expelled into the sac by their occlusion in relaxation of the lids.

It is as regards the action of the lacrimal sac, and especially of the pars lacrimalis or Horner's muscle in relation to it, that the most widely divergent views are expressed. The muscle is said either to compress the sac or to dilate it. It will be realized that, according to the above description of the sac and its relations, normally it is neither in a position nor in a condition to be compressed; its position is one of lying flattened in the lacrimal fossa covered over by the lacrimal fascia, and no part of the orbicularis muscle is actually inserted into it; the arc described by the pars lacrimalis muscle may be slightly flattened in contraction, but by so much less would the muscle press upon the sac since it lies behind the latter; the medial palpebral ligament, bound to the fore and upper part of the lacrimal fascia, will be straightened and pulled away from the sac on contraction of the orbicular fibres which arise from it (Schirmer, Fuchs). The normal condition of the sac, flattened from side to side with the walls in apposition so that its lumen is a mere cleft, also allows no further occlusion, though Ploman found its lower part compressed in forcible winking. On the other hand, the sac is both in a position and in a condition that readily admits of dilatation, as is seen in pathological conditions and by experimental injection. It has been pointed out above that the area of attachment of the pars lacrimalis muscle frequently encroaches upon the fascia covering the sac, and if the two be adherent, contraction of the muscle will dilate the sac; indeed, in one preparation of a horizontal section through the parts where the muscle fibres were so disposed, the slightest traction upon them made the cleft of the sac visible, and, moreover, in cases of epiphora due to paralysis of the orbicularis it has been observed that tears can be made to enter the sac by drawing the lids laterally. The importance of the pars lacrimalis muscle is further shown by the fact that the first symptoms of a facial paralysis may be a complaint of epiphora even before the puncta become everted (Kalt, 1903). Scimemi (1892) found experimentally that the capacity of the sac is increased by 2 c.mm. in each physiological closure of the lids, by 10 c.mm. with moderate effort, and by as much as 30 c.mm. by forced effort, and that an upward and outward traction of the upper eyelid will increase the capacity much more than the most energetic closure of the lids will do; further, clinical observations have shown that a drop of fluid at the entrance to a fistula of the sac is aspirated in winking (Kalt). On such grounds, therefore, the lacrimal sac must be considered to be *dilated* and not compressed in the act of winking. It then regains its normal and reduced capacity owing to the elasticity of the walls, and will drive its contents onwards into the wider channel, the naso-lacrimal duct, rather than backwards along the narrower canaliculi. That the sac must have some action is shown by the fact that in atony of this structure the proper conduction of tears is arrested (Fuchs).

Finally, as regards the passage of the tear-fluid down the duct into the nasal cavity, the flow is favoured by gravity, by the passage from narrow to wide channels, and perhaps by inspiration of air through the nose, since one sniffs in the act of weeping. Permeability of the duct is not a necessary condition for the absorption of tears, since the sac will fill in spite of complete obliteration of the passage below it. The so-called valves of the lacrimal passages play no part in occasioning the downward flow.

To sum up then, it is probable that the capillarity of the puncta primarily suffices to draw off from the conjunctival sac any slight surplus of fluid left over from evaporation; that the canaliculi, shortened and dilated in the act of winking, take the chief part in effecting the drainage; and that the lacrimal sac participates in the removal of excess fluid by aspiration, being slightly dilated in the act of winking, and more so in forced closure of the lids; the elastic reduction of the sac to its resting dimensions drives the fluid into the nasal duct, whence the flow into the nasal cavity is directed by gravity.

PART III

THE CONTENTS OF THE ORBIT

1. The Eyeball.
2. The Muscles.
3. The Fascia and Fat.
4. The Vessels.
5. The Nerves.

General Arrangement.—The eyeball is situated in the anterior and roomiest part of the orbit, of which it occupies about one-fifth the volume; from its posterior pole the optic nerve emerges and passes backwards to leave the apex of the space by the optic foramen; the nerve lies in the midst of the four recti muscles, which, arising close around the foramen, diverge as they pass forwards to be inserted on to the eyeball. Within the 'cone' so formed by the muscles lie, as well as the optic nerve, their own nerves (third and sixth), the naso-ciliary nerve and ciliary ganglion, the opthalmic artery and vein together with many of their branches. Outside the cone of muscles lie the superior oblique and levator palpebrae superioris muscles and the fourth, frontal and lacrimal nerves. Beneath the eyeball lies the inferior oblique muscle, above it is the reflected tendon of the superior oblique muscle, whilst on its upper and lateral side is placed the lacrimal gland. The rest of the orbital space, both within and without the cone of muscles, is completely and compactly filled with the orbital fat, from which the eyeball is separated by a fascial envelope, the fascia bulbi or capsule of Tenon. Clinically, a regional distinction is made between the ocular and retro-ocular contents and relations of the space.

1. **THE EYEBALL** (BULB OR GLOBE OF THE EYE, *BULBUS OCULI*)

The eyeball is here considered only as regards its external configuration and relations, that is, the organ as a whole is described; for the gross structure the reader is recommended to consult *The Anatomy of the Human Eye, as illustrated by Enlarged Stereoscopic Photographs*

(Thomson, Clarendon Press, 1912); the development and histological structure will be found fully described in the 'Organe de la Vision'

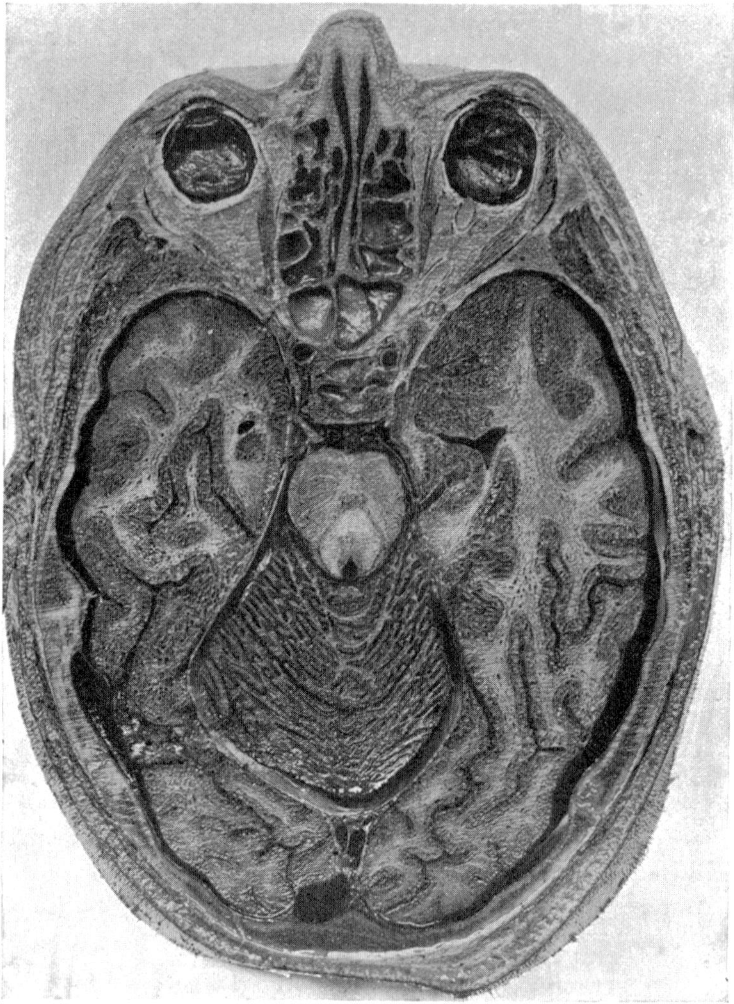

FIG. 136.—Horizontal section of Adult Male Head to show the relative position and relations of the orbits and their contents. Compare with Figs. 3 and 17. From the McGill Anatomical Museum. × $\frac{2}{3}$.

(Druault, in Poirier's *Traité d'anatomie humaine*, Paris, 1912); 'The Development of the Human Eye' by Ida Mann (1928), may be referred to; and Hereditary Anomalies of the eye have been described in a series of articles by M. Macklin (1926, 7). Articles by Merkel and Kallius and Pütter in the *Graefe-Saemisch Handbuch der Augenheilkunde,* or by Piersol in Norris and Oliver's *System of Diseases of the Eye,* should be consulted. Finally, the clinician interested in cases of 'blue sclera' associated with fragile bones and deafness will find the literature given by Singer (1923) and Oast (1928), and hereditary cases collected and described by Kunii (1930).

The eyeball has the form of a sphere, but not a strictly symmetrical one, being slightly flattened from above downwards (forming an 'oblate spheroid'), with the contour interrupted by the bulging of the cornea in front. It may be described as composed of segments of two spheres, an anterior corneal segment with a radius of 8 mm., and a posterior scleral segment of 12 mm. radius; the corneal area extends over about one-sixth of the whole bulbar surface. Where the two areas join, that is, at the sclero-corneal junction or *limbus corneae,* is a shallow groove, the *sulcus sclerae.* The central points of the curvatures of the corneal and scleral spheres are called the anterior and posterior **poles** (*polus anterior vel posterior*) of the globe; and a straight line joining them represents its **axis** (*axis oculi externa vel interna*); a line encircling the globe midway between the poles is the anatomical **equator** (*aequator*), and it divides the surface into two **hemispheres**, anterior and posterior; any line surrounding the surface of the globe and passing through both poles is a **meridian**, and crosses the equator at right angles; the antero-posterior length, height, and width of the globe are expressed by sagittal, vertical, and transverse or horizontal diameters respectively. The 'visual axis' (*axis optica*) passes from the *fovea centralis* of the retina through the 'point of rotation' of the globe, and cuts the cornea rarely at its

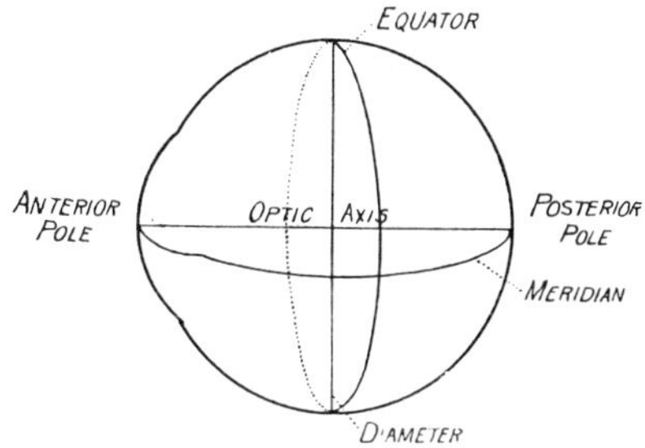

FIG. 137.—Scheme of the **Eyeball** seen from the side. × 1½. (Half a meridian only has been drawn.)

mid-point but usually slightly to its nasal side; it therefore does not exactly coincide with the axis of the globe. When the visual axis is directed straight forwards, the axis of the cornea presents a deviation from this line of 5° laterally and 2° below; the long axis of the orbital cavity deviates still further from the visual axis, being from 23° to 25° lateral and 15° to 20° below it (Druault, 1912); the visual axes are roughly parallel to one another, at most forming an angle of 10° (Testut), whereas the orbital axes enclose an angle of about 45°, as already noted on p. 97.

Most eyeballs vary slightly from the schematic form; the most important variations are those where the length of the optic axis is either less (hypermetropic globe) or greater (myopic) than the normal (emmetropic), since in such eyes, when accommodation of the lens is completely relaxed, the image of a distant object falls respectively either behind or in front of the retina instead of exactly upon it.

The **dimensions** of the normal globe vary (but not beyond one millimetre) according to the findings of different authorities. A full list has been compiled by Goldnamer (1923). The figures in millimetres given by Sappey (and quoted by Dwight, Druault, and Baker) from 26 adult eyes (14 male, 12 female) measured from 24 to 36 hours after death are as follows:

Diameter.	Antero-Posterior.	Transverse.	Vertical.
Maximum	26·4	27·1	25·8
Mean	24·2	23·6	23·2
Minimum	22·9	22·2	22·2

Each diameter is on an average five- to six-tenths of a millimetre greater in the male than the female. Merkel (1901) gives

24·3 (external axis) 22·5 (internal axis)	23·6	23·3

The last-named author gives, in addition, numerous other dimensions of the parts of the globe as found by himself, and also by Quain, Rauber, Schwalbe, and Weiss; the mean of the combined above measurements of all these authors is

Antero-Posterior.	Transverse.	Vertical.
24·15	**24·13**	**23·48**

The average diameters of the eyeball in the new-born, as found by Weiss (1897) in 14 observations, are

16·40	16·00	15·40

and he found the eyeball at this period to be hypermetropic in 92 per cent. Weiss (1897) gives the comparative diameters and also the circumferences at various ages from birth to adult; and Seefelder (1908) has made a very complete series of measurements of the embryo and infant eyeball; Scammon (1925) finds higher values in volume, weight, and diameters at birth than those of Weiss.

The cornea has an average horizontal diameter of 12 mm. and a vertical one of 11 mm. (Druault); its post-natal growth has been studied by Hymes (1929).

The weight of the eyeball varies in individuals, but may be considered to be about 7 grammes; Sappey estimated it as 7 to 8 grs.; Testut at 7·14 grs. on a mean of ten eyes taken from the cadaver; Weiss at 7·45 grs. on a mean of 5 adult emmetropic eyes, with weights of 2·29 grs. at birth, 4·05 grs. at one year, and from 5·87 grs. to 6·50 grs. at from 13 to 15 years; the volume is about 6·5 c.mm., with a specific gravity of 1·077 (Schäfer).

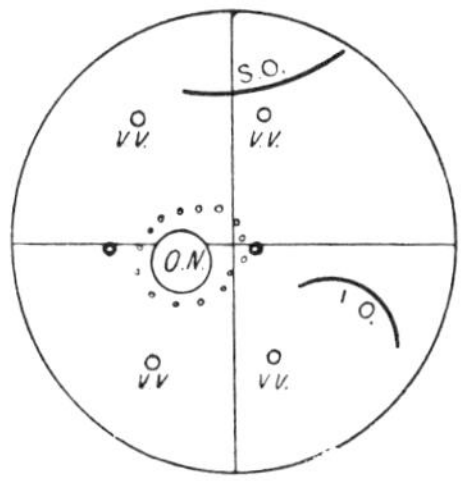

FIG. 138.—Scheme of the **Eyeball** seen from behind, to show the position of the various structures attached to it. (After Testut.) × 1½.

O.N. = the optic nerve, surrounded by a ring of ciliary nerves and vessels, with a long posterior ciliary artery on the horizontal meridian on either side of it; V.V. = the four *venae vorticosae*; S.O. and I.O. = the lines of insertion of the superior and inferior oblique muscles respectively.

The Structures attached to the Eyeball.—The optic nerve emerges not exactly from the posterior pole, but a little medial to and below it, the central point of the nerve lying 3 mm. medial to the vertical meridian and 1 mm. below the horizontal meridian (Fig. 138). Round the nerve exit in an irregularly scattered ring are the points of entrance of the minute ciliary nerves and short posterior ciliary arteries; the long posterior ciliary arteries pierce the globe one on each side of the horizontal meridian and a little distant from the optic nerve; four venae vorticosae emerge on or a little behind the equator, and the anterior ciliary arteries enter just behind the sulcus sclerae. The four recti muscles are inserted by tendons into the globe in front of the equator, and the two oblique muscles behind this line and on the lateral side. The fascia bulbi or capsule of Tenon is attached to the globe in front immediately behind the corneal margin, and at this point is fused with the bulbar conjunctiva, the two being closely applied to the anterior third of the bulb; the capsule closely envelops the entire scleral part of the globe, being united to it by a very delicate loose connective tissue, and terminates behind round the site of emergence of the optic nerve; it must consequently be pierced by all the above-named structures which are attached to the globe, and it becomes fused with the fascial sheaths of the muscles where they enter it.

The relation of the eyeball to the orbit is best studied in horizontal and vertical views, as shown in Figs. 136, 161, and Fig. 60, p. 128. As already noted, it occupies about one-fifth of the cavity, lying in its roomiest forepart and slightly nearer the upper and lateral sides, being nowhere in contact with the walls. The average distances from the walls are: 4·5 mm. from the lateral, 6·5 mm. from the medial, 4·5 mm. from the superior, and 6·2 mm. from the inferior (Goldnamer).

The summit of the cornea is normally just touched by a ruler held vertically against the superior and inferior orbital margins, but the prominence varies in individuals and according to age and state of health; it thus may vary from 12 mm. in front to 10 mm. behind this plane, and in exophthalmic goitre may even lie 24 mm. in front (Cohn); indeed, Ambialet (1905) finds that in 85 per cent. of living cases the protrusion is positive, that is, in front of the vertical axis; Hymes (1929) gives the projection in infants as 5·84 mm., at puberty as 15 mm., and finds that many adults exceed 17 mm. with relatively few less than 11 mm. The lateral orbital margin, on the other hand, is so recessed that half of the eyeball is here exposed, as is well shown in Fig. 139, and a straight line passing from this margin to the anterior lacrimal crest would traverse the globe behind the *ora serrata* of the retina on the one side and emerge at the junction of the ciliary body and iris on the other. A strong medial rotation of the globe will almost bring the *macula lutea* in line with this margin (Henke). The distances which separate the equator of the globe from the orbital walls are: 4 mm. from the roof, 4·5 mm. from the lateral wall, 6·8 mm. from the floor, and 6·5 mm. from the medial wall. The mid-points of the two pupils lie from 58 to 60 mm. apart. The posterior pole lies about 18 mm., varying from 14 to 24 (Weiss), or $\frac{3}{4}$ inch in front of the apex of the orbit.

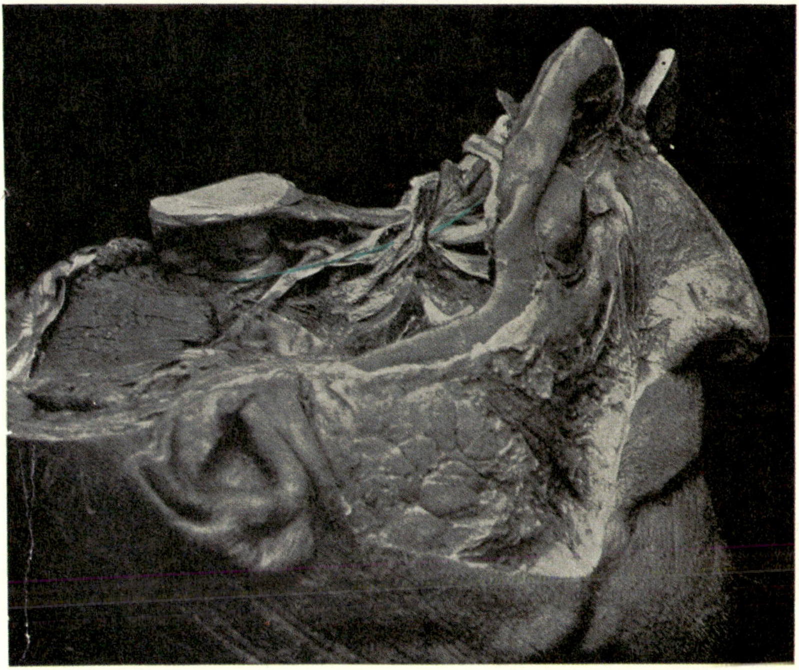

FIG. 139.—Dissection of the Head, to show the **relation of the eyeball** to the orbital margin, the course of the **optic nerve,** the position of the optic chiasma, the trochlear nerve in its whole course, the cavernous sinus, and the semilunar or Gasserian ganglion. The same preparation is viewed from above in Fig. 197, p. 379. × $\frac{2}{3}$.

2. THE MUSCLES (*MUSCULI RECTI OCULI*)

The muscles which move the eyeball (sometimes referred to as *extrinsic*, in contradistinction to the intrinsic muscles which are connected with the mechanism of the lens and iris within the globe) are six in number. There are four recti, distinguished as superior, inferior, medial, and lateral, and two obliques, the superior and inferior, all named from their positions relative to the eyeball. Each rectus muscle presents a narrow, tendinous origin from the apex of the orbit, an elongated strap-like fleshy belly, and a thin tendon of insertion onto the eyeball, all four being attached in front of the

FIG. 140.—The Left Eyeball from behind to show the four **recti muscles** spread out, and the oblique muscles in their respective positions. Natural size. Note the curved edge of the tendon of the lateral rectus on the left.

(An enucleated eyeball can always be relegated to the side to which it belongs if viewed from this aspect and if sufficient of the oblique muscles be left to distinguish the superior (tendinous) from the inferior (muscular); these cut ends point towards their origins on the nasal or medial side of the orbit, or, it may be said, away from the side to which the eye belongs.)

equator. The superior oblique arises with the recti from the apex of the orbit, and runs forwards above the medial rectus to the trochlea, where it becomes tendinous and is reflected backwards onto the globe. The inferior oblique arises near the orbital margin at its inferior medial angle, and passes laterally and posteriorly across the forepart of the floor of the orbit to reach the globe behind the equator, being attached, like the superior oblique, to the postero-lateral quadrant of the eyeball. Each muscle is enveloped in a fascial sheath, which is so thin as to be practically invisible posteriorly, but becomes thick and opaque where the muscle pierces the fascia bulbi (capsule of Tenon), at which point the two fascial structures blend.

The lateral rectus is innervated by the abducent or sixth nerve, the superior oblique by the trochlear or fourth nerve, and the superior, medial, and inferior recti, together with the inferior oblique, by the oculomotor or third nerve. Their blood-supply is derived from the ophthalmic artery, and is drained away by the ophthalmic veins into the cavernous sinus.

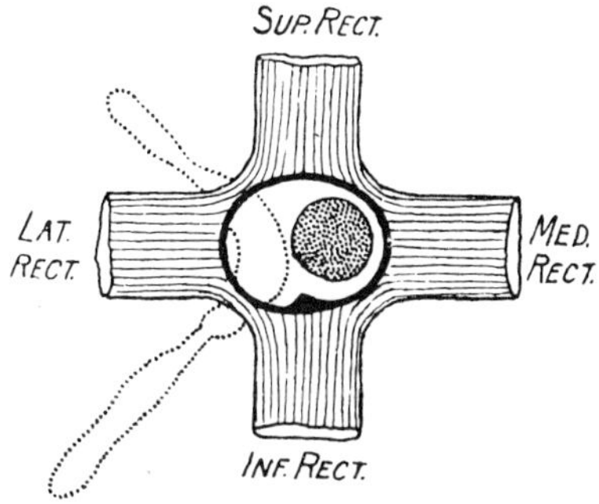

FIG. 141.—Scheme to show **origin of the four Recti Muscles** from the annulus of Zinn and its relation to the optic foramen and superior orbital fissure (in dotted lines).

The STRUCTURE of the ocular muscles has been specially studied by Schiefferdecker (1904), who found that the parallel fasciculi of muscle-fibres of which they consist are remarkable for the richness of their nerve-supply and the large amount of elastic fibres in the perimysium and intermuscular septa; the latter character is responsible for a passive contraction of the muscle after extension by its antagonizer, the action of which is, moreover, moderated by the same means; the provision ensures a delicate regulation of the eye movements. Woollard (1927, 1931) finds that it is possible to distinguish three groups of muscle-fibres according to size. First, a very thick fibre, not very numerous; second, a medium-sized type, composing the greater part of the muscle belly; third, thin, few in number, and mostly on the margins of the muscle. Broadly there are thick and thin types, each with a different nerve-supply (p. 336). Here it may be noted that the eye muscles are distinguished from those elsewhere in the body, not only by a peculiar innervation, but also by the fact that they give a prolonged tonic contraction to choline without previous denervation, as shown by Duke-Elder (1930).

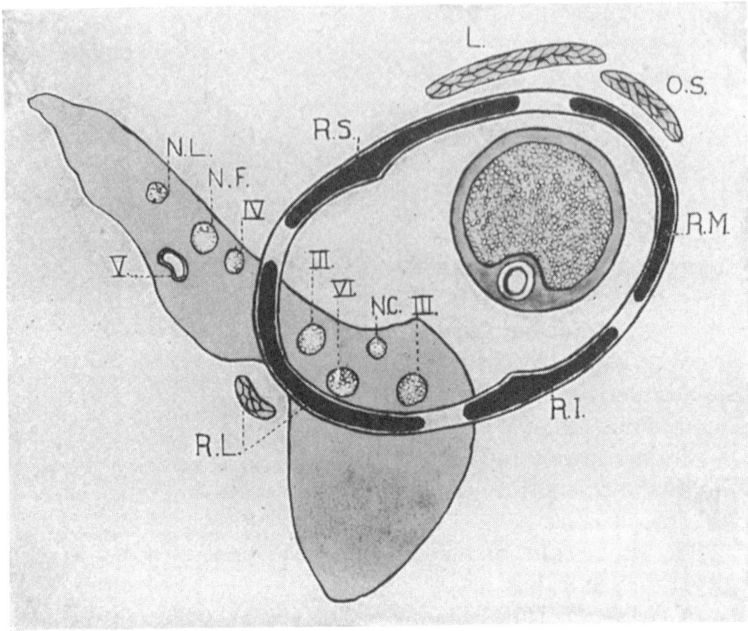

FIG. 142.—Scheme to show the position of the **annulus of Zinn,** implanted on the apex of the right orbit in such a way as to embrace both the optic foramen (in which is seen the optic nerve and ophthalmic artery) and the broad end of the superior orbital fissure. See Fig. 65, p. 136.

The annulus is resolved into its component muscle origins, and the relative positions of the nerves which enter the orbit through the superior orbital fissure both within and without the annulus are shown.

R.S., R.M., R.I., R.L., indicate the four Recti muscles, showing the real 'two heads of origin' of the last, the lateral rectus; L. is the levator, O.S. the superior oblique; N.L. the lacrimal nerve, N.F. the frontal, and N.C. the naso-lacrimal; V. is the superior ophthalmic vein. This scheme agrees with the representations of Merkel, Poirier, and Hesser (Fig. 167, p. 313).

The Recti Muscles are approximately equal in length, each being about 40 mm. or $1\frac{1}{2}$ inches long including the tendon of insertion, but there is a difference in the length of the tendons and also in the size of the muscle belly. The medial rectus is thickest, the lateral rectus is thicker than the inferior, and the superior is thinnest of all. The weights, as given by Volkmann, are:

Medial rectus,	0·747 gramme;	inferior,	0·671 gramme.
Lateral ,,	0·715 ,, ;	superior,	0·514 ,,

and the exact lengths in millimetres are:

Medial rectus,	40·8;	inferior,	40·0.
Lateral ,,	40·6;	superior,	41·8.

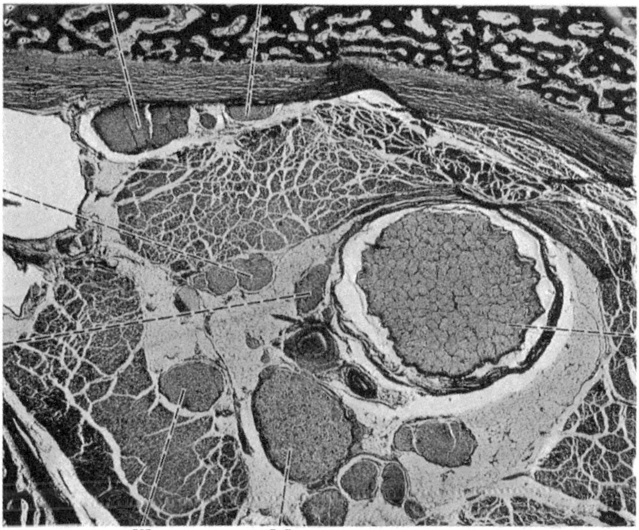

FIG. 143.—Section through the **Apex of the Orbit** (right) in a foetus at term, to show the close packing and relative positions of the contents.

F.L. = frontal and lacrimal nerves; IV = fourth nerve lying on superior rectus muscle; P = periorbita; O.N. = optic nerve in its sheath; M.R. = medial rectus muscle; C.G. = ciliary ganglion with the ophthalmic artery and vein lying above, and inferior branches of III nerve to the right, VI = sixth nerve lying against lateral rectus muscle; N.C. = naso-ciliary nerve; and III*a* the upper division of the third nerve. (Preparation and micro-photograph by E. P. Stibbe.)

The **origin** or posterior attachment of the recti at the apex of the orbit is complicated by the configuration of this region and by the presence of the many important structures which here enter the space. Looked at from the outside, after they have been stripped from the bone as in Fig. 145, the muscles are seen to have a common origin from a short tendinous tube, the annulus of Zinn (1755) (*annulus tendineus communis*). This tube is oval in cross-section, and is most simply considered as being implanted by its base onto the apex of the orbit in such a way as to embrace both the optic foramen and the medial extremity of the superior orbital (sphenoidal) fissure (Figs. 65, 142); that sector of the annulus which embraces the end of the fissure is fixed to the bone forming its anterior boundary, that is, to the tip of the lateral orbital wall, a site marked in 80 per cent. of cases by a spur (the *spina recti lateralis*, p. 35); in the specimen photographed on p. 36, the spur is crescentic in outline, indicative of the contour of this portion of the ring.

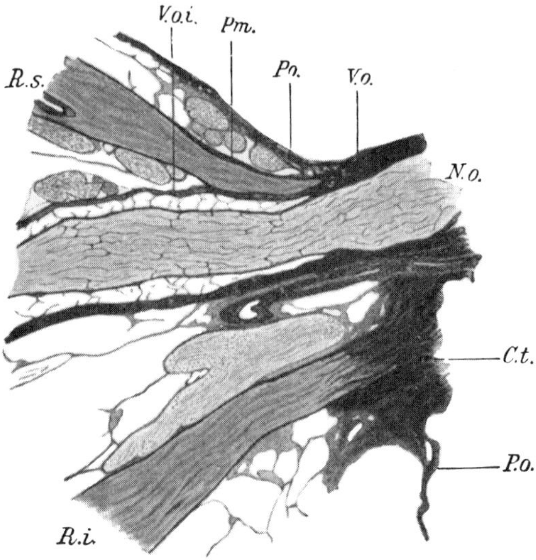

FIG. 144.—Sagittal section through the **Apex of Orbit** to show the **tendon of Zinn** (C. T.), and the connexions of the **optic sheath** (V.O.I.).

R.S. and R.I. = superior and inferior recti muscles respectively. V.O. = dura mater entering optic canal and splitting into P.O. the periorbita, P.M. the muscle sheath, and V.O.I. the sheath of the optic nerve (N.O.). (From Kiss, 1919.)

At its base the annulus is intimately related to the dura mater lining the middle cranial fossa of the skull, since this membrane extends into the orbit through its apertures to become continuous with the sheath of the optic nerve on the one hand and the periorbita on the other; the base of the medial half of the annulus is inserted into the cleft formed by this splitting of the dura (Fig. 144).

The muscles arise from the fore end of the annulus in their

respective positions; the superior and medial recti are most closely related to the optic nerve, being, indeed, adherent to its sheath, a fact that would explain painful movements of the eyeball in inflammation of the nerve, and owing to the great obliquity of the optic foramen their origins are on a plane anterior to the rest; just outside and above them lie the

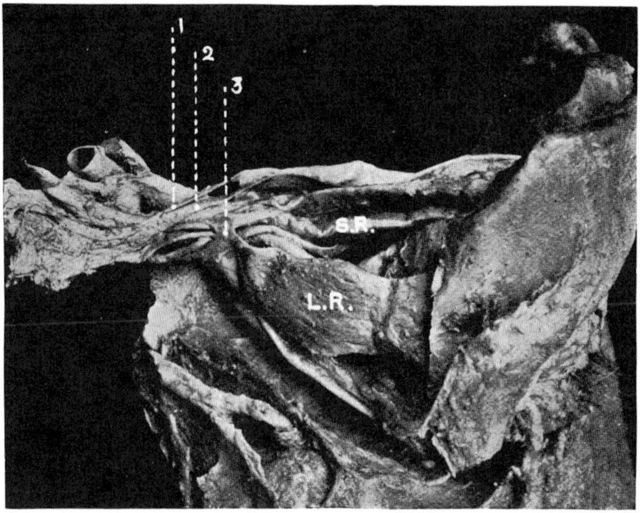

FIG. 145.—Dissection of Right Orbit in which all the bony walls have been removed and the margin alone left, viewed from the lateral aspect. The preparation shows the trochlear (1) and frontal (2) nerves as they enter the orbit outside the **annulus of Zinn** (3); entering within the annulus are seen the oculomotor, naso-ciliary and abducent nerves, in that order from above downwards. The continuity of origin of the lateral rectus (L.R.) with that of the superior rectus (S.R.) is shown. Natural size.

origins of the levator palpebrae superioris and superior oblique muscles; the inferior rectus arises below and a short distance away from the nerve.

The lateral rectus arises from that part of the annulus which spans the superior orbital fissure, a position that has led to the statement that it arises by 'two heads', one from each margin of the gap; but such description, like so many drawings of the parts, gives an exaggerated idea of the actual facts, for there is no break in the annular origin of the muscle belly, as can clearly be seen in Fig. 145; its upper edge is fused at this point with the lower edge of the superior rectus, and similarly below it blends with the inferior rectus. Merkel (1901), indeed, described and figured 'two heads' of origin of the lateral rectus, as actually there are, one from the spina recti lateralis and the other from the annulus next the optic nerve sheath, but they do not span the fissure, no structures enter the orbit through them, and they are not the same 'heads' as are commonly described in later works.

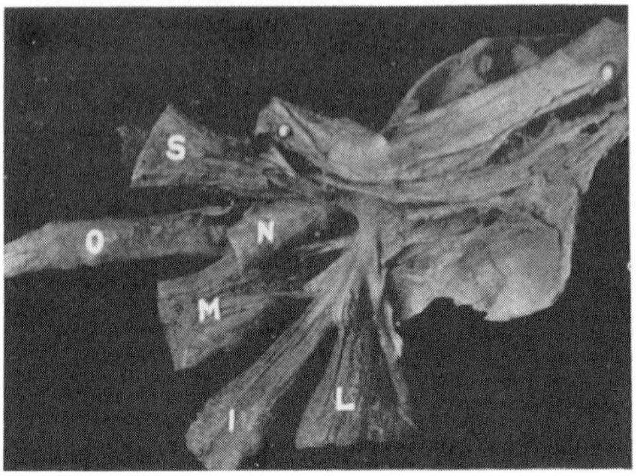

FIG. 146.—Dissection of the parts in relation to the **Apex of the Left Orbit,** lateral view. The annulus of Zinn has been split open between the superior (S.) and lateral (L.) recti to expose the **tendon of Zinn.** The inferior (I.) and medial (M.) recti and superior oblique (O.) are spread out, and in the centre of the muscles lies the optic nerve (N.). Between the base of the nerve and the root of the tendon is a round orifice, the 'oculomotor foramen'. Natural size.

If the annulus be now examined from the interior by splitting it open between two adjacent muscle origins, as has been done between the superior and lateral recti in Fig. 146, further complications of the muscle origins are seen. The ring on this inner aspect is not regularly smooth as on its outer side, but presents below a well-marked tendinous ridge, which gives origin to fibres of three of the muscles, namely, to part of the lateral, all of the inferior, and part of the medial rectus; this is called the tendon of Zinn, or *tendo orbitalis inferior* or *centrum tendineum*, and its attachment to the lower lip of bone below the optic foramen may be marked by a small pitting, but more often forms an 'infra-optic tubercle' (*tuberculum musculare*), as found present by Kiss in 87 per cent. of cases; it can be discerned in the figures on pp. 36, 91. On the upper half of the interior of the annulus a similar but much less definite thickening, the *tendo orbitalis superior*, was described by Lockwood (1886) as giving origin to part of the lateral, all of the superior, and part of the medial recti. Rouvière (1914), in a special study of this region, describes a single tendon, that of Zinn, arising from the bone as above noted but dividing into six parts from which the muscles arise. Kiss (1919) likewise considers the chief origin to be from a *centrum tendineum* (Fig. 144) attached to the infra-optic tubercle, and he holds the description of an annulus to be unfounded; he would ascribe 'two heads' of origin to the superior rectus, but agrees with Merkel's description of the lateral rectus having a second origin from the spina recti lateralis.

FIG. 147.—Dissection of the **Structures in the Apex of the Right Orbit** seen from in front. The four recti muscles have been spread out and occupy their respective positions, and the nerves are seen entering the orbit within the cone of muscles. In the centre is the optic nerve in its sheath with the ophthalmic artery on its infero-lateral side. Below the artery is the inferior division of the third nerve, the upper division of which lies above the vessel. The sixth nerve lies lateral to the artery. The cut ends of the levator (left) and superior oblique (right) appear between the superior and medial recti. Natural size.

Since the annulus of Zinn surrounds the optic foramen and the medial end of the superior orbital fissure, it follows that all the structures which enter the orbit through the foramen and this part of the fissure will lie at first within the cone of muscles. Through the optic foramen there enter the optic nerve enveloped in its dural sheath, and the ophthalmic artery embedded in the sheath and accompanied by branches from the carotid plexus of the sympathetic nervous system; through the medial end of the fissure there pass the oculo-motor or third nerve, the abducent or sixth nerve, the naso-ciliary branch of the trigeminal or fifth nerve, the sensory root of the ciliary ganglion, and usually the superior ophthalmic vein (p. 313).

These latter structures are all closely packed together in a narrow space, as is seen in Figs. 143 and 147, and also in the fourth section in Fig. 198, p. 381, and their relative positions are shown in Fig. 142. The space through which they pass is formed into a smooth and glistening tendinous foramen bounded medially by the optic nerve sheath and laterally by the loop of the annulus bridging the fissure; it is termed the 'oculomotor foramen' (*foramen oculomotorium*), and its contained structures are better described as entering the orbit through this foramen, or 'within the cone of muscles', than 'between the two heads of the lateral rectus muscle'.

The **insertions** or anterior attachments of the recti muscles onto the globe are effected by means of thin, flat, glistening tendons, formed by parallel rectilinear fasciculi of connective tissue, without anastomoses but weakly united by transversely placed and easily separable fibres. The fasciculi are several layers deep, those of the medial rectus being thickest, but form a single layer at the edges, which are recurved to form a little cul-de-sac (Fig. 140). Some of the fasciculi leave the main layer to be inserted separately as far back as from 1 to 5 mm. behind the remainder, and consequently, like the recurved edges, may escape the tenotomy hook in operations. The tendons penetrate the sclera to unite most firmly with it, causing a slight thickening of the surface as they spread out within it in fan-like manner. They are accompanied by the anterior ciliary arteries, and lie beneath the anterior part of the fascia bulbi, which, with the overlying bulbar conjunctiva, must be removed to expose them from the front, as shown in Fig. 148; so closely are they applied to the sclera and so much do they resemble it in colour, that the student may not easily recognize them.

The antero-posterior length of the tendons in millimetres is given by Merkel (1901) as follows:

Rectus Medialis.	Rectus Inferior.	Rectus Lateralis.	Rectus Superior.
3·7	5·5	8·8	5·8

The breadth of the actual line of insertion is given by Fuchs as

10·3	9·8	9·2	10·6

By Weiss (in adult):

Rectus Medialis.	Rectus Inferior.	Rectus Lateralis.	Rectus Superior.
10·76	10·35	9·67	10·75

By Weiss (in new-born):

Rectus Medialis.	Rectus Inferior.	Rectus Lateralis.	Rectus Superior.
7·35	6·25	5·85	6·95

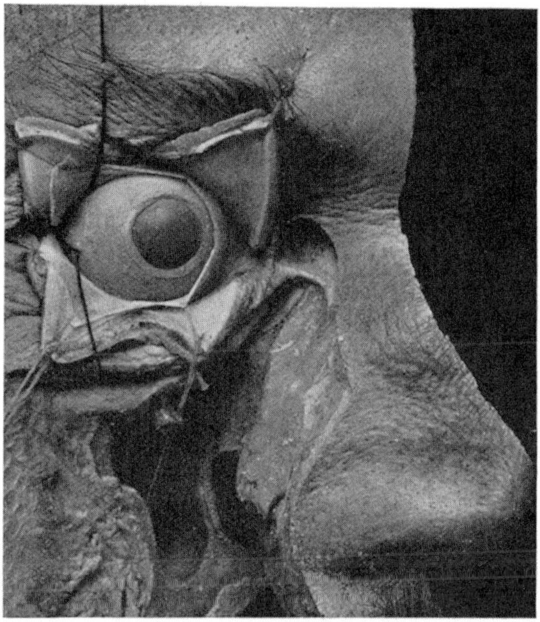

FIG. 148.—Dissection of the Right Eyeball from in front. The eyelids as a whole have been crucially incised and the flaps pulled aside. The bulbar conjunctiva and **anterior part of the** *fascia bulbi* or **capsule of Tenon** have been cut just wide of the cornea, and pulled away from the globe on the lateral side to expose the **tendon of the lateral rectus muscle,** beneath which a bristle has been placed.

The lacrimal sac and naso-lacrimal duct have been exposed from in front in their entire extent. Natural size.

The distance of the mid-point of this line from the corneal margin varies slightly according to the observations of different authors. The measurements made by Fuchs (1884) were taken from the largest series of eyes, and are most widely accepted; they are

Rectus Medialis.	Rectus Inferior.	Rectus Lateralis.	Rectus Superior.
5·5	6·5	6·9	7·7

and a mean of the figures given by Howe, Adachi, Weiss, Fuchs, Testut, Gerlach, Sappey, Merkel, Motais, Tillaux, Macalister, and Krause is:

Rectus Medialis.	Rectus Inferior.	Rectus Lateralis.	Rectus Superior.
5·84	**6·65**	**7·18**	**7·88**

The corresponding distances in the new-born are given by Weiss as

3·6	5·0	4·9	5·8

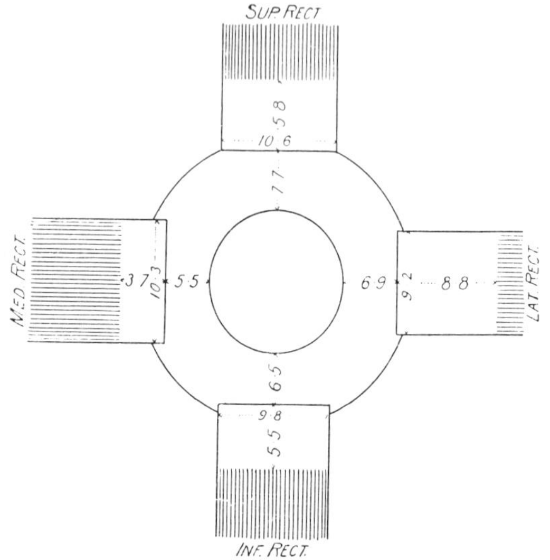

FIG. 149.—Scheme to show the **anterior attachments of the Recti Muscles** to the Eyeball, with the various measurements given in the text. × $1\frac{1}{2}$.

The above measurements of the length, breadth, and distances from the corneal margin are shown diagrammatically in Fig. 149, and the more exact positions of the lines on the globe in Fig. 150. The oblique muscles are about 10 mm. broad at their insertions (see pp. 276, 279).

It is to be noted, however, that these lines of insertion are not straight, but are curved or sometimes even wavy; they lie neither parallel to the corneal margin nor at right angles to the respective meridian either of the muscles or of the cornea, though those of the medial and lateral recti are approximately so placed (Fig. 151). The line of insertion of the medial rectus is straight or feebly convex forwards, and may either cut the horizontal meridian symmetrically or cross it obliquely and unequally, with the greater length lying above; in the case of the inferior rectus, the line is oblique and markedly convex forwards, and in two-thirds of

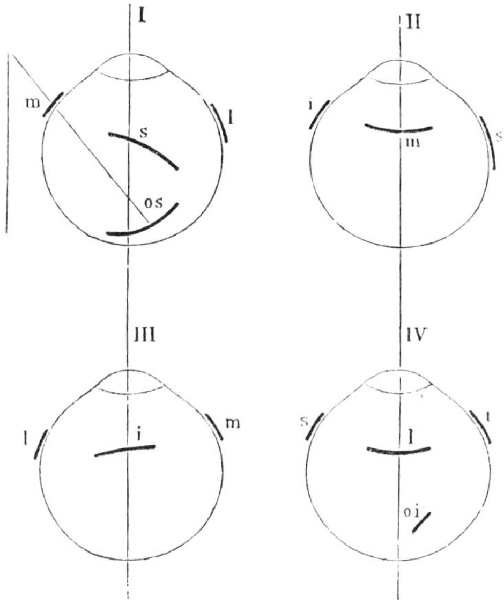

FIG. 150.—Schematic representation of the lines of the **Muscle insertions** on to the Eyeball (from Merkel, 1901). Natural size.

I. = seen from above, II. from the medial side, III. from below, IV. from the lateral side. *s.* = superior rectus, *i.* = inferior rectus, *m.* = medial rectus, *l.* = lateral rectus; *os.* = superior oblique, *oi.* = inferior oblique.

cases the greater moiety lies lateral to the vertical meridian; the line of the lateral rectus is usually straight and lies perpendicular to the horizontal meridian, though in abnormal cases it presents an asymmetry the reverse of that of the medial rectus; the line of the superior rectus is convex forwards and is most often asymmetrically placed relative to its meridian, sloping obliquely upwards and lateralwards with the greater length on the lateral side.

It follows, therefore, that the above distances, measured from the mid-points of the lines of insertion, are not the shortest nor the mean distances from the cornea, and Motais suggests taking the nearest and farthest points, which he finds are respectively as follows, in millimetres:

Rectus Medialis.	Rectus Inferior.	Rectus Lateralis.	Rectus Superior.
5·5	5·5	6·7	6·5
7·0	8·0	7·0	11·0

and he maps out the lines as in Fig. 151, from which it is seen that the so-called spiral line on which the insertions lie is actually a very irregular one.

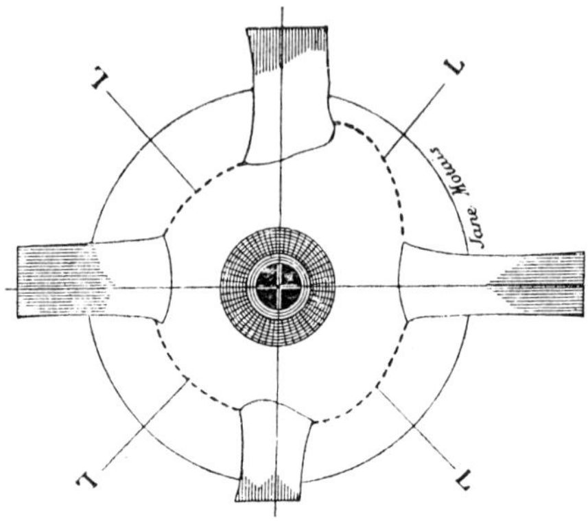

FIG. 151.—The **insertions of the Recti Muscles** onto the Left Eyeball according to Motais (in Poirier, 1912). The line joining them (L) describes a very irregular spiral. The medial rectus lies to the reader's left.

Finally, Weiss gives the distance of the insertions from the optic nerve as follows:

16·95	18·80	20·60	18·51

That there is much individual variation in the position, length, and form of these lines of insertion is seen from Fig. 152, as found by Howe on 20 eyeballs, and is illustrated also by Fuchs (1884).

On the whole, the medial rectus is the largest and heaviest muscle, and is most advantageously placed nearest the corneal margin, whilst the superior rectus is the weakest and is farthest away, especially at the lateral edge of its insertion.

The above described tendon is sometimes termed the 'primary or principal insertion' of the muscle, the following being then considered as 'secondary insertions' (the *adminculum tendonis* of Virchow):

(i) The thin recurved strands at the edges of the tendons.
(ii) The few fasciculi which are inserted on to the globe separate from and behind the remainder.
(iii) Certain attachments of the muscle sheaths to the walls of the orbit, which form the so-called 'check ligaments', described on p. 295.

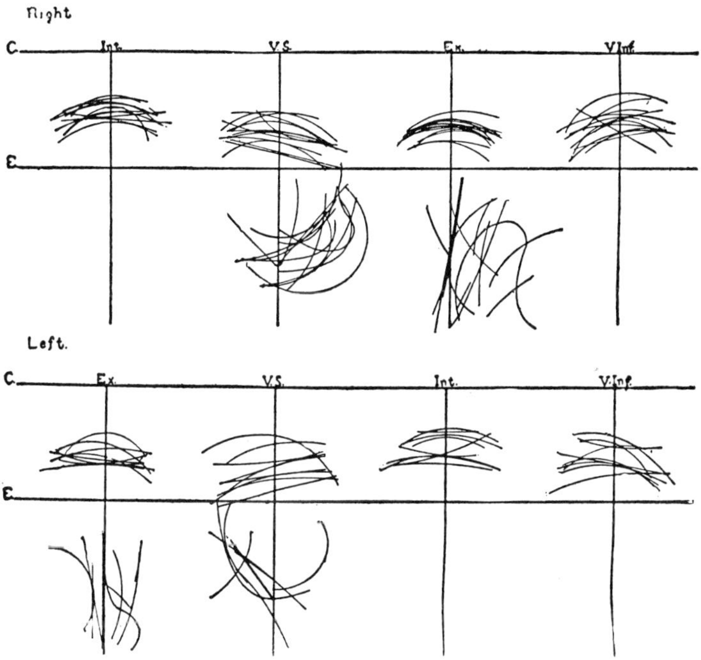

Fig. 152.—Lines of insertion of the Muscles on several Eyeballs plotted together in order to show the difference in the form and place of the insertion. All the lines of the right eyes are grouped together in the upper portion of the figure, and all of the left eyes in the lower portion. In both, C. represents the line of the cornea, E. the equator of the globe. (From Howe, 1907.)

The 'arc of contact' which the muscle tendon makes with the globe prior to its insertion has been studied by Weiss (1894), especially as regards the lateral and medial recti, and he points out the increase of the arc of the lateral muscle and decrease of that of the medial consequent to the greater divergence of the orbits during growth; the changes may explain cases of the spontaneous cure of convergent strabismus which existed in childhood (see also p. 285). The point of contact is, as Howe remarks, the physiological insertion of a muscle of the globe.

Having considered the recti muscles together as regards their origin and insertion, the relations and certain other points of interest remain to be described for each muscle.

The **medial rectus** (*m. rectus medialis*) from its origin, where it is closely applied to the medial side of the optic nerve sheath, runs forwards and inclines a little upwards and medialwards; at first it lies in contact with the corresponding wall of the orbit, separated by the thin os planum from the ethmoidal air-cells, but in its anterior third it inclines from the orbit wall to reach the globe, the intervening space being filled by a compact mass of fat (Fig. 136). Above it lies the superior oblique muscle, closely applied posteriorly but separated by the naso-ciliary nerve, ethmoidal branch and terminal part of the ophthalmic artery anteriorly. A branch from the oculomotor nerve enters its substance in several strands on the ocular side about 15 mm. from its origin, and its blood-supply is provided by the inferior muscular branch of the ophthalmic artery.

The **inferior rectus** (*m. rectus inferior*) runs along the floor of the orbit closely related to the palatinal air-cell near its origin, but separated by a mass of orbital fat from the roof of the sinus maxillaris or antrum of Highmore in front (Fig. 178, p. 332); its infero-lateral margin lies above the infra-orbital nerve in the hinder part of its course. Anteriorly and a little in front of the equator of the globe the inferior oblique muscle passes laterally and obliquely beneath the inferior rectus, the fascial sheaths of the two being fused at this point and forming the suspensory ligament of Lockwood. The lower division of the oculomotor nerve runs along its ocular or superior aspect, and the branches of supply enter about the junction of its posterior third and anterior two-thirds; the nerve to the inferior oblique is closely bound to its lateral margin (Fig. 145). The blood-supply is derived from the inferior muscular branch of the ophthalmic artery and also from the infra-orbital artery.

The **lateral rectus** (*m. rectus lateralis*) is attached at its origin to the apex of the lateral orbital wall by means of its lacertus and the annulus; between the muscle origin the optic nerve is the fibrous oculomotor foramen through which there enter the orbit the oculomotor, the abducent, the naso-ciliary branch of the trigeminal nerves, and the ophthalmic vein, these structures being consequently here closely related to the muscle (Figs. 145, 147). A little in front of this point (5 mm. or about ¼ inch) lies the ciliary ganglion (*ganglion ciliare*), and further forwards the ophthalmic artery and the naso-ciliary nerve lie likewise between the muscle and the optic nerve (Fig. 191, p. 362). In its forepart it is separated laterally by orbital fat from the wall of the temporal fossa (Fig. 136, p. 254); above, it is in contact with the lacrimal gland, the nerve and artery to which run along its upper border; whilst between its lower border and the inferior rectus lies the nerve to the inferior oblique; at the globe it crosses over the lateral side of this muscle. The sixth nerve enters it on the medial aspect, about 15 mm. beyond the oculomotor foramen, and its blood-supply is derived from the lacrimal artery.

The **superior rectus** (*m. rectus superior*) lies beneath and in intimate contact with the levator palpebrae superioris, separated by it from the frontal nerve and roof of the orbit (Fig. 69, p. 141); it is united to the levator along its medial border by the adherence of their fascial sheaths, and also by the passage through the latter of the superior division of the oculomotor nerve, which supplies both muscles. The ophthalmic artery and the naso-ciliary nerve lie beneath it posteriorly, between it and the optic nerve; in front it passes over the tendon of the superior oblique muscle. The direction of its long axis from the apex of the orbit to the globe inclines laterally, and so does not coincide with the sagittal axis of the latter, but forms an angle of about 25° with it. It does not lie directly above the inferior rectus, but is placed slightly lateral to it.

The **superior oblique muscle** (*m. obliquus superior*) lies along the angle between the medial wall and roof of the orbit (Fig. 69, p. 141). It arises by a short, slender tendon anterior and medial to the optic foramen, from which it is separated by the medial rectus, in the angle between the annulus of Zinn and the periorbita. It may further be attached to the medial orbital wall for a few millimetres from its origin. The belly of the muscle, more rounded than that of the recti, runs forwards, parallel and close to the nasal wall above the medial rectus, but separated from the latter by the naso-ciliary nerve and ophthalmic artery, and crossed by their ethmoidal branches; a few millimetres behind the orbital margin at its supero-medial angle the muscle becomes tendinous and passes through a cartilaginous ring, the pulley or *trochlea* (which point is the physiological origin of the muscle); it then bends abruptly backwards, slightly down, and lateralwards at a 'trochlear angle' of about 54°, and the round tendon broadens out, pierces the fascia bulbi and passes beneath the superior rectus, to be inserted into the back and lateral side of the eyeball. The first part of the muscle from its origin to the trochlea is referred to as the 'direct' portion, and the tendinous part between the trochlea and the globe as the 'reflected' portion. The tendon measures 20 mm., or ¾ inch, in length from pulley to globe, and is enveloped by a prolongation of the fascia bulbi as far as the pulley.

In most mammals this reflected portion is also fleshy, and indeed in lower vertebrates the whole muscle is represented by the reflected portion only, which arises from the medial angle of the orbit near the inferior oblique. The insertions of the tendons of the oblique muscles differ from those of the recti in being placed on the posterior instead of the anterior hemisphere of the eyeball; they also show much more variation in position, as is shown in Fig. 152.

The line of insertion of the superior oblique tendon crosses the vertical meridian obliquely and is curved with a backwardly directed convexity. The breadth as measured by Fuchs on 31 emmetropic globes was on an average 10·7 mm., with extremes of from 7·5 to 12·7 mm., but was found narrower in myopic globes, in 20 of which it averaged 9·6 mm. with extremes of from 6·8 to 14 mm. The distance from the corneal margin to where the line crossed the meridian is from 16 to 17·9 mm. (Fuchs), but Weiss, in 5 adult emmetropic eyes, found the anterior edge of the obliquely placed line to lie 13·85 mm. from the cornea and 14·66 mm. from the optic nerve, whilst the posterior end was 18·8 mm. from the cornea and 7·56 mm. from the nerve; the anterior end also lies on the same meridian as the lateral end of the line of attachment of the superior rectus, but distant 4·6 mm. (with extremes of from 0·5 to 8 mm.) from it, though the two tendons are often connected by connective tissue strands. The line of the superior oblique is similarly connected with that of the inferior oblique, but is separated from it by 15 mm. at the anterior and 11·9 mm. at the posterior end (Fig. 138). Fuchs described two types of insertion: one in which the line is broad with a strongly marked anterior concavity, and lies in an equatorial direction across the vertical meridian with the greater part on its lateral side (Fig. 150), as more commonly found in emmetropic and hypermetropic eyes; the other narrower and flattened, and lying entirely in the lateral quadrant almost parallel to the vertical meridian, as found in myopic (though occasionally in normal) globes.

The trochlear or fourth nerve crosses over the muscle from the medial side to enter its substance at about 12 mm. from its origin; the blood-supply is derived from the superior muscular branch of the ophthalmic artery.

The *trochlea* or pulley of the superior oblique muscle consists of a finely grooved and curved plate of hyaline cartilage, 4 mm. long by 6 mm. broad; its edges are attached to the *fovea trochlearis* of the frontal bone (p. 25) by fibrous retinacula, so completing the tube through which the muscle tendon slides, as shown on p. 209. The tendon here, as it actually passes through the pulley, is invested by very lax areolar tissue with an imperfect endothelial lining in its clefts, but there is scarcely a true synovial membrane lining the trochlea (Macalister, 1889). The fascial sheath, which envelops the tendon from its point of entry into the capsule of Tenon, extends up to the pulley, and is fixed to and terminates at the latter. The trochlea was primitively a specialized part of the sheath of the muscle at its origin, which was situated near the orbital margin, but the need for a longer muscle belly in mammals has led to the development of the pulley and the retrogression of the origin backwards towards the apex of the orbit. Maddox (1907) suggests that the reason why the superior oblique has a pulley is that since the speed with which a muscle's point of insertion moves is proportional to its length, it is necessary for the superior oblique, the virtual origin of which as regards its action is from the pulley, to have a long, fleshy, reflected portion extending to the back of the orbit if it is to keep pace with the movements of the recti; and as regards this advantage over the inferior oblique, the reason may be that prolonged looking downwards is more important for daily work than looking upward, and therefore the former excursions of the eye are more amply provided for than the latter. It may be pointed out, however, that the disparity between the lengths of the contractile parts of the two muscles is not great, since the muscle belly of the superior is approximately 40 mm. long, and the total length of the inferior, which is fleshy throughout nearly its whole extent, is about 37 mm.

The **inferior oblique** (*m. obliquus inferior*) muscle is the only one of the ocular muscles which does not arise from the apex of the orbit. The origin is from a small shallow depression on the orbital floor, just within the inferior orbital margin lateral and usually close to the opening of the naso-lacrimal canal (Fig. 19, p. 43); some fibres often arise from the lower part of the fascia covering the lacrimal fossa, and so are closely related to the lower part of the lacrimal sac (Fig. 135, p. 250).

The origin, however, has been found by the writer to be abnormally situated in several instances. Out of 100 orbits examined, it occupied the usually described position in 45 cases; it lay from 2 to 5 mm. away in 47, most often in the left orbit; and it lay from 6 to 7 mm. distant in 8 cases. Of the last, the muscles were examined on three eyeballs and found to be inserted higher up than usual, though the total length was normal. Such position resembles that found in certain fishes (Whitnall, 1921). As noted on p. 106, the site of origin lies in the same vertical plane as that of the pulley of the superior oblique muscle, and the distance between them is not affected by the shape of the orbit.

The muscle passes laterally and backwards as a flat band

beneath the inferior rectus, to which it is closely connected by the fusion of their fascial sheaths (Fig. 73, p. 147); it is separated from the floor of the orbit by a compact mass of fatty tissue and inserted, like the superior oblique, onto the posterior hemisphere of the eyeball; its course intersects the axis of the orbit at an angle of 75°; it is fleshy in nearly its whole extent of about 37 mm., the tendon being only 2·6 mm. long or, indeed, often non-existent, so that the muscle fibres pierce the sclera (Fig. 140). The line of insertion, usually 10 mm. broad, lies obliquely in the infero-lateral quadrant of the posterior hemisphere, farther back than that of the superior oblique, with which it makes an angle of 25°, and nearer the optic nerve, from which it is distant about 5·2 mm. (Fig. 138) (from 3·8 to 7·5 mm.) in normal but 7·1 mm. (from 4·8 to 11·5 mm.) in myopic globes (Fuchs); it lies 9·5 mm. away from the insertion of the lateral rectus and almost on the same meridian, though at right angles to the line of the latter, and is only 2·2 mm. from a spot corresponding to the *fovea* of the retina. The distance from the corneal margin is from 17·3 mm. to 19·1 mm. (Fuchs). The variations in its position are seen in Fig. 152. Salzmann (1912) notes that the line of its insertion often shows gross irregularities, angular serrations, or even dehiscences. The branch from the inferior division of the oculomotor nerve enters the muscle belly about the middle of its hinder border, and its blood-supply is derived from the infra-orbital artery as well as the inferior muscular branch of the ophthalmic artery.

The Action of the Muscles.—The movements of the eyeball have been resolved by Fuchs (1917) into components which correspond to three primary axes, perpendicular to each other and intersecting at its 'centre of movement' or 'point of rotation'; this point in a emmetropic eye is situated not at the mathematical centre of the globe, but a little behind the mid-point of its antero-posterior axis, or 14 mm. behind the summit of the cornea and 10 mm. in front of the posterior pole; in myopic globes the distance is 14·52 mm. and in hypermetropic 13·22 mm. behind the anterior surface of the cornea (Donders). The three primary axes are:

The vertical axis, around which lateral movements occur to the right and left, that is, abduction and adduction or divergence and convergence of the cornea on the horizontal plane;

The frontal axis, running from right to left, and related to movements of elevation and depression of the cornea;

The sagittal axis, passing from before backwards, corresponding to the visual axis, and correlated with medial and lateral rotation, or intorsion and extorsion of the cornea.

As regards the actions of the muscles, it is first to be emphasized that the eyeball is delicately poised in the fascia and fat of the orbit; the capsule of Tenon forms its primary socket, and the two, globe and fascia, move together in any but the smallest excursions upon the bed of fat which completely fills the surrounding space up to the bony orbital wall. The globe undergoes no change of place as a whole, but simply rotates about the centre of movement; since, however, this point is not the mathematical centre of the eyeball, it is more correct to describe the movement as *oscillation* rather than rotation. A state of equilibrium is maintained by the tonicity of all the ocular muscles, and individually it is impossible for any one muscle to act without the others participating both in the movement and in the maintenance of the new position of the globe; in every motion from three to five muscles participate, one or two acting directly and the others helping by steadying the course.

Incidentally there is to be noted the extent to which the head itself participates in all except the most narrow ranging of the visual regard, a co-ordination which serves to relieve the ocular mechanism of wide excursions and favours concentration of effort upon the finer movements; the prevalence of this unconscious act is well realized when the observer is suffering from a stiff neck, for example, and is a noticeable feature in the muscular rigidity of paralysis agitans.

As a whole, the four recti may be regarded as retractors of the eyeball; they are antagonized by the two obliques, which from their origins and directions are protractors (apart from their pull nasalwards on the globe), and this is probably the active element in maintaining the equilibrium of the globe

(Theobald, 1918); the medial and lateral recti are nearly perfect antagonists in movements of the cornea upon a horizontal plane; the superior and inferior recti are opposed in producing elevation and depression and in torsion, but both act as adductors; the two oblique muscles rotate the eyeball in opposite directions in elevation and depression of the cornea and in torsion, but both produce abduction. Thus the equilibrium of the globe is assured by the anatomical dispositions of the muscles. In movements, the action of a contracting muscle, which shortens in full extent about a quarter of its length, is moderated by its antagonizer, which elongates to the requisite degree. A rise of intra-ocular pressure in consequence of the muscle action, especially in convergence of vision, has been assumed, but Levinsohn (1910) considers the action to be too slight to affect the pressure; Lederer (1912), on the other hand, found that a rise of pressure equal to 5 mm. of mercury occurs in marked lateral motions of the eye, due, however, to the compression of the globe against the orbital fat. It may be noted that Ochi (1919) ascribes juvenile progressive myopia to the pressure of the muscles on eyeballs with weak sclerotic coats.

The individual action of each muscle can be described as determined by its anatomical disposition as follows, but it is to be noted that, except for the medial and lateral recti, each has a main and a subsidiary action.

The medial rectus is the adductor of the eye, turning the cornea medialwards, and the movement is sensibly direct, with a possible excursion of 45°; the simultaneous contraction of the two muscles produces convergence of vision.

The lateral rectus has a directly opposite action; it is the abductor of the eye and directs the cornea lateralwards; the motion is direct and has an excursion of nearly the same extent as that of the medial rectus.

The superior rectus is an elevator of the cornea, but since its direction does not coincide with the sagittal axis of the eyeball but forms an angle of 23° with it, and since its insertion falls on the anterior hemisphere in front of the centre of movement of the globe, it will not only elevate but adduct, and can, moreover, incline the upper end of the vertical axis medialwards, that is, rotate the globe in this direction. The action then is to elevate, adduct, and intort the cornea.

The inferior rectus is a depressor of the eye, but since, like the superior rectus, its long axis inclines laterally forwards and so makes an angle with the visual axis of the globe, it also possesses a slight action of adduction, directing the cornea medially; it can also cause a small degree of extorsion, and its action, therefore, is to depress, adduct, and extort the cornea.

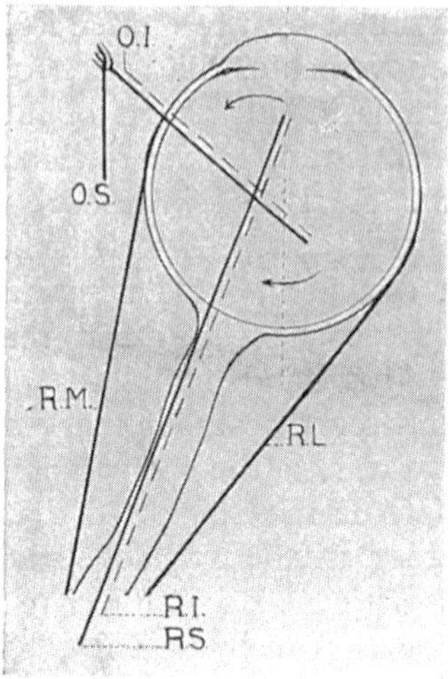

FIG. 153.—Scheme to illustrate the **Action of the Eye Muscles.**

O.S., O.I. = the superior and inferior obliques; R.S., R.I. = the superior and inferior recti; and R.M., R.L. = the medial and lateral recti, respectively.

The superior oblique must be considered to act relative to the globe from its physiological origin at the pulley. Primarily it will raise the posterior hemisphere of the globe, to the upper aspect of which it is fixed, and so depress the cornea, but since it is also inserted behind the centre of movement of the globe at an angle of 39° to the optic axis (Fig. 150, I), it will at the same time draw the posterior pole medially, so abducting the cornea; finally, from its insertion lateral to the sagittal meridian, it can rotate the eyeball slightly medialwards. Its action, therefore, is to depress, abduct, and intort the cornea.

The inferior oblique similarly can be understood from its position beneath the globe and its insertion on the posterior hemisphere to the lateral side of the sagittal meridian, primarily to elevate, secondarily to abduct and extort the cornea. It would appear that were it not for the forward divergence of the orbital axes and the consequent angular medial pull of the superior and inferior recti, there would be no need for post-equatorial insertion of the oblique muscles to counteract this secondary action of the recti (see Fig. 153).

The movements of the cornea are caused by combinations of these individual muscular actions; thus direct

Elevation results from the combined action of the superior rectus and inferior oblique (both supplied by the third nerve);

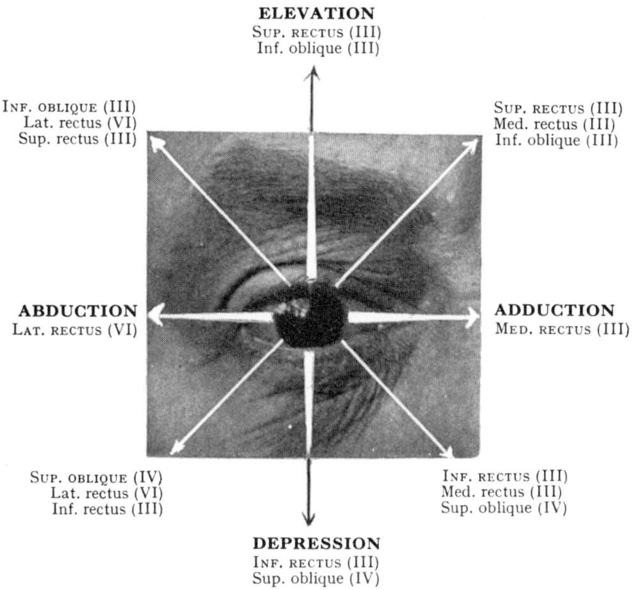

FIG. 154.—Scheme of **movements of the Cornea** from the resting position with the muscles that effect them. The predominant muscles are in capital letters; the roman numerals after the names indicate the nerve supply. (Adapted from Testut, 1914.)

Depression is effected by the inferior rectus (third nerve) and superior oblique (fourth nerve);

Abduction by the lateral rectus (sixth nerve), assisted by the two obliques (third and fourth nerves), especially towards the end of the movement;

Adduction by the medial rectus assisted towards the end of its excursion by the superior and inferior recti (all supplied by the third nerve);

Further movements:—

Up and lateralwards are effected by the superior and lateral recti and inferior oblique muscles;

Up and medialwards by the superior and medial recti and inferior oblique;

Down and lateralwards by the inferior and lateral recti and superior oblique;

Down and medialwards by the inferior and medial recti and superior oblique;

Intorsion by the superior oblique and superior rectus;

Extorsion by the inferior oblique and inferior rectus.

The actions of the muscles are illustrated by the schematic plan in Fig. 153, and are in agreement with the ideas of Fuchs, from whose classical text-book the above description is largely taken, and the movements of the cornea are shown in Fig. 154. More detailed descriptions of the action of these muscles and the movements of the eyeballs, of which the above is but a brief summary of the main facts, will be found in the works of Stevens (1887), Maddox (1907), Howe (1907), Peter (1927), and in any of the current text-books of ophthalmology.

From a clinical point of view, the position of the insertion and action of the muscles in the *myopic* or elongated eyeball are of interest. Fuchs states that the distances of the recti from the corneal margin are not increased, the elongation only affecting the posterior hemisphere as regards axial length. The oblique muscles, however, are thereby affected, passing more transversely and less antero-posteriorly to their insertion, so that the angles which they make with the medial plane are increased; their abducting power is greater, and the adducting power of the superior and inferior recti diminished, conditions favouring the condition of divergent strabismus (Fisher, 1904). As regards the range of movement of the insertions, Landolt finds an equal excursion of 47° from the position of rest in all directions, and Charpy gives ranges of 45° (medial rectus), 50° (inferior), 46° (lateral), 45° (superior).

As quite subsidiary actions of the muscles, it may be added that by means of the connexions of their fascial sheaths the superior rectus pulls up the superior conjunctival fornix and thus assists the levator in raising the upper lid, the inferior rectus slightly depresses the lower lid, the lateral rectus draws the corresponding commissure slightly backwards, and the medial rectus causes marked recession of the lacrimal caruncle.

Development.—It is commonly assumed on morphological grounds that the orbital muscles are derived from the anterior head or cephalic mesoblastic somites, though the latter have never been observed in man or mammals. Their mode of development in lower forms is described in some detail by Ryder (1900). The first somite gives rise to the levator, superior, medial, and inferior recti and inferior oblique, supplied by the third nerve; the second somite gives origin to the superior oblique, supplied by the fourth nerve; and the third somite to the lateral rectus, supplied by the sixth nerve. Neal (1918), however, produces evidence from comparative anatomy to support the contention of Dohn that the lateral rectus is derived in part also from the second somite; in his paper will be found the literature upon the subject. Marshall states that the lateral rectus has nothing whatever to do with the first head cavity, though it ultimately reaches the eyeball. Lewis (in Keibel and Mall, 1910) found that in man all the eye muscles arise from a common pre-muscle mass, first recognizable in an embryo of 7 mm. in length, and connected with the third nerve only; the fourth and sixth nerves entered at the 9 mm. stage, wherein the muscle mass began to show signs of cleavage; at 14 mm. all the muscles are to be distinguished, and have nearly the same position relative to the eyeball as in the adult, though the inferior oblique does not completely separate off from the inferior rectus until later (cf. also Ida Mann, 1928). In a reconstruction by Ask (1910) of the muscles of the eye in an embryo 55 mm. long, the levator is only just beginning to separate off from a well-developed superior rectus. Poyales (1917) has also studied the muscles in the human embryo and finds that the lateral rectus shows inferiority as compared with the medial rectus in age, length, and number of fibres; he considers that the consequent greater power of the medial rectus would be the cause of the congenital strabismus of the new-born (see also p. 274).

Abnormalities.—It is probable, to judge from the writer's individual experience in finding quite a number of gross abnormalities of the ocular muscles in his series of dissections, that such are by no means as excessively rare as would appear from the number recorded in the literature; dissecting-room conditions do not favour their identification, and in life some may be unrecognizable through compensatory action of the other muscles. They are explained by developmental errors in cleavage from the common pre-muscular mesoblastic mass. Congenital anomalies are considered by Hardy (1914), Posey (1923, 24), and Natale (1929), and those especially affecting abduction by Gifford (1926) and Weill and Nordmann (1927); absence of each one of the ocular muscles has been recorded. The following are additional and personal observations.

The superior rectus has been found to give off a muscular slip 15 mm. long, which arose from the same origin from the annulus of Zinn and passed downwards and forwards across the lateral face of the optic nerve to join the inferior rectus about its mid-point; the nerve-supply came from the inferior division of the third nerve (Aubaret). See also p. 192.

The medial rectus has been found fused in its posterior third with the inferior rectus, or absent, in some cases of divergent strabismus (Ledouble, Krause). A bifid sclerotic insertion by two tendons, 16 mm. in length, is recorded (Wicherkiewicz).

The lateral rectus has likewise been found undeveloped in some cases of convergent strabismus (Ledouble, Krause), and a case of atrophy of the muscle has been noted on operating for strabismus in the living (Bourgeois). A fasciculus may pass from it to the inferior rectus, as is normal in certain ruminants, or to the lateral wall of the orbit. A lateral rectus with two extra fasciculi which passed forwards to end on the inferior tarsal plate and lateral wall is recorded (Curnow). In the specimen photographed in Fig. 42 there was a well-marked, fleshy bundle 7 mm. long and 2 mm. in diameter, passing from the lateral rectus across the posterior third of the orbit beneath the optic nerve to fuse with the belly of the medial rectus; no nerve could be traced to it.

The inferior rectus in the preparation photographed on p. 296 gave off a large muscular bundle which passed lateral to the optic nerve and joined the superior rectus; it was innervated by the lower division of the third nerve.

The superior oblique may be closely accompanied by an offshoot from the levator palpebrae superioris, sometimes called the *comes obliqui superioris* (see p. 149). In the preparation shown on p. 379 there were present two long muscle bundles, arising in common with the levator and ending anteriorly, the one upon the fascia bulbi between the superior oblique and the globe, the other on the orbital margin beneath the pulley; the nerve supply came from the fourth nerve; the superior oblique was broader than usual. Ledouble has found supernumerary fasciculi accompanying the reflected tendon; and has further recorded a case where the direct or normal fleshy part of this muscle was absent, the reflected or usually tendinous part being muscular and arising from the site of the pulley, recalling the type normally found in non-mammalian vertebrates.

The inferior oblique may have an abnormally placed origin (p. 278). An abnormal muscle bundle (*musculus obliquus accessorius inferior*) has been found by Rex passing from the apex of the orbit to the inferior oblique, but also sending a slip to join the inferior rectus; it was found in both orbits, and was supplied by the third nerve.

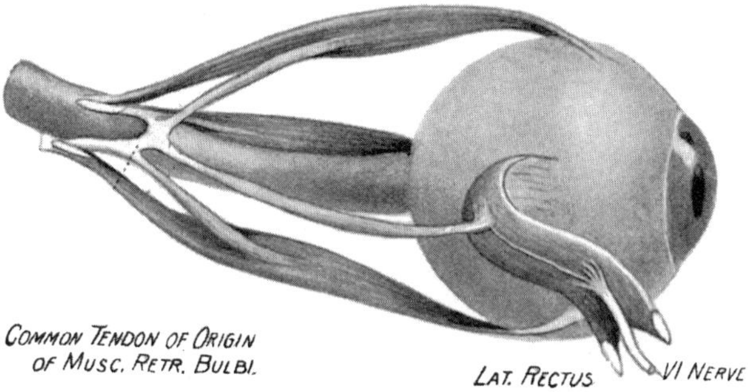

FIG. 155.—Sketch of an instance of the **Retractor Bulbi Muscle** in Man. There were four muscular bundles passing forwards towards the globe, but each fused with a rectus muscle before reaching it; one bundle was innervated by the sixth nerve as shown in the sketch, but the others were supplied by the third nerve. For description see the paper by the writer, 1911.

Vestiges of the *musculus retractor bulbi* (*retractor oculi, superior oculi, rectus posterior, choanoid*) muscle have on rare occasions been recorded in man, and what would appear to be a well-marked example is illustrated in Fig. 155. The muscle is first met with in amphibians and certain reptiles, and exists in the greater number of mammals, reaching its highest development in the ruminants; vestiges occur in some monkeys (e.g. macacus rhesus). The muscle arises from the apex of the orbit, lies wholly within the cone of muscles formed by the recti, and is inserted onto the posterior hemisphere of the globe; it may exist either as a cone-shaped mass cleft only by a passage for the optic nerve, as in the pig, or be subdivided into two or more parts; in the cat, for example, the muscle is divided about 2 mm. from its origin into four slips, which diverge and pass forwards to be inserted onto the globe behind the line of attachments of the recti and opposite the intervals between them. It is to be regarded as a derivative of the lateral rectus (Corning, 1900; Johnson, 1901), and is innervated in the domestic animals by the sixth nerve (Hopkins, 1916), though in the writer's example in man the third nerve shared. Its action is antagonized by the *membrana orbitalis musculosa* or protractor of the globe. In the macaque monkey it is found reduced to a single bundle placed between the superior and lateral recti, in which position Ledouble found delicate muscle bundles in two cases in man. Other instances have been noted by Nussbaum, and Fleischer (1907), and according to Lewitsky (1910) it is always represented in man (thirty specimens examined) by a well-marked strand of connective tissue (*fascia retro-bulbaris*) lying between the lateral rectus and the optic nerve, and attached to the back of the fascia bulbi anteriorly; it does not appear, however, to be readily identifiable. Cases of involuntary retraction of the globe in life have been reported by Axenfeld, The literature will be found in a paper by the writer (Whitnall, 1911).

For the comparative anatomy of the oblique and recti muscles the reader interested should consult the work of Motais (1887), and papers by Struthers (1849), Ottley (1879), Corning (1900), Prangen (1928); Poole (1905) discusses the variation in the trochlear angle in the domestic animals.

3. THE FASCIA OF THE ORBIT (*FASCIAE ORBITALES*)

This comprises:

(i) A condensation of connective tissue containing both collaginous and elastic fibres surrounding the eyeball, and forming a primary socket for it called the *fascia bulbi* or Tenon's capsule.

(ii) The fascial sheaths of the muscles, with extensions to the orbital walls known as 'check ligaments'.

(iii) The connective tissue supporting the orbital fat.

The muscle sheaths blend with the fascia surrounding the globe, and both are connected to the stroma of the orbital fat, but for descriptive purposes the three parts of the fascia are best considered apart.

(i) The **fascia bulbi (Tenon's capsule,** Bonnet's capsule, *aponeurosis orbito-ocularis, tunica vaginalis oculi*, &c.), from the great attention that has been paid to it and the conflicting views regarding its exact constitution, deserves a brief historical note. Mention is vaguely made of it under the name of 'tunica adnata' by Galen, Colombo, Casserius, Riolon, and others, but Tenon, of Paris, in 1806 gave the first good description of the structure, which bears his name. It then appears to have been forgotten until about 1840, when

Stromeyer's operation of strabotomy raised the question of the more exact relations of the muscles of the eye to the orbital fascia, and investigations by Malaigne again brought

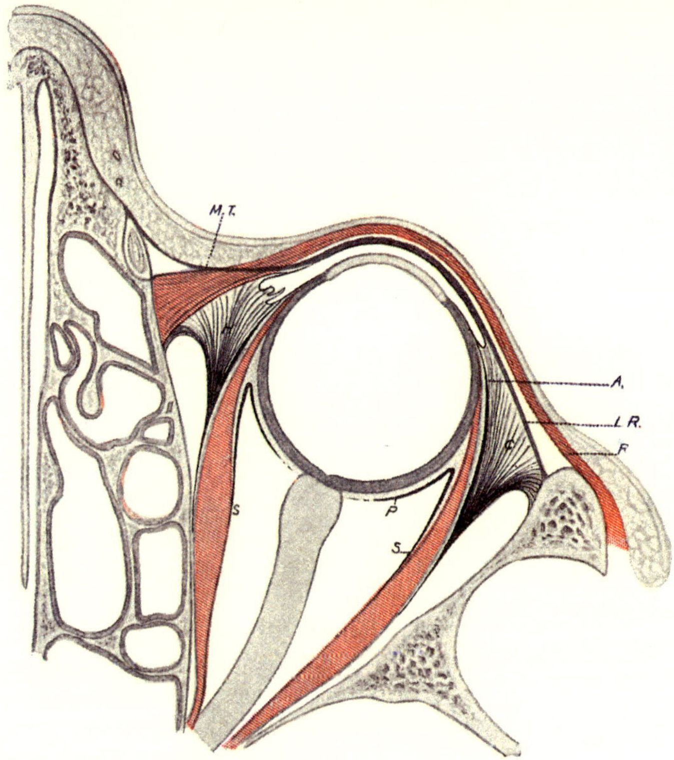

FIG. 156.—Schematic view of a horizontal section through the Right Orbit to illustrate the **Fascia of the Orbit.** Compare with Fig. 161, p. 296. Fig. 73, p. 147, shows the arrangement in vertical section.

The *fascia bulbi* or capsule of Tenon is shown by A., its anterior part, and P., its posterior part. The fascial sheaths of the muscles are marked by S., and their offshoots form C., the lateral, and H., the medial 'check ligaments'.

The drawing also illustrates certain points in the anatomy of the eyelids, M.T. being the medial palpebral ligament with its two limbs passing in front and behind the fossa for the lacrimal sac, L.R. being the lateral palpebral ligament, and R. the lateral palpebral raphe.

the capsule into notice under the name of the 'orbital aponeurosis'. Then Bonnet in 1841 pointed out the surgical importance of the relation of its posterior part in connexion with enucleation of the eyeball, and Ferrall gave a very clear description of the 'tunica vaginalis oculi' about the same time. Sappey in 1867 first described the presence of smooth muscle fibres in the check ligaments of the lateral and medial recti, and showed that these ligaments were not formed by the muscles themselves but by their fascial sheaths. As regards later accounts, like fascial dispositions elsewhere in the body such as the pelvic fascia, the capsule has received most elaborate attention, and is still to this date the subject of reinvestigation and laboured description. Dwight, indeed, remarks that 'the complications of this membrane are limited only by the perverted ingenuity of those who describe it'. Motais, for example, devotes nearly fifty pages to its description in Poirier's *Anatomy*, and deals with it from the standpoint of comparative anatomy in his book on the motor apparatus of the eye (1887); his method of demonstrating layers of fascia by pressing the connective tissue between sheets of hot blotting-paper to absorb the fat within its meshes appears liable to form artificial conceptions of its disposition. Schwalbe (1887), Lockwood (1886), Howe (1907), Lewitsky (1910), Gallemaerts (1899), Virchow (1902), and Charpy (1910), the latter of whom has studied the effects and extent of experimental chemosis by gelatine injections into the capsule and beneath the bulbar conjunctiva, have also written papers on it. Hesser (1913) has compiled what is practically a monograph upon the subject; an account is also given by Sutton (1920), who has, however, little to add to previous descriptions. It has been suggested that the capsule is the degenerate representative of the retractor bulbi muscle (Bland Sutton), a supposition that the writer cannot agree with. Its embryology is the subject of a paper by Goldstein (1923), who claims it is recognizable in embryos of 12 mm. long, though Mann (1928) found it to appear first posteriorly at 5 months.

The writer has studied the capsule and muscle sheaths not only by repeated dissections of material preserved in the usual way, but by examining practically fresh preparations in the post-mortem room, and feels convinced that connective tissue strands, hardened by alcohol or formalin and isolated by dissection from fat and neighbouring structures, assume in many regions an unwarrantable distinction as membranes, ligaments, or capsules. It should be realized that the fascia of a muscle is the remains of a mesoblastic tissue mass undifferentiated into muscle fibre; also that if a firm body such as the eyeball revolves against a cushion of fatty tissue, the fat will be pressed away and the connective tissue mesh-work supporting it condensed into a membrane-like surface. Such appears to be the case in the formation of the capsule of Tenon: the surface applied to the globe is shown definitely membranous after enucleation of the latter (hence meriting the term capsule, which for clearness of description will be used in this section), but the posterior surface is connected with the mesh-work of the orbital fat, and even the most careful dissection from this aspect leaves it ragged and perforate.

The following account of the capsule is such as is demonstrable by ordinary dissection.

The capsule of Tenon is a thin envelope of connective and elastic tissue completely surrounding the eyeball from the circumference of the cornea in front to the entrance of the optic nerve behind; it is so closely applied to the globe that in sections of the orbital contents it is not readily identified, but by pulling the structures apart posteriorly a space between it and the globe, the 'episcleral space' or *spatium interfaciale*, can be formed. Careful scrutiny will show, however, that the interval is really occupied by a felting of extremely fine connective tissue, the tissue of Tenon, episcleral tissue or *tunica adventitia*, comparable to the arachnoidea of the brain, though finer and denser. There is no sign of an endothelial lining, and the second or visceral layer of the capsule, as sometimes described (Schwalbe, Parsons), is merely that part of the adventitia which remains adherent to the sclera after separation of the capsule (Fig. 160); the space, according to Charpy and Hesser, is neither a true serous nor a formed lymph cavity, as described by some authors (Schwalbe); Leber (1876) also denies that it communicates with the perichorioidal space of the eyeball or contains lymph. The eyeball can move within the capsule, the adventitia being lax enough to allow of slight excursions, but in wider movements both globe and capsule move together as a whole upon the bed of orbital fat which is loosely connected to the capsule behind. It is obviously impossible for the eyeball to rotate freely within a socket that is fixed to and terminates at the circumference of its anterior sixth, that is, the cornea, as well as attached to the muscles at their points of entrance.

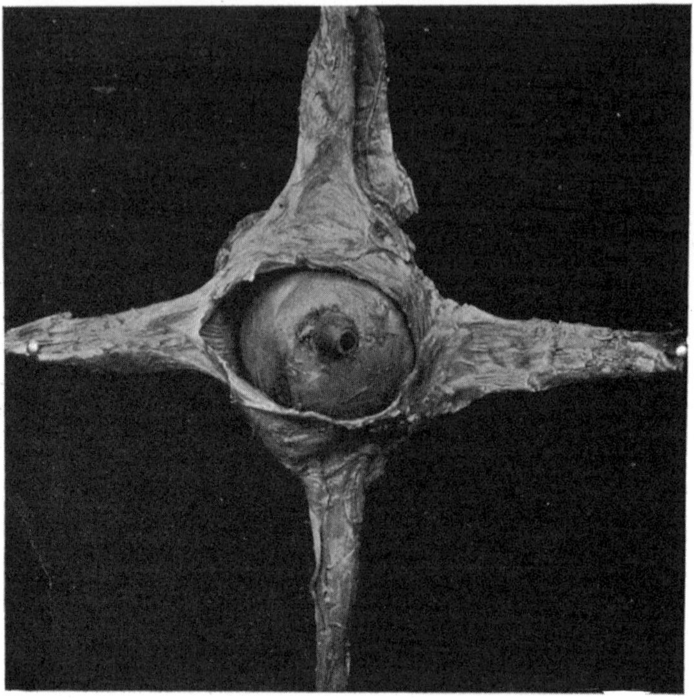

FIG. 157.—Left Eyeball seen from behind. The four recti have been spread out to show the continuity of their fascial sheaths with the *fascia bulbi*, or **capsule of Tenon,** which has been cut away on the left side to expose the medial rectus. Such a hardened and dissected-out preparation conveys an exaggerated idea of the density of the capsule. Natural size.

Since the capsule envelops the globe from the cornea in front to the optic nerve behind, it follows that it is pierced by all the structures which are attached to that organ, namely, the ocular muscles, the venae vorticosae, and the ciliary arteries and nerves. The recti muscles enter the capsule in front of the equator of the globe and serve to demark a smaller anterior from a larger posterior part. The anterior part of the capsule, in extent about one-third of the whole and lying in front of the muscle attachments, is thinner and more closely applied to the globe, being fused to the sclera for 1 to 2 mm. wide of the corneal margin (Fig. 103, p. 198); on its external face lies the bulbar conjunctiva, and the two membranes are raised together in exposing the tendons of insertions of the recti, as is well illustrated in Fig. 148, p. 269; the episcleral space here is extremely narrow and the intervening tissue greatly reduced; it is easily separable from the globe in growing subjects, but is fused with the conjunctiva in later years.

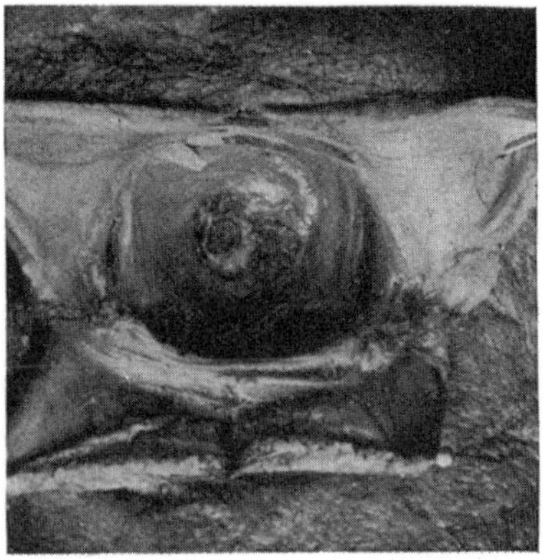

FIG. 158.—The Left Eye-Socket. The eyelids have been crucially incised and pulled aside and the eyeball removed to show the *fascia bulbi* or **capsule of Tenon** forming a barrier to the orbital contents behind it. Natural size.

The posterior part of the capsule is thicker and more yellow than the anterior, save just round the attachment of the optic nerve to the globe; here it is so broken up by the entrance of the posterior ciliary arteries and nerves that it is difficult to define, so much so that Merkel considers it to cease altogether a few millimetres from this point; in any case it is quite clear, as both Merkel and Virchow point out, that there is no continuation of the membrane backwards along the optic nerve to the apex of the orbit to form a so-called and often figured 'supravaginal space' around it and continuous with the episcleral space (as described by Schwalbe, Piersol), the optic nerve in its movements occasioned by the excursions of the globe has naturally necessitated a looser packing of the surrounding orbital fat.

The practical importance of the capsule lies in the fact that it forms a barrier between the globe and the other contents of the orbit, and after enucleation of the eyeball it helps to form a socket for an artificial eye, as is shown in Fig. 158, where the slit-like openings of the muscle sheaths into the capsule are seen, and also in Fig. 160.

(ii) The **muscle sheaths** (*fasciae musculares*) are so thin and transparent in their posterior third (and in the case of the fleshy part of the superior oblique in the whole extent) that they are practically invisible; traced forwards, however, they become thick and opaque, and at the points where the muscles pierce the capsule of Tenon, they blend and become directly continuous with the latter. So uniform is the connexion that some authors consider the sheaths to be formed by backward diverticula from the capsule, as the body of a shirt is continuous with its sleeves along the arms (Testut, Charpy, Virchow), and on the other hand the posterior part of the capsule has been considered to be formed by the fusion of backward extensions of the sheaths (Motais); it is much clearer to describe, as Hesser does, the sheaths and their offshoots as separate formations which fuse with the capsule when they come into contact with it. Just before this union with the capsule, the adjacent margins of the sheaths are continuous with one another, with a wide sweep between them when laid out as in Fig. 157, but in the natural position forming a backwardly directed 'inter-muscular membrane', which is especially well marked between the superior and lateral recti, where it passes beneath the lacrimal gland; the backward extension of the membrane is greatly exaggerated in many

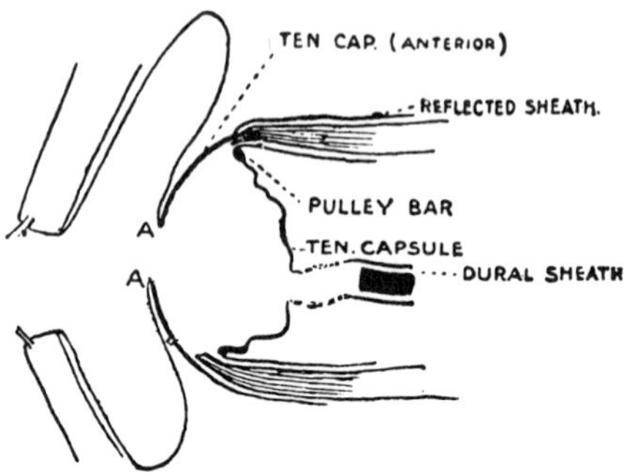

FIG. 159.—Diagram to show the *fascia bulbi* or **capsule of Tenon** after excision of the Eyeball. The cut margins of the conjunctiva at A A heal and complete the posterior wall of the new socket which will carry the artificial eye; the recti muscles, by their connexion with the anterior part of the capsule which is blended with the bulbar conjunctiva, retain an attachment to the new socket. (From Fisher, 1904.)

illustrations (e.g. those of Motais), and practically exists only a short distance behind the globe.

Where the sheath is reflected backwards on the inner or ocular surface of a rectus muscle to become continuous with the posterior part of the capsule it is thick and naturally its folding forms a lip, but is not expressly developed into an 'intracapsular ligament' (Lockwood) or a 'falciform fold' (Charpy), and the appearance of such a thickening as shown in Fig. 160 is caused by a gathering together or rucking of the fascia, and disappears when the muscle sheath is outstretched. Nor does there appear to be specially developed any 'pulley bar' (Fisher, 1904) at this point, over which the muscle tendon turns; the muscle makes no angle at the beginning of its arc of contact with the globe such as would require such a contrivance, and the idea that there are so formed slings connected through the medium of the external part of the sheath or capsule with the orbital wall to prevent muscle pressure on the globe is also erroneous. The muscles are firmly connected with the margins of the openings through which they enter the capsule; moreover, their sheaths are adherent to their muscle bellies along the whole extent, a fact of considerable importance, since the attachment prevents their extensive retraction after severance in enucleation of the globe (Fig. 159).

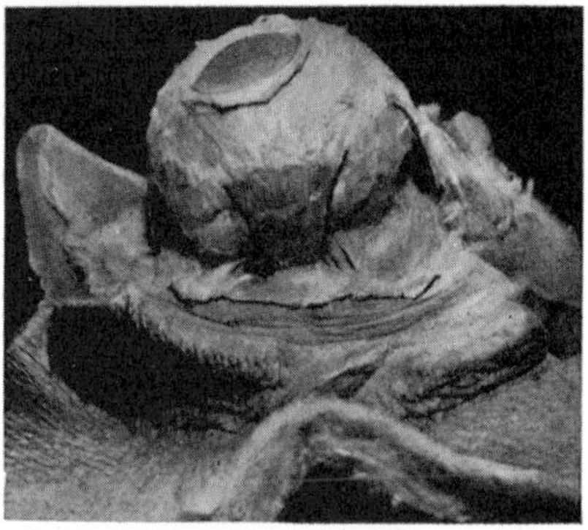

Fig. 160.—Dissection of Eyeball to show the Muscles entering the *fascia bulbi* or **capsule of Tenon,** and the episcleral tissue adherent to the sclera. The eyelids were first cut and turned down, then the bulbar conjunctiva and anterior part of the capsule cut round the cornea and reflected from the forepart of the globe. Natural size.

The muscle sheaths give offshoots from their external surfaces to the orbital walls, known as 'check ligaments', *Fascienzipfel*, or *ailerons*; they are best developed in the case of the medial and lateral recti (muscles whose excursions are greater and more exercised than are those of the others), and are therefore most clearly observed in horizontal sections through the orbit (Figs. 156, 161). That part of the sheath which clothes the lateral rectus on its external aspect becomes greatly thickened anteriorly and divides into two lamellae

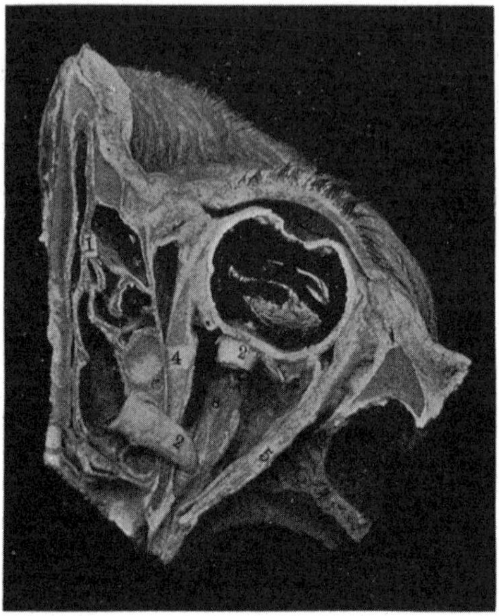

FIG. 161.—Horizontal section through Right Orbit seen from above, to show the **fascial sheaths** of the medial (4) and lateral (5) recti with their expansions which form the 'check ligaments'. 2 is placed on the optic nerve, which has been cut and turned aside to show 3, the inferior rectus, which in this instance presents a large abnormal bundle, noted on p. 286. Natural size. Compare with Fig. 156, p. 289.

one, the outer and larger, radiates forwards and laterally as a fairly dense triangular mass of tissue as broad as its muscle, and is mainly inserted into the periosteum over the orbital tubercle, just within the mid-point of the lateral orbital margin, but is also attached to the hinder face of the lateral palpebral ligament, which is attached to the same tubercle and to the lateral conjunctival fornix; this mass is the 'check ligament' or *lacertus* of the lateral rectus. The inner lamella is thinner, and closely follows the tendon of the muscle forwards; it unites with the corresponding continuations of the other recti to form the anterior part of the capsule of Tenon, lying beneath the bulbar conjunctiva and extending up to the sclero-corneal junction. The check ligament of the medial rectus presents a similar formation, its anterior fixation being to the lacrimal bone just behind the posterior crest, to the septum orbitale behind the pars lacrimalis muscle, and to the caruncle and medial fornix. Ledouble states that it is not rare to find in man a muscle fasciculus in either of these expansions, presumably an offshoot from the muscle itself, but Sappey first described smooth or involuntary muscle fibres in them, the presence of which has been verified by later authors (Motais, Groyer, Hesser). That these expansions do act as check ligaments to over-contraction of the muscles can readily be verified by cutting one of them and noting the greatly increased range of excursion of the globe on pulling the appropriate muscle; their attachments to the commissures explain the recession of the parts, notably the caruncle, seen on strong movements of the eyeball in the one direction or the other. The checking of the action of the superior or inferior rectus is not so apparent. The sheath of the former blends with that of the overlying levator palpebrae superioris, and the fixed mass between them is attached to the back of the superior conjunctival fornix; the two muscles act harmoniously, and the superior rectus may be limited in range by the same fascial mechanism as has been described for the levator on p. 148. The fascial sheath of the reflected tendon of the superior oblique clothes it as far as, and is attached to, the pulley. A definite check ligament is described by Motais, Charpy, and Maddox for the inferior oblique, in the form of a fascial strand passing laterally and backwards from its sheath to the floor of the orbit, and called the 'septal bridge' of the lower eyelid by Virchow; an example is seen in Fig. 59, p. 127, but it may be pointed out that this is the only instance in many dissections that the writer has found of a strand which could reasonably possess any such action. Distal to the point where the inferior oblique crosses the inferior rectus, their blended sheaths send forwards thin lamellae, one of which follows the tendon of the latter muscle to help form the anterior part of the capsule of Tenon, whilst the other can with difficulty be traced into the lower eyelid, as shown in Fig. 73, p. 147.

These expansions of the sheaths draw the fornices of the conjunctiva backwards, exactly as muscle fibres adjacent to joint capsules draw the slack of synovial membranes out of the way in extension, but a more important function is that by their fixation to the orbital walls they anchor the whole fascial apparatus of the globe in position. In this respect special note is to be made of the sheath of the inferior rectus which is thickened on the lower aspect and blended with the sheath of the inferior oblique. As previously noted, the sheaths of the muscles become confluent with each other by their adjacent margins immediately prior to the fusion with the capsule; the sheath of the inferior rectus is no exception to the rule, and the thickening of its sheath can be traced laterally and medially up to the sheaths of the lateral and medial recti, and since these are fixed to the orbital walls by expansions there is formed a continuous band about one-tenth of an inch thick beneath the globe, supporting it as in a hammock. The band was first described by Lockwood (1886) and named the *ligamentum suspensorium oculi*; it is well displayed in Figs. 60, 73, and 178; and its function as a suspensory ligament to the globe is manifest after operations in which the maxilla is removed for disease, since it then prevents the eyeball from sinking. It may be added that the medial and lateral check ligaments, blended with the extremities of the suspensory ligaments, meet the horns of the aponeurosis of the levator and are attached to the orbital margin at the same mid-points, and the conjoint masses of tissue so formed have been named the *retinaculum oculi laterale* and *mediale* respectively (Fig. 60, p. 128).

A final point of interest in regard to the capsule lies in the fact that there is found in close relation to it a scattered and ill-defined area of smooth muscle fibres. Hesser (1913) has recently investigated this **peri-bulbar musculature** in the foetus and the adult by means of serial microscopical sections, and finds it to extend round the anterior half of the globe in a nearly continuous but very thin layer, from 3 to 7 mm. wide in antero-posterior extent, but broken on the lateral side, and he terms the whole the *musculus capsulo-palpebralis.* Anteriorly it extends into the eyelids, constituting the palpebral involuntary muscles (of Müller), and designated by Hesser *pars superior* and *pars inferior* respectively; these are the only parts of the musculature which form definite isolated lamellae, the *pars medialis* being an extension of the latter (Fig. 162). Posteriorly, behind the conjunctival fornix, it is

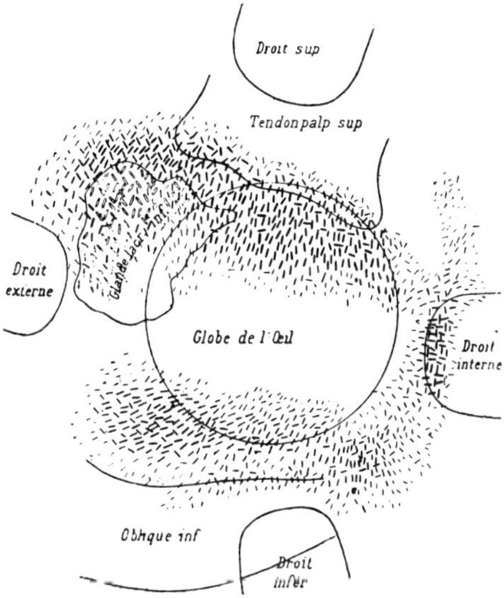

FIG. 162.—Diagram showing the position of **smooth Muscle Fibres** in the Fascia round the eyeball of an adult man, as described by Hesser (1913) and called by him the *musculus capsulo-palpebralis.* (From Hesser's figure, reproduced by Rousseau, 1916.)

quite indefinite. The whole sheet, which is innervated by the sympathetic system (from the cavernous plexus through the ciliary ganglion and long ciliary nerves, according to Starling), is of interest in that its contraction was considered by Landström (1908) to be the cause of the exophthalmos of goitre although it lies mainly in front of the globe. Further, Adler (1924) confirmed the findings of Starling and others that contraction of this unstriped musculature can compress the eyeball and raise the intra-ocular pressure; in some experiments on cats and dogs he removed the fascia bulbi containing this musculature before stimulation of the cervical sympathetic, with negative results; he found no evidence of vaso-dilator fibres to the eyeball. Groyer's morphological derivation of this involuntary musculature (p. 146) may here be referred to.

(iii) The **orbital fat** (*corpus adiposum orbitae, capsula adiposa bulbi*), which in life is no doubt semi-fluid, compactly fills all the space not occupied by the other structures, as is seen on p. 254; in formalin-hardened specimens, indeed, it forms so firm a mass that excellent sections of the orbital contents can be cut with a razor by hand, as shown by Fig. 198, p. 381. It varies in consistency in different regions, and the difference between the amount of connective tissue in the central and peripheral regions can be made out in that figure, the looseness of the former obviously facilitating the movements of the optic nerve and vessels in excursions of the globe. Within the cone of muscles at the back of the globe and along its whole extent round the optic nerve (central part of the orbital fat) it appears as large and loosely connected lobules which can easily and cleanly be picked away with the forceps in dissections of the orbit; between them the ciliary nerves and arteries make their way forwards, and it is by the passage of these structures that the fat is broken up; the inter-lobular tissue is loose and can be made oedematous by artificial injections, and it is along these channels that blood extravasated in fractures reaches the deeper parts of the orbit; Birch-Hirschfeld (1909) thought that the spaces between the lobules represented the unknown orbital lymph channels. Outside the cone of muscles, especially beneath the globe, behind the lacrimal gland, and even between the muscles anteriorly, the fat is collected in larger and firmer masses, which are much more difficult to remove owing to the quantity of connective tissue they contain and their close connexion thereby with the muscle sheaths; in this region it may be distinguished as the peripheral part of the orbital fat. Non-medullated nerve-fibres which may regulate the activity of the cells have been traced to the orbital fat in the rabbit by Wilkinson and Burkitt (1926).

As regards the relations of the capsule and orbital fat to the conjunctiva and eyelids, it is first to be noted that if the eyelids, including the septum orbitale, be completely dissected away from the front of the orbit, the capsule and muscle sheath expansions form an interrupted barrier to the orbital fat between the globe and the walls of the orbit; the gaps through which the fat is visible are called the 'orifices adipeux' of the base of the orbit by Charpy (1908), and are shown in Fig. 60, p. 128; it is through one such orifice that the lobule of fat ('palpebral hernia') illustrated in Fig. 54, p. 119, protrudes.

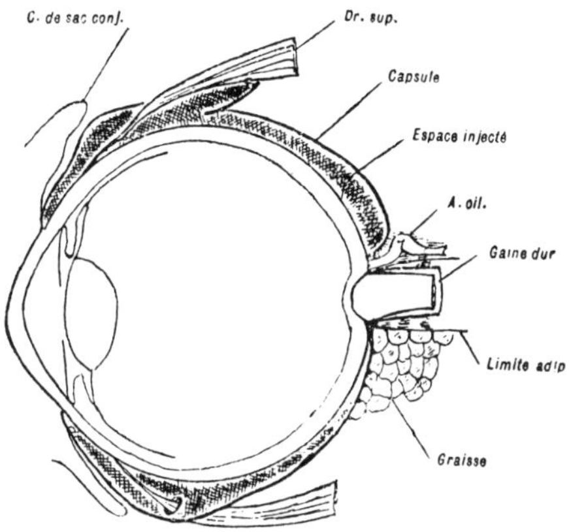

FIG. 163.—Diagram of a sagittal section through the Eyeball, illustrating the confinements of various injection masses, as made by Charpy (1910).

Charpy (1911) further studied the relations between the eyelids and the orbit by the injection of coloured gelatine between various muscle and fascial planes in the orbit, and his results may be summarized as follows:

(i) Injections within the capsule of Tenon can distend and are confined to the whole episcleral space, from the corneal margin to the optic nerve, as shown in Fig. 163; they produce anteriorly a regularly disposed tumefaction round and limited by the cornea, and lying beneath the bulbar conjunctiva and the anterior part of the capsule.

(ii) Injections into the central part of the orbital fat form a retro-ocular mass, which is confined anteriorly within the muscle cone by the conjoint anterior edges of the sheaths, the so-called inter-muscular membrane, but in the posterior half of the orbit extends between the muscles into the peripheral region; there is no communication with the eyelids, since the injection is blocked by the globe and the muscle cone in front; hence moderate invasions of fluid into the central region neither invade the conjunctiva nor reach the lids, but produce an exophthalmos and immobility of the globe.

(iii) Injections into the peripheral part of the orbital fat, outside the cone of muscles, are chiefly confined to the lower part of the orbit, and lie between the orbital fat and the periosteum lining the orbital walls. They can reach both the conjunctiva at the fornices and the eyelids behind the septum orbitale by passing through any of the 'adipose orifices' left between the attachment of the fascial sheaths of the muscles to the orbital walls, and this is the ordinary path along which fluids pass from the depths of the orbit to the eyelids. They may further transude through the venous or vascular apertures in the septum orbitale or through that membrane itself, and so reach the superficial parts of the lids, but Heerfordt (1904) has shown that the septum will stand a pressure of at least from 40 to 50 mm. of mercury before allowing an intra-orbital injection to pass into the eyelids. The limits of injection masses in the superficial parts of the eyelids have been described above on p. 142. (See also p. 42.)

Finally, sub-periosteal injections beneath the periorbita easily strip it off from the orbital walls and are confined to the space so formed, being arrested at the margin; they may rupture the membrane, especially if beneath the roof, and then behave like an invasion into the peripheral orbital fat.

Pathological observations support the results of such experimental injections, and emphysemata, abscesses, and haemorrhage have been shown to follow the paths outlined above; the orbital course of haemorrhage in fractures of the base of the skull is determined by the same structures in reaching the eyelids (Liebrecht, 1906).

4. THE VESSELS OF THE ORBIT

The **arteries** supplying blood to the orbital contents are chiefly the ophthalmic, with minor additions from the infra-orbital and middle meningeal branches of the internal maxillary.

The ophthalmic artery (*arteria ophthalmica*) arises from the internal carotid, being the last branch given off before it divides into its terminal branches, the anterior and middle cerebral arteries; it usually springs from the medial side of the fourth part of this vessel's course just as it ascends out of the cavernous sinus, and pierces the dura between the anterior clinoid process and the body of the sphenoidal bone, though sometimes it arises just within the sinus. The artery leaves the cranial cavity through the optic canal, embedded in the dural sheath of the optic nerve below and to its lateral side. Having entered the orbit, it pierces the nerve sheath, and lies at first within the apex of the cone of recti muscles, between the optic nerve on the one side, and the ciliary ganglion, third, fifth and sixth nerves, and lateral rectus muscle on the other (Fig. 147); it ascends and winds round the optic nerve, passing over it (in 93·5 per cent., according to Adachi, 1928) about half-way along its course beneath the superior rectus, and in company with the naso-ciliary nerve (Fig. 197); it then runs forward in a sinuous course towards the medial orbital wall between the superior oblique and medial rectus, and at the medial angle of the upper eyelid ends by dividing into its terminal branches, the frontal and dorsal nasal arteries. These pierce the septum orbitale, and the latter vessel, passing between the trochlea and the medial palpebral ligament along with the infra-trochlear nerve, becomes directly continuous with the angular part of the facial artery.

As regards the origin of the vessel, the writer has often observed a marked reduction in the calibre of the internal carotid artery just after it has given off the ophthalmic branch, a narrowing out of proportion to the consequent diminution of its blood-stream as judged by measurement of cross-sections of the vessels (see Fig. 172, p. 323); on an average of ten specimens, the diameter of the internal carotid on entering the cavernous sinus was 5·4 mm. (cross-section area = 22·89 sq. mm.), but at a point 5 mm. distal to the origin of the ophthalmic, the width was 3·8 mm. (area = 11·33 sq. mm.), the corresponding dimension of the ophthalmic artery being 1·5 mm. (area = 2·009 sq. mm.). This conformation may be related to the intra-cranial blood-pressure, or be a means of reducing the shock of arterial pulsations to the brain (in conjunction with the curved course of the vessel as it traverses the carotid canal and cavernous sinus, a similar bending being found in the case of the vertebral arteries before they enter the cranial cavity), but it is also of interest to speculate whether the reduction in calibre of the main vessel, by forming a slight damming up of the blood-stream, does not favour the passage of the side current down this important ophthalmic branch. In connexion with this suggestion it is to be noted that the branch not only comes off the main stem at right angles, but at once bends downwards to lie almost parallel with the latter at the commencement of its course, and it is not unreasonable to imagine that in such case the main stream would tend to draw blood out of the side branch rather than to guide fluid into it (analogous to the action of a spray producer, wherewith the air is blown across the end of the fluid-containing tube), were there not some such provision as apparently exists here.

The exact relation of the ophthalmic artery to the optic nerve in the intra-cranial part of its course appears subject to slight variation, the vessel lying more to the medial or lateral underside of the nerve according as it arises nearer or farther away from the optic foramen (Fawcett, 1896).

The branches of the ophthalmic artery are small but numerous, and like the parent stem are characterized by a tortuosity of course, which obviates tension in ocular movements; anastomoses are formed with neighbouring extra-orbital vessels, though not as freely as in the case of the ophthalmic veins. They may be classified in various ways, either:

(i) Topographically into groups supplying the eyeball, the other contents of the orbit, and extra-orbital regions, i.e., ocular, orbital, and extra-orbital groups; or

(ii) From their sites of origin from the main vessel relative to the optic nerve, into groups arising lateral, superior, and medial to that structure, together with terminal branches; or

(iii) According to their sequence of origin.

The eyeball is supplied by the central artery of the retina, two long posterior ciliary, and seven short anterior ciliary arteries; there are groups of branches to the muscles, and an artery to the lacrimal gland; and, passing on to supply parts beyond the orbit, are the posterior and anterior ethmoidal arteries, the supra-orbital, and finally the terminal branches, the frontal, nasal and palpebral vessels. Relative to the optic nerve, there lie laterally the central artery of the retina and the lacrimal, superiorly the supra-orbital, posterior ciliary and muscular arteries, medially the ethmoidal and palpebral. The sequence in which the branches arise from the parent stem is, according to Meyer (1887), the retinal and medial posterior ciliary, the lateral ciliary, the lacrimal, the upper and lateral muscular twigs, the supra-orbital and posterior ethmoidal, the lower and medial muscular twigs, the anterior ethmoidal, and the terminal branches.

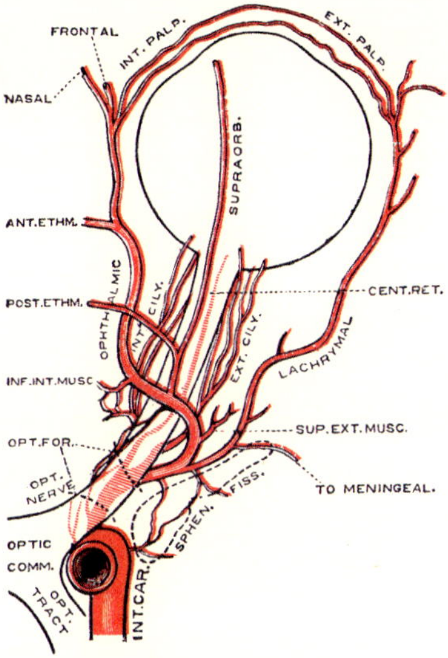

FIG. 164.—Diagram of the branches of the **Ophthalmic Artery.** (From Quain, 1909, after Meyer.)

It is also to be noted (i) that an artery lies in close relation to each of the walls of the orbit, namely, the supra-orbital to the roof, the lacrimal to the lateral wall, the ethmoidal vessels and the terminal part of the ophthalmic artery itself to the forepart of the medial wall, and the infra-orbital to the floor; (ii) that they anastomose with vessels outside the orbit; and (iii) that each is accompanied for at least a part of its course by a nerve: namely, the frontal, lacrimal, naso-ciliary, and infra-orbital respectively. The various branches are now to be described in detail.

The central artery of the retina (*arteria centralis retinae*), the first to be given off from the parent stem and the most important of its branches, arises from the latter whilst it still lies beneath the optic nerve, and runs forward to pierce the nerve sheath at a distance variously given from 5 to 15 mm. behind the posterior pole of the eyeball; the site of entry is variously given as either on the lower, the medial, or the lateral aspect of the nerve, but Deyl (1896), after a special examination of twenty-one specimens, found the position in each case to be infero-medial, and since this is the site of the optic fissure in the embryo where the mesenchyma invaginated the primary optic vesicle to form the vitreous body, it appears that the vessel maintains its primitive position, and the supposition that the nerve and eyeball have rotated through an angle of 90° during development (Vossius, in Keibel and Mall, 1912) is without foundation. The artery, accompanied by a central vein and a sympathetic plexus derived from the ciliary nerves (the so-called Tiedemann's nerve) is enveloped as it enters the nerve substance by a prolongation of the pial sheath, and it courses along the centre of the nerve to supply the inner layers of the retina. The intra-neural course is noted in conjunction with that of the vein on p. 314; each vessel sends a recurrent branch backwards within the nerve almost up to the optic foramen (the *arteria vel vena centralis posterior*). The finer branches of the network into which it ultimately breaks up in the retina make no connexion with any other artery supplying the eyeball, though the main stem of the vessel may anastomose with a short posterior ciliary artery just after its entrance into the eyeball, such a cilio-retinal vessel being present in about 17 per cent. of cases (Lang and Barrett). Sometimes the central artery may be replaced by a posterior ciliary artery, and indeed Meyer (1887) states that normally it arises in common with the medial long posterior ciliary, an origin found in a special study by Beauvieux (1924), however, to occur only in 6 out of 20 orbits.

The ciliary arteries, which are also destined to supply the

eyeball (chorioid coat, ciliary processes, and iris), are variable in number and origin; they are distinguished as posterior and anterior groups. The posterior (*arteriae ciliares posteriores longae et breves*) arise from the ophthalmic artery as it crosses the optic nerve, and may appear as two single trunks or as six to eight vessels; in either case they subdivide near the

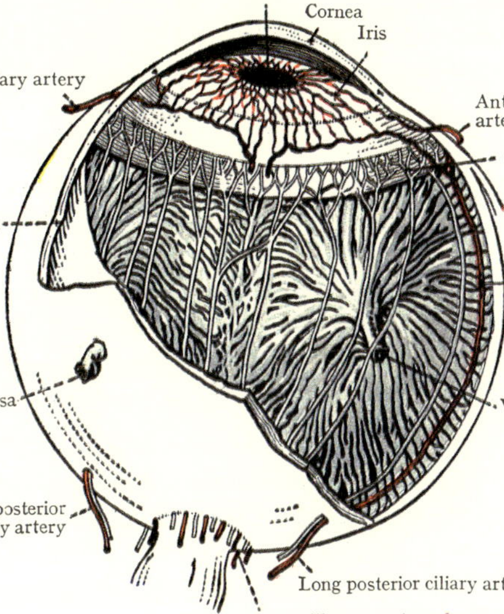

FIG. 165.—Dissection of the Eyeball showing the Vascular Tunic and the Arrangement of the **Ciliary Nerves and Vessels**. (From Cunningham's *Textbook of Anatomy*.)

globe into fifteen or twenty branches arranged round the optic nerve, and pierce the sclera to supply the chorioid. Two of them are larger than the others, and under the name of long posterior ciliary arteries enter the globe one on either side of the optic nerve (Fig. 138, p. 257); their ultimate destination is the ciliary muscle and the iris, where they anastomose with the anterior ciliary vessels to form the *circulus arteriosus iridis*. The anterior ciliary arteries (*arteriae ciliares anteriores*) arise usually from the branches of the ophthalmic artery which supply the muscles of the globe; they pierce the muscle bellies about their mid-points and accompany the tendons of the recti (two to each, save the lateral rectus which has one), pierce the sclera just anterior to the line of tendon insertion, and anastomose with the long posterior ciliary vessels, as just noted; their main stems pass onwards toward the corneal margin to supply its periphery and the adjacent bulbar conjunctiva (see Fig. 109, p. 204, where the clinical significance of this is noted); an anterior ciliary vessel may arise from the lacrimal artery. Fuchs (1929), in a special study of the ciliary vessels, finds that normally the arteries have a thicker wall than others of the same sized lumen elsewhere in the body, and describes the senile changes that occur in them.

The lacrimal artery (*arteria lacrimalis*) arises from the ophthalmic artery on the lateral side of the optic nerve, and in company with the lacrimal nerve passes forward along the upper margin of the lateral rectus muscle through the orbital fat to reach the gland. Early in its course it gives off a recurrent meningeal branch, which passes through the lateral end or pierces the bony margin of the superior orbital fissure and anastomoses with the middle meningeal artery in the middle cranial fossa. It then supplies the medial and superior recti muscles, and sometimes provides an anterior ciliary artery; it gives off temporal and malar branches which traverse the zygomatic bone in company with the zygomatic nerve and anastomose with the anterior deep temporal and transverse facial arteries in the temporal fossa. Finally, after passing outside or through the lacrimal gland, which it supplies, it bifurcates into the two *lateral palpebral* vessels and ends in the conjunctiva and skin of the eyelid. Since it also anastomoses with the infra-orbital artery, it forms an extensive series of unions with other vessels.

The muscular branches (*rami musculares*) of the ophthalmic artery are usually found as two main stems, with a variable number of smaller twigs; when distinct, one stem arises inferiorly close to the lacrimal artery and supplies the inferior and medial recti and inferior oblique muscles, the other stem, the upper, supplies the superior and lateral recti, the levator and superior oblique muscles. As noted above, the muscular arteries provide the anterior ciliary twigs.

The supra-orbital artery (*arteria supraorbitalis*) arises from the ophthalmic above the optic nerve, and runs forwards at first medial, then superior to the levator, and between it and the roof of the orbit, in company with the corresponding nerve; it leaves the orbit through the supra-orbital notch, and ascends in the scalp over the forehead towards the vertex of the skull. Besides supplying the scalp, it sends twigs to the levator, the periorbita of the orbital roof, the diploë and periosteum of the frontal bone, and the upper eyelid, and it anastomoses with the superficial temporal artery (Fig. 91, p. 172).

The ethmoidal arteries, two in number, leave the ophthalmic as it passes along the medial wall of the orbit between the superior oblique and medial rectus. The *arteria ethmoidalis posterior*, smaller and less constant, leaves the orbit, according to Adachi, more often (18 to 8) by passing above the superior oblique, to enter the foramen of the same name and end in the mucous membrane lining the neighbouring air-cells and upper part of the nasal fossa, anastomosing with the spheno-palatine branch of the internal maxillary; it may be accompanied by a nerve filament (of Luschka, p. 53), and possibly lymph vessels. The *arteria ethmoidalis anterior* is a constant and larger vessel; it leaves the orbit through the anterior ethmoidal canal in company with the nasal branch of the naso-ciliary nerve and a vein, enters the anterior cranial fossa, crosses the cribriform plate of the ethmoidal bone, and enters the nasal cavity by traversing a slit at the side of the crista galli; it then descends along a groove beneath the nasal bone to the tip of the nose. It supplies a meningeal branch to the dura of the anterior cranial fossa, twigs to the anterior ethmoidal cells, the frontal sinus, the nasal mucosa, and the skin of the dorsum of the nose.

The palpebral branches (*arteria palpebrales mediales*) arise from the ophthalmic just below the trochlea; one enters each eyelid to form the superior and inferior tarsal arcades, which anastomose laterally with the palpebral branches of the lacrimal artery and also with the supra-orbital and infra-orbital arteries. The details of this palpebral distribution are illustrated on p. 172.

The frontal terminal branch of the ophthalmic artery

(*arteria frontalis*) leaves the orbit by piercing the septum orbitale in company with the supra-trochlear nerve, and ascends over the forehead to supply a small part of the eyelids, but chiefly the mesial region of the scalp; it anastomoses freely with the supra-orbital artery and with its fellow of the opposite side (Fig. 91, p. 172).

The nasal terminal branch (*arteria dorsalis nasi*) is the real termination of the ophthalmic artery, and, as already noted, it leaves the orbit by piercing the septum orbitale between the trochlea and the medial palpebral ligament, and becomes directly continuous with the angular part of the facial (external maxillary) artery; it supplies the lacrimal sac and the skin of the root of the nose, and anastomoses with its fellow of the opposite side and with the lateral nasal branch of the facial artery.

The infra-orbital artery (*arteria infraorbitalis*), although not a branch of the ophthalmic but a terminal part of the internal maxillary, itself a branch of the external carotid, deserves notice here from its relation to the floor of the orbit beneath the periosteum of which it lies. It originates in the pterygo-palatine fossa, enters the orbit through the posterior end of the inferior orbital fissure, and runs along the infra-orbital sulcus and canal to emerge on the face by the infra-orbital foramen; in the sulcus it gives branches to the inferior rectus and inferior oblique muscles, orbital fat, and lacrimal gland, besides the alveolar branches, and anteriorly it supplies the lower eyelid and lacrimal sac (Fig. 91, p. 172, and see p. 246); on the face it divides into palpebral, nasal, and labial branches, and anastomoses with various branches of the facial artery, with the transverse facial branch of the superficial temporal, and with the dorsal nasal branch of the ophthalmic.

The anastomoses which the ophthalmic artery, a derivation of the internal carotid artery, forms with the internal maxillary and facial, which are branches of the external carotid, have been given in the above description; the chief communications are the supra-orbital with the transverse facial, the lacrimal with the middle meningeal, the nasal with the angular arteries.

Henderson (1903) gave a description of the ophthalmic arteries in the rabbit and dog, which, with a paper by Parsons (1903), will be found of value in studying the intra-ocular circulation. The development of the vessels is given by Dedekind (1909).

Variations in the branches of the ophthalmic artery are common. In 15 per cent. of cases the main vessel passes beneath instead of over the optic nerve (Quain), and then may lie entirely to its medial side; it has been found buttonholed by the optic nerve (Poirier). The lacrimal artery may be reinforced by the anterior deep temporal; its communication with the middle meningeal may be unusually large, and, indeed, sometimes the latter vessel provides the lacrimal branch instead of the ophthalmic; as a further development, the whole of the ophthalmic arterial system has been found in very rare cases to be provided by the middle meningeal (a derivative of the internal maxillary branch of the external carotid), as in the specimen photographed in Fig. 191, p. 362, where there was no trace of an ophthalmic artery proper on either side. Conversely, the ophthalmic artery has been seen to give off the middle meningeal (Quain). Such anomalies are due to the embryonic anastomosis between the supra-orbital branch of the stapedial and the ophthalmic vessels, and will be found illustrated in the monograph on the arteries of the body by Adachi (1928). The supra-orbital as well as the posterior ethmoidal is inconstant; the latter may be a branch of the former instead of coming directly from the ophthalmic. The nasal frequently supplies a deficiency of the upper part of the facial artery. Types of such abnormalities are figured in a paper by Meyer (1887). As Parsons (1903) states, the general tendency as we ascend the animal scale is to find the principal ophthalmic artery, which in the lower members is supplied by the external carotid, to be derived from the internal carotid; in intermediate forms there is usually one vessel derived from each source, commonly with an anastomosis between the two systems; hence individual variations may be expected.

The Veins.—The blood is drained from the orbit by three main routes; firstly and in greater part, backwards by the ophthalmic veins into the cavernous sinus of the intra-cranial venous system; secondly, forwards by the free communications of the ophthalmic veins with the angular vein into the facial system; thirdly, and in least extent, downwards through the inferior orbital fissure into the pterygoid plexus.

The ophthalmic veins are two in number, the superior and inferior. The superior (*vena ophthalmica superior*) is the chief vein of the orbit; it is formed behind the medial corner of the upper eyelid by the union of two branches from the supra-orbital and angular veins of the face, which enter the orbit by passing through the septum orbitale respectively

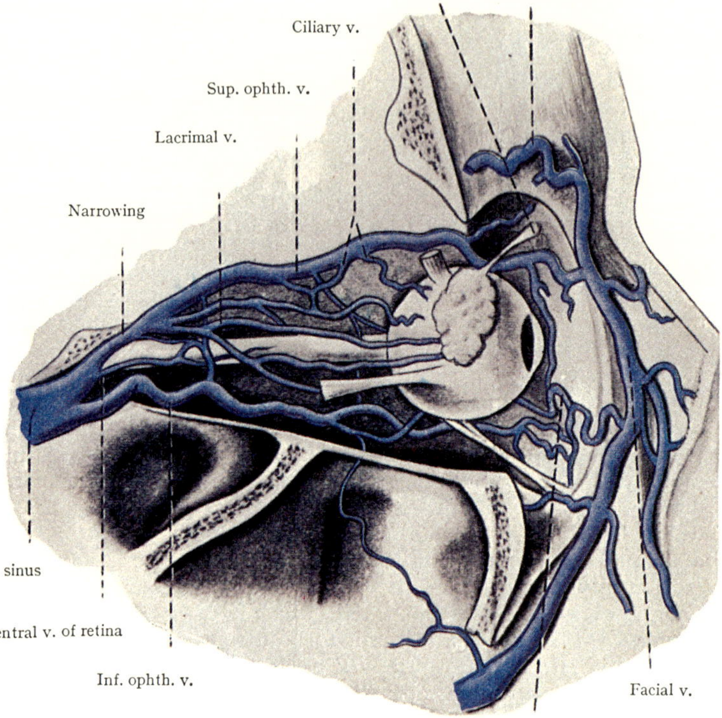

FIG. 166.—The **Veins of the Orbit** and their connexion with those of the Face. (After Sesemann.)

above and below the trochlea; both these veins are seen in Fig. 91, p. 172, and Fig. 166 above, and the communication with the angular vein has been dissected out in Fig. 59, p. 127.

According to Zinn (1755), it was Vesalius who first remarked the ophthalmic vein as beginning at the medial angle of the eye, and Fallopius its anastomosis with the facial; Walter

(1755) published and beautifully illustrated the first complete work on these vessels, and Festal (1887), like Sesemann (1869), devised and studied them by means of mass injection.

After its formation, the superior ophthalmic vein passes backwards in the orbit, at first very superficially and tortuously, and separate from the ophthalmic artery; accompanied by an anastomotic satellite-vein, it crosses the optic nerve beneath the superior rectus muscle to reach the apex. Here it generally receives the inferior ophthalmic vein and bends downwards to enter the cavernous sinus; before its entry it is stated to become markedly constricted by some authors (Sesemann, Dwight, Festal, Krauss, Hesser), as seen in Fig. 166, but enlarged by others (Henle, Luschka, Sappey, Merkel). It must pass through the superior orbital fissure either above, within, or below the annulus of Zinn, as shown in Fig. 167 (on which point neither Sesemann nor Festal made any particular observations), the last position being held especially important as exposing it to compression by the involuntary orbitalis muscle, which may occur to a separate inferior ophthalmic vein (p. 317). In two-thirds of cases, Gurwitsch (1883) found it to communicate near its termination with the 'ophthalmo-meningeal' vein of Hyrtl, which is a middle cerebral vein coming from the fissure of Sylvius and ending in either the spheno-parietal sinus (p. 324) or the orbital system. It frequently presents varicosities in its general course, the widest dilatation found by the writer being nearly 4 mm. in diameter; Krauss considers that these varices, if voluminous, may produce a pulsating exophthalmos.

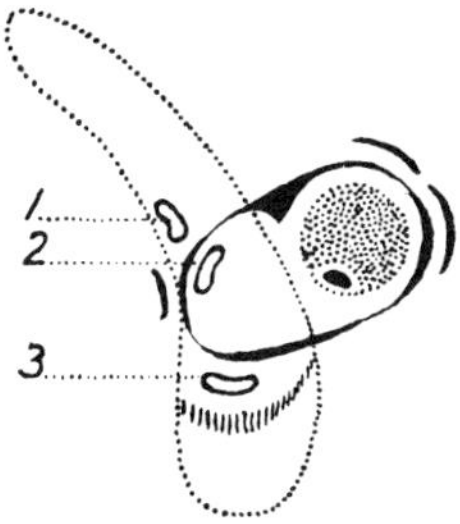

FIG. 167.—Diagram to show the various positions of the **Superior Ophthalmic Vein** relative to the annulus of Zinn as it leaves the orbit through the superior orbital fissure, as described by different authors. (After Hesser, 1913.)

1. According to Zinn, Festal, Merkel, Charpy, Hovelacque, Kiss, and most others.
2. According to Schwalbe, Hesser.
3. According to Fründ. (This is the position of the inferior ophthalmic vein at this point on the rare occasions when it enters the cavernous sinus separate from the superior vessel, according to Sesemann, Charpy, Hesser, and Hovelacque.)

The shading below 3 indicates the position of the orbitalis muscle of Müller.

Its tributaries are from the eyeball and other contents of the orbit, the latter corresponding mostly to the arterial vessels; just at its origin it occasionally receives the anterior frontal diplöic vein and generally the lacrimal vein posteriorly; from the eyeball it receives the upper *venae vorticosae,* the lower ones joining the inferior ophthalmic vein. These veins, usually four in number and each draining a quadrant of the chorioid coat, ciliary body, and iris of the globe (Fig. 165), emerge obliquely after a passage of about 3 mm. through the sclera at nearly equidistant points relative to the vertical meridian, though the lateral ones are nearer, and about 6 mm. behind, the equator (Weiss); anastomoses are formed between them. They are sometimes thought to be compressed by the muscle tendons, but whilst the movements of the recti have no action upon them, those passing beneath the oblique muscles, that is the upper and lower lateral pair, certainly appear to lie in such a position that they could be pressed upon, especially in convergence. Variations in number up to five or seven are not infrequent, generally on the nasal side, and they may empty into the veins of the muscles (Fuchs).

The *vena centralis retinae* accompanies the artery of the same name in its intra-neural course, but here Fry (1930), from a study of serial sections of thirty-six nerves, describes a new point in variations. Whilst it is known that both vessels have a certain length of course backwards in the subarachnoid space between nerve and sheath, at a variable distance behind the lamina cibrosa both vessels make an abrupt turn to reach the periphery of the nerve; they may do this in company and cross the intra-vaginal space to pierce the nerve sheath together, but more often (in 17 out of 21 nerves) they part at right angles to one another within the nerve and reach its periphery at different levels, with the vein usually nearer the globe; in the intra-vaginal space the contrary occurs, the vein having a longer course and always piercing the nerve sheath farther from the globe; thus it may obviously be affected by any rise of pressure to which the fluid in the space is subjected (Fig. 168). The site of the point of emergence of the vein from the nerve sheath exhibits the same variations as does that of the artery, described on p. 306; the details are given by Deyl (1896), Hovelacque (1920), and Beauvieux (1924). The central vein then passes backwards to open most frequently separately into the cavernous sinus, rarely into the superior or even inferior ophthalmic vein; sometimes a delicate plexus surrounding the optic nerve may receive the vein and transmit it to the sinus; it has always at least one side connexion, commonly with the former vein, an important provision that ensures an escape for the blood of the retina in cases of thrombosis of the cavernous sinus.

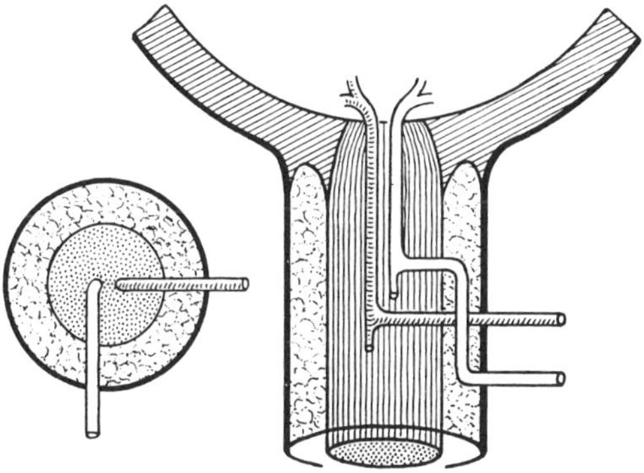

FIG. 168.—Scheme to show the intra-neural course of the **Central Vessels of the Retina.** Transverse and longitudinal sections of the Optic Nerve and its Sheath. The divergence of the Artery (cross-hatched) and Vein from one another is shown in the first drawing, and the longer course of the vein through the intra-vaginal subarachnoid space in the second. (Modified from Fry, 1930.)

The ciliary veins (*venae ciliares*) accompany the corresponding arteries, the anterior ones, through the muscular veins, draining the ciliary body and great circle of the iris; it is these anterior ciliary veins which become engorged and appear as dark serpentine vessels on the forepart of the sclera in cases of chronic glaucoma, the intra-ocular pressure occluding the obliquely running vorticose veins through which the bulk of the blood of the eyeball is normally drained (Fisher). Sesemann would omit the term 'posterior ciliary veins' since they are only continuations of the venae vorticosae.

The ethmoidal veins (*venae ethmoidales*), anterior and posterior, drain part of the blood from the sphenoidal sinus, ethmoidal cells and superior meatus of the nose, where they communicate with the veins of that cavity; they enter either ophthalmic vessel. The multiple veins from the upper and medial ocular muscles call for no special note.

The inferior ophthalmic vein (*vena ophthalmica inferior*) is the vein of the orbital floor; it originates in a plexus of small vessels in the orbital fat beneath the globe, and, indeed, may exist as a network throughout its whole course; it receives venules from the lower and lateral ocular muscles, the two inferior vorticose veins, and also helps to drain the conjunctiva of the lower eyelid and the lacrimal sac. It runs backwards above the inferior rectus and ends either by joining the superior ophthalmic vein (as was always the case in Hesser's observations, generally so in those of Krauss and Festal, found in two out of three by Gurwitsch, but rarely by Sesemann), or by entering the cavernous sinus separately by passing beneath the annulus of Zinn, where it lies in contact with the orbitalis muscle (Figs. 166, 167); it may even enter the ophthalmo-meningeal vein. Anteriorly it anastomoses with the (anterior) facial vein by branches which pass over and sometimes groove the orbital margin; posteriorly it always communicates with the superior ophthalmic vein; inferiorly it establishes communication through the inferior orbital fissure with the pterygoid plexus, in which it may terminate entirely; the connexion, sometimes called the 'ophthalmo-facial vein', may become of great importance. Found present usually by Festal, in 57 per cent. by Gurwitsch, but very seldom by Krauss, this vein is really a nasal one, since it arises in the mucous membrane of the nose, passes through the spheno-palatine foramen to the infratemporal fossa and runs parallel with the inferior orbital fissure, where it drains the maxillary antrum and makes the above noted connexion. The pterygoid plexus, buried in the fatty tissue of the pterygoid fossa, communicates also with the cavernous sinus, and by the deep facial vein with the facial system; it receives blood from the field of distribution of the internal maxillary artery, and ultimately condenses to form the vein of the same name which in junction with the superficial temporal becomes the temporo-maxillary vein, a part origin of the external jugular. In its passage through

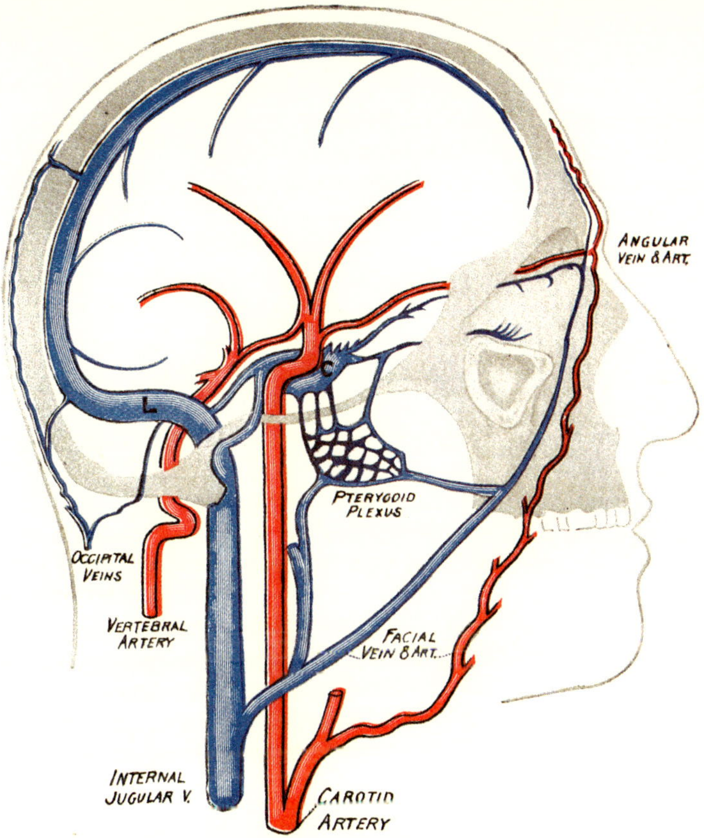

FIG. 169.—Scheme to show the **connexions of the Vessels** of the Orbit with those of the Face on the one hand and with those of the Intracranial System on the other. (The external jugular vein is not shown.)

C. = the cavernous sinus, traversed by the internal carotid artery. L. = the lateral or transverse sinus.

the inferior orbital fissure the communication between the inferior ophthalmic vein and the plexus is subject to compression by the involuntary orbitalis muscle, which fills the fissure, and the vein itself is related to and may be affected by contraction of the same muscle if it should pass beneath the annulus of Zinn to enter the cavernous sinus. By these relations an explanation has been sought for the exophthalmos of Basedow's disease, a point that has already been referred to on p. 88.

The lacrimal vein (*vena lacrimalis*) is large enough to be considered as the third main vessel of the orbit; it originates in the lacrimal gland, but receives blood also from the neighbouring conjunctiva and anastomoses with the temporal veins round the margin and by venules passing through the lateral orbital wall; it accompanies the artery in its course backwards, and usually opens into the superior ophthalmic vein, but sometimes into the cavernous sinus separately or even into the ophthalmo-meningeal vein. Gurwitsch found the lacrimal usually joined by veins from the superior and lateral recti and by a vorticose vein, the common trunk entering the superior ophthalmic vein farther back than any other tributary; in only 10 per cent. of cases did the vessel enter the cavernous sinus separately.

The **anastomoses** of the orbital system of veins are of the greatest importance from a clinical point of view, and have been specially studied in this respect by Krauss (1910) and from the anatomical point of view by Sesemann (1869), Gurwitsch (1883), and Festal (1887). They communicate anteriorly with the superficial veins of the face and posteriorly with the intra-cranial system through the cavernous system, inferiorly with the deep veins of the pterygoid plexus, medially with the veins of the nasal cavity through the ethmoidal vessels and plexus round the lacrimal sac, laterally with the temporal veins by venules passing through the lateral wall of the orbit; no venules pass through the roof. The superior ophthalmic vein is the largest channel in the adult between the veins of the face and those inside the skull.

As regards the *direction* in which the blood of the orbit flows, if, as is generally stated, no valves exist in any of the branches of the ophthalmic veins, the blood can pass in either direction forwards or backwards, and they form important emissary veins connecting the cavernous sinus with the facial system; Merkel, Gurwitsch, Dwight, Krauss, and Festal, however, state that valves do occur in these branches of junction with the facial veins just outside the orbit, and are so disposed as to prevent the flow of injections from the face backwards into the orbit; Sesemann, however, made his injections through this route; Festal, like Merkel, found that the temporal and pterygoid plexus could be engorged with an injection mass with but little penetration into the orbit. The valves, as elsewhere in the body, are doubtless inconstant, and tend to become inefficient in old age. Probably the orbital blood passes both forwards into the facial system along with the blood from the eyelids, although contraction of the orbicularis would tend to impede its passage, and backwards into the cavernous sinus, this last being the normal direction according to most authors, a conclusion supported by clinical and pathological observations, though the posture of the head may be a factor in determining its course; incidentally, the movements of the ocular muscles must aid the circulation. But it is to be noted that the ultimate channel formed by the ophthalmic vessels on reaching the sinus is small compared with the size of the numerous radicles they collect, and, moreover, usually exhibits a marked narrowing at this point; also that the cavernous sinus is not only largely occupied by the internal carotid artery, but such space as is left is filled by trabecular tissue, so that the blood-flow through it must be sluggish. On the other hand, as noted on p. 170, the common appearance of a prominent frontal vein during laughter or coughing when the intra-cranial pressure is raised would point to a normal flow of at least part of the contents of this vessel into the orbital system and backwards to the sinus. It seems reasonable to conclude, therefore, that the orbital venous system consists of a network of vessels which can drain either forward and freely into the facial vein, as Sesemann considers the greater part does, or backwards into the cavernous sinus, or inferiorly and least freely into the pterygoid plexus and through other minor channels; the result, thus attained by the anastomoses between the vessels of the orbital system itself and by the multiplicity of paths along which the blood can drain away, being to facilitate the ocular circulation.

It may be noted that the bulk of the veins lie in the section of the orbit behind the globe, a disposition that accounts for the general formation of angiomata in the posterior rather than in the anterior part of the orbit, as illustrated by a case of Byers (1924).

The development has been studied by Dedekind (1909), from whose reconstruction of the embryonic network in the orbit which gives rise to the orbital veins it is not surprising to find such variations and anastomoses as do occur in the adult.

The **variations** and anomalies of the orbital veins are numerous and, as elsewhere in the body, less important than those of the arteries, but some are of clinical interest. The inferior may alone form the chief vein of the orbit and anastomose with the angular; the superior has been found to join the middle cerebral (Sylvian) vein of the brain (Hyrtl); the ethmoidal veins may enter the cavernous sinus (Sesemann), or communicate through the lamina cribrosa with the veins of the olfactory tract (Zuckerkandl). Stanculeanu (1902) states that he has sought in vain for the venule described by Gaillard as draining from the antrum through the orbital floor into the ophthalmic system. See also p. 32.

The **cavernous sinus** (*sinus cavernosus*), to which reference has been made, here calls for description not only from its connexion with the ophthalmic veins, but from the fact that, apart from the optic nerve itself, all the nerves entering the orbit pass either through or in close proximity to it. It is generally considered to be formed, like the other intracranial blood sinuses, by a local separation of the two layers of which the dura mater lining of the cranium is composed, the outer endosteal layer closely following the contour of the bone, the inner meningeal layer forming the lateral wall of the sinus, whilst the space between them is lined by an endothelium continuous with that of the veins with which it communicates. Langer (1885), in a special study of the sinus, distinguishes two regions, a superficial more reticular one containing the third and fourth nerves, and a deeper more spacious one nearer the bone. Ferron (1913), however, has more particularly studied the constitution of the lateral wall of the sinus, and considers it to be formed by two layers, a thin deeper one, which is a fascial continuation and enlargement of the carotid sheath, and a thicker superficial layer continuous with the dura lining the middle cranial fossa; between the two the nerves pass, carrying with them sheaths formed by prolongations of the dura as they pierce it. Lockhart (1925, 1927), in describing the relations of the Gasserian ganglion, finds that Meckel's cave, in which it is lodged, sends a prolongation of its dural wall into the cavernous sinus to surround the ganglion and the three divisions of the nerve and separate them from the other contents of that space; the lateral wall of this extension, however, fuses with that of the sinus, but can be separated from it, as indeed it is behind, by the entrance of the superior petrosal sinus. Lockhart points out that an incision through these fused lateral walls could give access to the ganglion without tapping the sinus. An interesting point is the closer relation of the maxillary and mandibular nerves to the sphenoidal sinus as contrasted with that of the ophthalmic nerve. The description in some respects is comparable to that of Ferron, and is illustrated in Fig. 171.

The cavernous sinus is situated on the side of the body of the sphenoidal bone, and is therefore related on the one hand to the air-sinus which hollows out that bone (with the hypophysis cerebri or pituitary body resting above it, as seen in Fig. 170), and on the other to the middle cranial fossa (Fig. 177), in which is lodged the temporal lobe of the brain; it extends from the medial and widest part of the superior orbital fissure in front to the apex of the petrous portion of the temporal bone behind, where it is in close relation to the medial edge of the semilunar (Gasserian) ganglion of the fifth nerve, and has an average length of 20 mm. or $\frac{3}{4}$ inch and a width of 10 mm.; it is wider in front than behind, and narrower below than above. Its lumen is traversed by numerous interlacing trabeculae and loose fringes of connective tissue derived from its walls, so that it appears in section filled with a spongy tissue, and it is still further reduced in size by the winding passage through it of the internal carotid artery and abducent nerve. The artery enters the sinus below, on emerging from the carotid canal; it first ascends towards the posterior clinoid process, passes forwards along the side of the body of the sphenoidal bone, and finally curves upwards and pierces the dura forming the roof of the

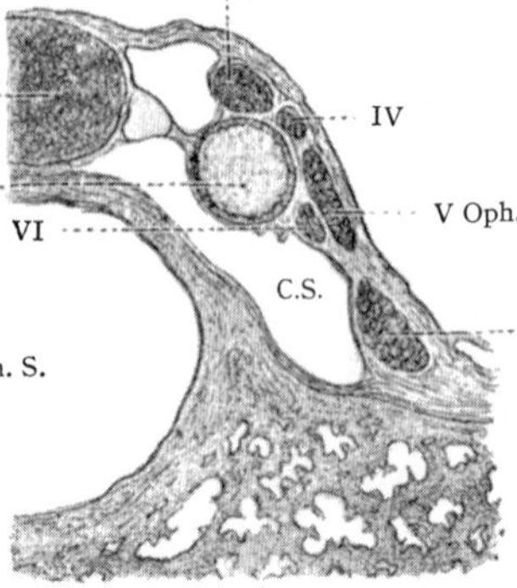

FIG. 170.—Transverse section through the **Cavernous Sinus** (C.S.), the Hypophysis Cerebri or Pituitary Body (H.), and the Sphenoidal Air Sinus (Sph.S.). (From Langer, 1885.) C.A. is the internal carotid artery.

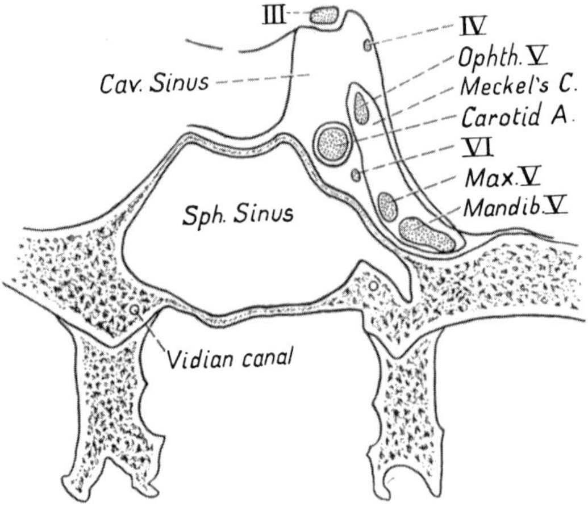

FIG. 171.—Coronal section through the **Cavernous Sinus** to show the prolongation in outline of the Cave of Meckel round the trigeminal divisions. The outer position of the III nerve is explained on p. 323. (Modified from Lockhart, 1925.)

cavernous sinus on the medial side of the anterior clinoid process; in this cavernous part of its course it is surrounded by a sympathetic nerve plexus and gives off minute branches to the walls of the sinus, the hypophysis and the semilunar ganglion. The abducent nerve crosses the lateral side of the artery, lying free in the sinus in the first part of its course but being adherent to the lateral wall towards its exit; in a minority of cases it is united to the wall in its whole course by a short fibrous 'mesentery'. The variations in position of artery and nerve are figured by Testut.

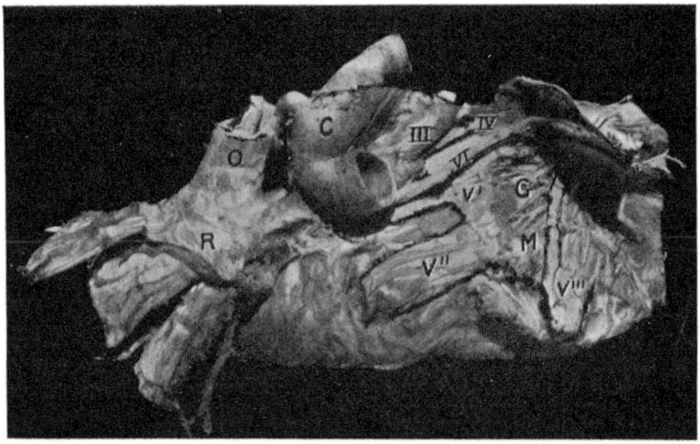

FIG. 172.—Dissection of the **Lateral Wall of the Cavernous Sinus** and its contents, viewed from the medial or inner side. The soft parts were stripped from the side of the body of the sphenoidal bone and the cavernous tissue cleared away. × 1½.

O. = optic nerve; R. = origin of the recti muscles; C. = the internal carotid artery, giving off the ophthalmic artery; III, IV, V', V''. V''', VI = the nerves; G = semilunar ganglion; M lies just to the left of the motor root of V nerve.

Embedded in the lateral wall of the sinus are the oculomotor (third) and trochlear (fourth) nerves, and the ophthalmic and maxillary divisions of the trigeminal (fifth) nerve, in that order from above downwards (Figs. 170, 172, 178, and 185). It is to be noted that at the level taken in Lockhart's drawing (Fig. 171) the third nerve is excluded from the sinus, since it does not enter until after it has dipped beneath the anterior clinoid process, as is well displayed in Fig. 177.

The cavernous sinus is joined to its fellow on the opposite side of the sphenoidal bone by a pair of transverse inter-cavernous sinuses placed anterior and posterior to the sella turcica; the anterior is the larger, and the posterior is often absent (twenty-six times out of forty-four cases, Knott, 1882); these, together with the portions of the cavernous sinus between their ends on either side, form a circular or coronary sinus round the pituitary body, into which its veins pass. Farther back, the two cavernous sinuses are connected by the basilar or transverse occipital sinus or plexus.

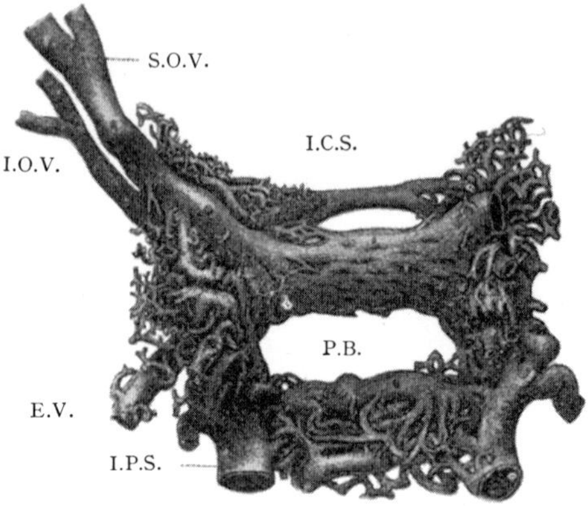

FIG. 173.—Corrosion preparation of the **Cavernous Sinuses** of a child. (From Langer, 1885. Note that the vessels are represented hollow, which would indicate a drawing and not a photograph of the actual injection mass.)

S.O.V. = superior ophthalmic vein, and I.O.V. the inferior. I.C.S = anterior inter-cavernous sinus. P.B. = position of pituitary body. E.V. = emissary vein, and I.P.S. the inferior petrosal sinus.

The tributaries and emissaries of the cavernous sinus are numerous. Besides the ophthalmic veins, there enter it the spheno-parietal sinus of Breschet (which runs along the margin of the lesser wing of the sphenoidal bone and sometimes joins the superior ophthalmic vein), veins from the neighbouring dura, and some of the veins from the muco-periosteal lining of the sphenoidal air-sinus which have pierced its bony wall. The middle cerebral vein or vein of the Sylvian fissure enters either the cavernous sinus or its tributary, the spheno-parietal sinus. The blood is drained away posteriorly by the two petrosal sinuses into which the cavernous sinus divides at the apex of the petrous bone; the superior petrosal sinus is the smaller, and runs along the upper border of that bone to enter the transverse (lateral) sinus, where it bends downwards to become the sigmoid sinus; the inferior petrosal sinus passes along the petro-occipital suture to the jugular foramen and enters the internal jugular vein; part of the blood also drains into the basilar sinus and thence into the anterior spinal plexus of veins. As emissary veins proper of the cavernous sinus are certain small veins leading from its lower part; they pass through (1) the foramen ovale, (2) the foramen lacerum (medium), (3) the foramen of Vesalius when this is present, (4) the foramen rotundum, though very rarely (Knott); all these pass into the pterygoid plexus; and (5) a venous plexus (carotid plexus or sinus), which completely invests the carotid artery as it traverses the carotid canal, and is prolonged downwards to empty into the internal jugular vein.

The **lymphatics** of the orbital cavity are imperfectly known in man. As elsewhere in the body, the system no doubt consists of closed spaces which are peri-vascular in position and into which lymph exudes from the blood, closed capillaries into which the fluid is then absorbed, and vessels. There are no lymphatic nodes, however, in the orbit, although the very rare occurrence of lymphangiomata (Birch-Hirschfeld, 1930) might indicate the presence of such tissue even in minute quantities. Definite lymph vessels can be demonstrated in the eyelids, conjunctiva, and lacrimal gland, as described on p. 174, but as regards the orbit itself, although Schwalbe described Tenon's capsule as enclosing a definite lymph space lined with endothelium and continuous with a supra-vaginal space around the optic nerve (denied by subsequent observers; see p. 291), and others have considered the inter-lobular spaces of the orbital fat to be likewise lymph channels (Birch-Hirschfeld, 1909), such intervals do not show any special structural development, though, like similar spaces elsewhere in the body, they may contain plasma exuded from the blood-vessels. Leboucq (1914) has studied the orbital lymphatics in the rabbit, and concludes that the sub-conjunctival space and that of Tenon's capsule are independent of the lymph circulation of the globe. The lymph from the latter passes out round the anterior ciliary and vorticose veins and the central vein of the retina, and eventually into the jugular lymph trunks (Bartels, 1909). The lymph vessels of the orbit are supposed to pass through the inferior orbital fissure to the internal maxillary nodes, and thence to those of the upper deep cervical groups, and others may pass through the superior orbital fissure; communications may also be assumed to exist between the orbital lymphatic system and those of the nasal cavity and accessory sinuses through the ethmoidal canals, though they have not been demonstrated in man. Channels also pass from the infra-orbital canal through the superior alveolar canaliculi to the teeth (p. 84, and p. 8, Fig. 5), and lymph vessels may accompany the vessels and nerves contained therein (Dewey, 1920).

It may be of interest to add that a paper on the technique of injecting lymph vessels has been written by Gerota (1896).

5. THE NERVES

The nerves which enter the orbit are as follows:

The optic nerve, the special nerve of sight;

The oculomotor, trochlear, and abducent (third, fourth and sixth) nerves, which are motor to the muscles of the eye and the levator of the upper lid;

The ophthalmic division of the trigeminal (fifth) nerve, which is the nerve of common sensation to the eyeball, and supplies also the lacrimal gland, conjunctiva and eyelids, but which in greater part traverses the orbit to reach facial areas beyond;

The maxillary division of the trigeminal, which runs beneath the periorbita of the floor and gives off one branch into the orbit, but is likewise mainly distributed to extra-orbital regions;

The sympathetic plexus of nerves.

As in the case of any other cerebral nerve, an orderly consideration must include these points:

(i) The 'nucleus of origin'; that is, the mass of nerve cells from which the axons forming the nerve trunk arise, sometimes called its 'deep origin'. In the case of a motor nerve, the nucleus lies buried in the mid-brain, where it is connected with other nuclei and parts of the brain by fibre tracts; in the case of a sensory nerve, such as the trigeminal, the nucleus of origin of its fibres (in this case the semilunar or Gasserian ganglion) lies outside the brain and is connected to it by centripetal fibres, which enter and terminate in a 'nucleus of reception'. After a short intra-cerebral course, the nerve emerges at its

(ii) 'superficial origin' on the surface of the brain; from this point, namely its superficial origin in the one case or point of entry in the other, the nerve has an

(iii) intra-cranial course inside the skull between the brain and dura mater. In the case of the orbital nerves other than the optic nerve itself they then enter the

(iv) cavernous sinus, and pass through the superior orbital fissure (comparable to the point of emergence from the skull of other nerves) to enter upon the

(v) orbital part of their course, and

(vi) final distribution.

Fig. 174 represents schematically the course of a typical orbital nerve showing these points; and it will be found that in each part of the course so outlined a nerve is related to structures an alteration in the disposition of which can become of great clinical importance.

It will be convenient to describe the optic nerve last (p. 372), so that its connexions may be followed into the cerebral visual tracts to which the last section of this work is devoted. In that section there will also be found an account of the nuclei of origin and the intra-cerebral connexions of the nerves. Here, in dealing particularly with the orbit, the nerves will be described from their superficial origin from the brain surface onwards.

The **Oculomotor** or third nerve (*nervus oculomotorius*) supplies all the extra-ocular muscles with the exception of the lateral rectus and superior oblique; it also sends fibres through the ciliary ganglion to supply the ciliary muscle and sphincter pupillae within the eyeball.

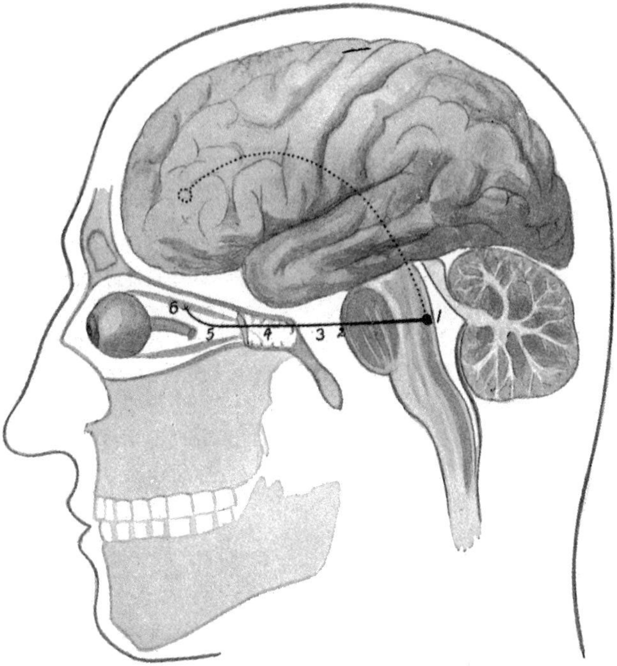

FIG. 174.—Scheme to show the **course of a typical Orbital Motor Nerve.** The fibres arise from a deeply-seated nucleus of origin (1), and after a short intra-cerebral course the nerve emerges from the surface of the brain stem at its superficial origin (2); it then traverses the cranial cavity (3), passes through the cavernous sinus (4), and after entering the orbit (through the superior orbital fissure) passes along it (5) to reach its final distribution in a muscle (6). The possible cortical connexion of the nucleus of origin is indicated by the dotted line.

The superficial origin on the surface of the brain is from the shallow oculomotor sulcus on the medial surface of the cerebral peduncle, in front of the pons and in the interpeduncular space, on the floor of which is the posterior perforated substance or spot (Fig. 175); there are usually two groups of fibres at the origin, separated very often by an arteriole derived from the posterior cerebral artery.

In the intra-cranial part of its course, from this origin to where it pierces the dura to enter the cavernous sinus, the nerve lies in the posterior cranial fossa, and is about 25 mm.

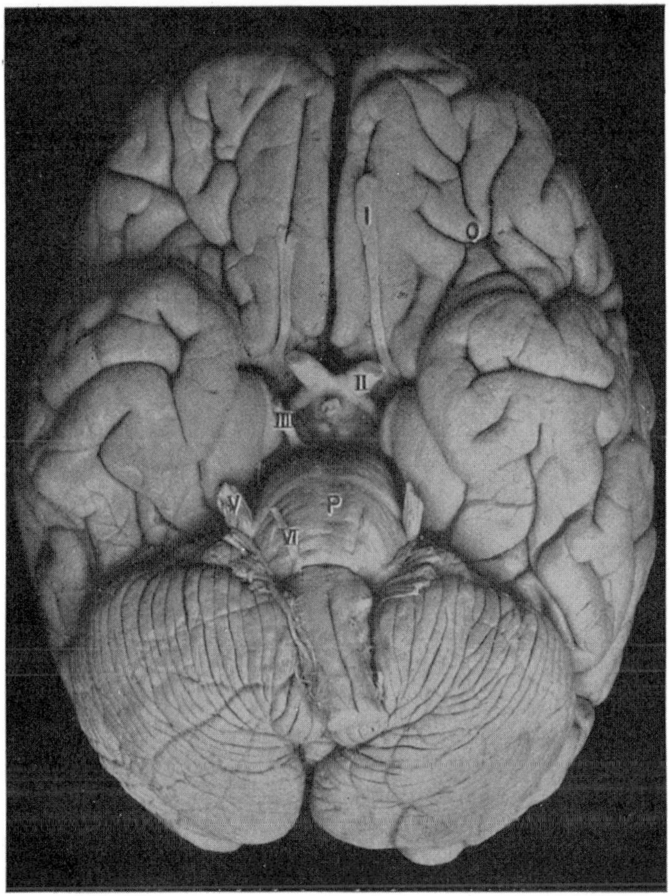

FIG. 175.—The Brain, inferior aspect, to show the **superficial origins of the nerves** (as numbered). The left (III) nerve has been removed; the VII, VIII, and *pars intermedia* are seen just below V, followed caudally by the IX, X, XI; the fourth nerve is seen above V on the reader's left-hand side.

O = orbital sulcus on base of frontal lobe; P = pons.

long; it is closely invested by the pia mater, and traverses the sub-arachnoid space, which is enlarged in this region between the peduncles into a cisterna interpeduncularis, an imperfect subdivision of the larger cisterna basalis; the nerve curves outwards and forwards, between and in close contact with two arteries, the posterior cerebral in front, the

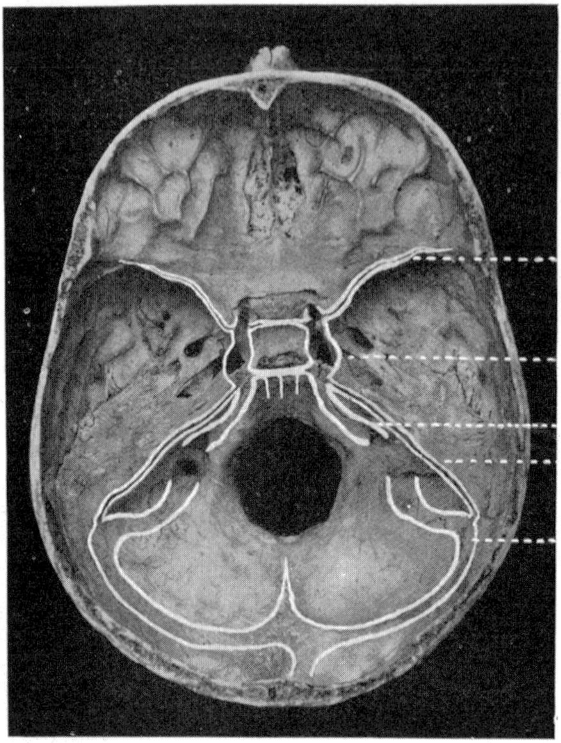

FIG. 176.—The **Base of the Skull** viewed from within, to show the position of the cavernous sinus and its connexions. Compare with Fig. 177.

superior cerebellar behind; the posterior communicating artery then runs parallel with it on its medial side, and it lies against the uncus of the temporal lobe of the brain, as seen in Fig. 182, p. 340; it pierces the arachnoid membrane forming the roof of the cisterna about 20 mm. from its origin.

The nerve enters the cavernous part of its course by piercing the dura to the lateral side of the posterior clinoid process in the small triangular area formed by the free and attached borders of the tentorium (Fig. 177); it here for the first time comes in contact with the base of the skull, lying against the anterior clinoid process, and is now situated in the middle cranial fossa. Its position in the lateral wall of

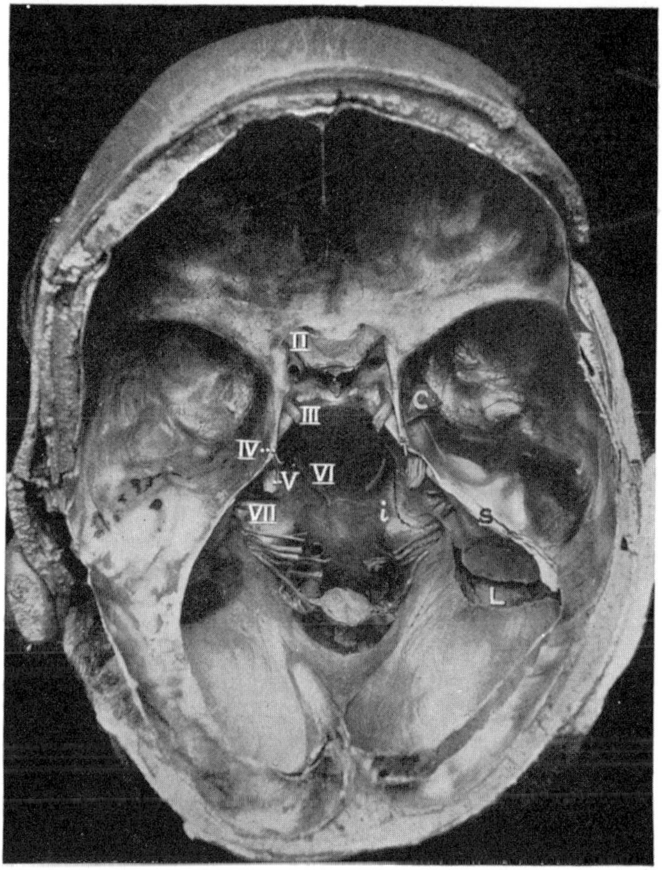

FIG. 177.—The Base of the Skull with the Dura *in situ* after removal of the Brain, to show the **nerve-exits and position of sinuses.** On the right side the lateral wall of the cavernous sinus (C) has been turned down as a flap. The superior (S) and inferior (*i*) petrosal and lateral or transverse (L) sinuses have been opened up. Compare with Fig. 176. × $\frac{2}{3}$.

the cavernous sinus has already been referred to, and its relation to the other nerves is well displayed in the dissection photographed in Fig. 178, as also in Figs. 170 and 185, p. 348.

It is to be noted, however, that the order of the nerves

changes in the forepart of the sinus: the ophthalmic nerve becomes superior, the oculomotor nerve passes below and medial to the trochlear, and the abducent nerve is now in intimate contact with the lateral wall of the sinus (Fig. 178).

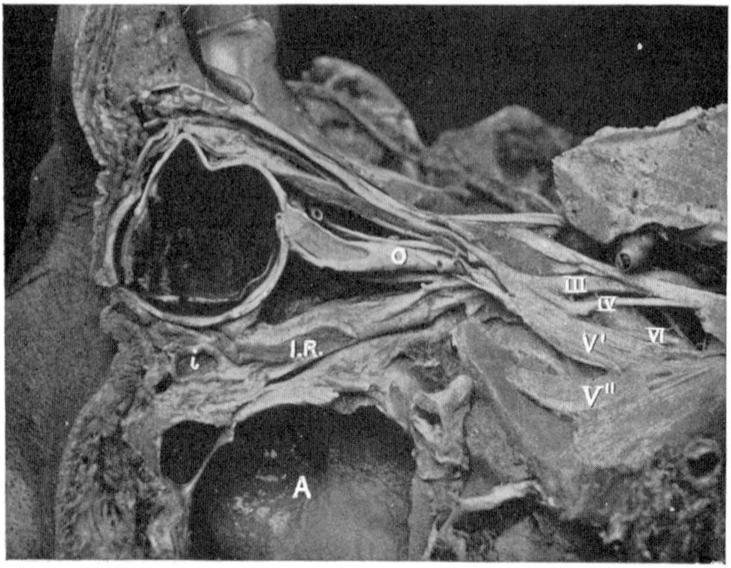

FIG. 178.—Sagittal section through Left Orbit; the lateral wall of the **cavernous sinus** has been removed to expose the **nerves**. The maxillary sinus (A) is cut open, and presents a well-marked pre-lacrimal recess in its fore and upper corner; the ostium is also visible. Natural size. Compare with the scheme shown in Fig. 73, p. 147, and the next Fig. 179 of the nerves.

O = optic nerve, cut to show its sheath; I.R. = inferior rectus; *i* = inferior oblique, with the 'suspensory ligament of Lockwood' immediately behind it.

The oculomotor nerve leaves the forepart of the sinus and enters the orbit through the superior orbital fissure. At or about this point it divides into two branches, superior and inferior, which enter the orbit through the oculomotor foramen, that is, inside the annulus of Zinn and within the cone of muscles, in close company with, but separated by, the naso-ciliary branch of the trigeminal, with the abducent nerve lying below. The position of these nerves relative to one another at this point, upon which considerable and unwarrantable stress is laid in the text-books, is illustrated on

p. 262, which should be compared with Figs. 141, 143, 147, and the fourth section in Fig. 198, p. 381, from which it will be realized that practically they are all crowded together here, compactly filling a narrow space certainly not more than 3 mm. in diameter.

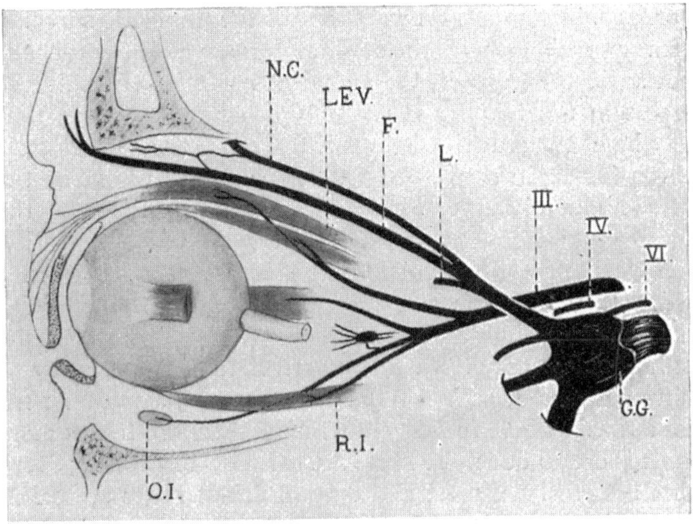

FIG. 179.—Drawing of the **Ophthalmic and Oculomotor Nerves,** composed from the dissection in Fig. 178.

N.C., F., L. = the Naso-ciliary, Frontal, and Lacrimal nerves respectively. LEV. = Levator muscle, R.I. = inferior rectus, O.I. = inferior oblique. The Oculomotor Nerve (III) is obviously seen in its two divisions, the upper supplying the superior rectus and levator, the inferior supplying the medial and inferior recti and inferior oblique (from which last a twig passes as the 'motor-root' to the ciliary ganglion). G.G. is the Gasserian or semilunar ganglion.

Within the orbit, the branches of the oculomotor nerve diverge; the superior and smaller of the two (*ramus superior*) passes upwards over the lateral aspect of the optic nerve to enter the ocular surface of the superior rectus, after supplying which it ends in the overlying levator palpebrae superioris muscle. The inferior and larger branch (*ramus inferior*), after supplying the medial and inferior recti, passes forwards beneath the optic nerve, between the inferior and lateral recti and closely bound to the former, to reach the posterior border of the inferior oblique, in which it ends. It contributes a short thick ganglionic branch, sometimes double, which forms the motor root of the ciliary ganglion (p. 364), whereby fibres are conveyed to the intrinsic muscles of the eyeball. The course and branches of the nerve are illustrated in Fig. 179.

As regards the CONSTITUTION and nature of the fibres of the oculomotor nerve, it is noteworthy that its size is remarkable in comparison with that of the muscle it supplies, a feature it shares with the other motor nerves of the eyeball; indeed, in this respect, the abducent nerve is relatively the largest in the body, and the only other motor nerve that may be compared with it in this respect is the hypoglossal; in the full-time foetus, the sixth is actually as large as the inferior gluteal nerve, whereas the relative weights of the muscles supplied is as 1 : 74, and there seems to be a correlation between the size of nerves and the precision of movement for which they are responsible (Stibbe, 1929). In the ocular muscles the function appears to be correlated not so much with rapidity as with a finely graded range of delicacy of movement; and, as Wood Jones points out, a muscle need not act as a whole to perform a minimal action, but as more effort is demanded more fibres are brought into action; every fibre acts to its utmost and only sufficient fibres are called into action to perform the demanded task. The oculomotor contains 15,000 fibres distributed to about 40,000 muscle fibres (that is, a proportion of 1 : 2·7; for the sixth nerve it is 1 : 1·8, and for the fourth 1 : 1; Quain, Macalister, 1889); its fibres are generally large, but it also contains smaller ones destined for the ciliary ganglion.

Whilst the great majority of its fibres are motor in function, it also contains a certain number of *sensory* fibres derived from the ophthalmic division of the trigeminal nerve, by an anastomosis usually considered to occur whilst the nerves are in the cavernous sinus. Barratt (1901) in a study of transverse sections of the nerve failed to find the communication, and Bischoff (1865) denied its existence. Stibbe (1929), however, has identified by dissection and serial sections this communication from the ophthalmic not only to the third but also to the fourth and sixth nerves (Fig. 180). Further, Sherrington and Tozer (1910) have shown experimentally

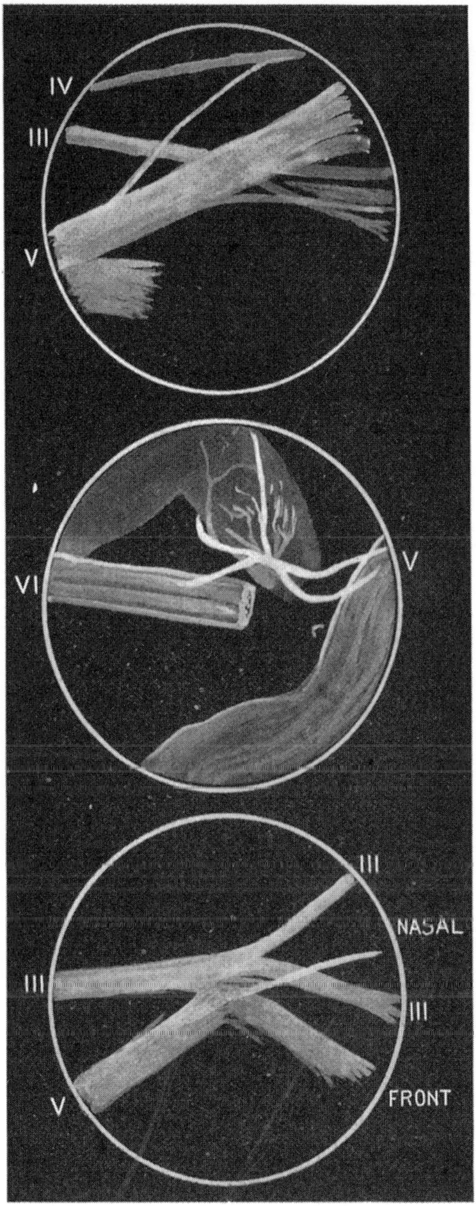

FIG. 180.—Micro-dissection drawings of the **Communications of Fifth Nerve** to the fourth, sixth, and third nerves within the cavernous sinus. (Original drawings by E. P. Stibbe.)

that all three motor nerves of the eyeball contain some sensory fibres as well as the motor ones, not derived from the fifth nerve but incorporated with the eye-muscle nerves themselves, and the former discusses the sensual role of the *proprioceptive* nerve supply of the extrinsic ocular muscles relative to the adjustment and maintenance of postures of the eyes in a later paper (1918). Woollard (1931), who regards the mode of innervation of the eye muscles as a peculiarity not found elsewhere in the trunk or limb muscles, found two types of nerve fibres supplying the former: (i) thick and medullated, with motor terminal plates connected with thick muscle fibres, and (ii) thin mostly non-medullated, with claw-like endings related to thinner muscle fibres; the latter were considered by Boeke (1915, 1927) to be sympathetic nerve fibres derived from the superior cervical ganglion, but Woollard, like Wilkinson (1929) and Hines (1931), considers them sensory (proprioceptive) in function, and definitely traces their origin from the mesencephalic root of the fifth nerve (Fig. 212, p. 415). The presence of these afferent fibres explains the fact that after total desensitization of the eyeball by means of cocaine, the ocular movements, the precision of which requires the co-operation of afferent impressions from the muscles, are carried out as accurately as in the normal animal (Starling, 1920). In primitive animals these nerves are clearly mixed, containing both motor and sensory fibres.

Gaskell (1889), like Thomsen, described the remains of a degenerate ganglion in the root of the oculomotor as well as of the trochlear and the abducent nerves, which from a study of ancestral development he regards as representing an anterior root ganglion; Tozer (1912) and Nicholls (1915) made similar observations, and Nicholson (1924) confirms these facts in the human foetus, demonstrating in the third nerve four ganglia (30 cells), three of which are on the branch to the inferior oblique, one on that to the medial rectus, but none on the branches to the inferior and superior recti, and on the sixth nerve one ganglion of 30 cells (Fig. 181). Stibbe (1929) found ganglion cells on the communications between the ophthalmic and the third and sixth nerves. Woollard considers that these cells might be scattered

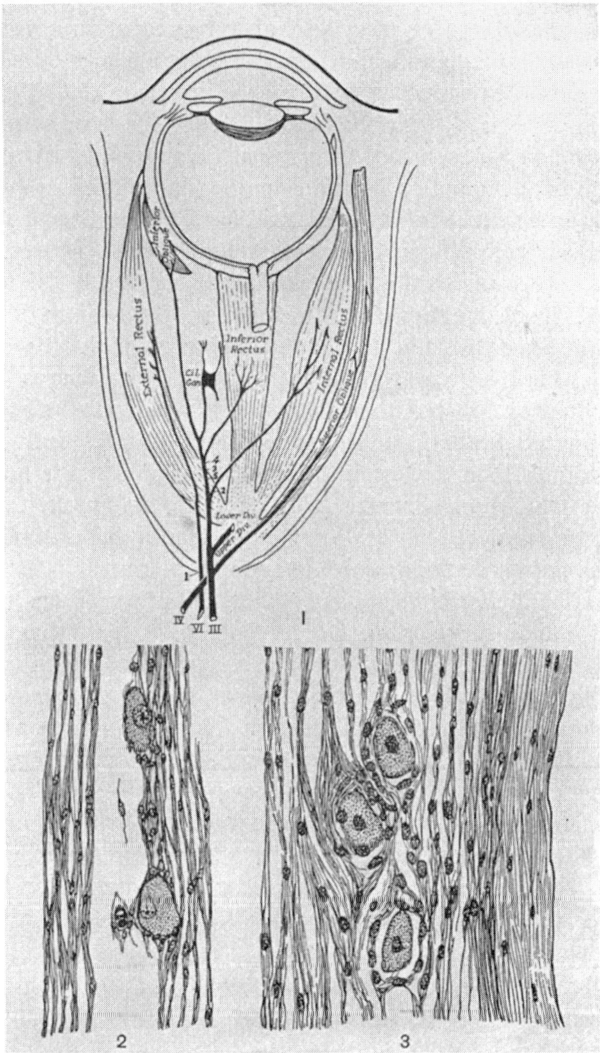

FIG. 181.—Schematic projection of the lower half of the human orbit, showing (1) the relative positions of the **Ganglia on the Eye-muscle Nerves,** 1, 2, 3, 4, 5;
(2) ganglion cells on root of III nerve in relation to the fasciculi;
(3) ganglion 3 in figure 1 on III nerve, showing the ganglionic group amongst the nerve fibres surrounded by a network of richly nucleated connective tissue. (From Helen Nicholson, 1924.)

members of the autonomic ganglia stranded in their peripheral migration, or more probably represent in a variable degree the ganglia of somatic motor ocular nerves present in the embryo but tending to disappear in the adult.

The third nerve, like the fourth and sixth, is said to receive *sympathetic* fibres from the cavernous plexus whilst traversing the sinus, although Wilkinson denies that there is a sympathetic innervation of striated muscle. The opinion of Boeke has been referred to above. Bischoff (1865) figures such fibres as passing to the third and sixth, but not to the fourth nerve. Koch (1916) identified by gross dissection in dog and man at least two fine branches passing from the cavernous sympathetic plexus to the third nerve, but microscopical examination showed them to leave it again; he traced unmyelinated fibres from the same source to the sixth nerve and found them present in the fifth, but could trace none to the fourth, though Boeke (1913) stated that in the cat the superior oblique muscle does receive sympathetic fibres.

As regards the *localization* of eye-muscle fibres in the trunk of the oculomotor nerve, it is known that the fibres to the levator and superior rectus run on the surface, thus being easily affected by a basal meningitis with consequent ptosis; and lately Ingvar (1928) has shown that in the macaque monkey the fibres from the inferior oblique run on the ventral surface of the nerve, and those from the medial rectus and ciliary ganglion in the centre. It would seem, since in some other nerves the surface fibres are oldest in development, that the original function of the oculomotor nerve was to raise the eyeball upwards, and the arrangement of fibres corresponds with the theory that accommodation and convergence are late acquisitions.

The *nerve-endings* of the motor nerves in these muscles, which differ by the richness of their ramifications from those of the skeletal muscles, were studied by Levinsohn (1901), and those of the sensory nerves by Dogiel (1906); Boeke (1915) found the sympathetic nerve-endings to be the same in smooth as in striated fibres; the observations of Woollard have been referred to above. The *myelination* of the fibres is described by Lucas Keene (1931). The *morphology* of the nerves was studied by Neal (1914), and that of the motor nuclei by Davidson Black (1917–22).

A note upon the possible LESIONS of the oculomotor nerve and their effects may be added as emphasizing its anatomical relations and distribution. A paralysis may result from injury to the nerve at its origin or in any part of its course, from cerebral disease, embolism, periarteritis, gumma or tumour, pressure of inflammatory exudate, periostitis, thrombosis of the cavernous sinus, and from fracture of the base of the skull or of the orbit. It may be affected by an aneurism or syphilitic periarteritis of the internal carotid, basilar, posterior cerebral, or superior cerebellar arteries, to all of which the nerve is closely related in its intra-cranial and cavernous course. In this respect the following notes on the arteries of the pons and medulla may be of interest, taken from Stopford's paper (1917), in which an extensive bibliography will be found.

Aneurism of either the posterior cerebral or superior cerebellar arteries may easily cause oculomotor paralysis, since the nerve lies in the angle formed by these two vessels at their origin from the basilar, a relation, it is to be noted, shared, though not so intimately, by the fourth nerve. It is also possible, as Cushing has shown, for the posterior cerebral, which has a more constant anatomical disposition, to indent the third nerve in cases of cerebral tumour accompanied by any considerable alteration in the position of the brain stem. The basilar artery itself has caused a paralysis of the nerve. Beadles, in his collection of 555 cases of aneurism of the larger cerebral arteries, found that the oculomotor was the nerve most frequently compressed by aneurisms of the posterior communicating artery (the vessel which joins the posterior cerebral to the internal carotid and so completes laterally the arterial circle of Willis, as seen in the figure following), and that dilatation of this vessel clinically produced mechanical symptoms more frequently (47 per cent.) than any other. Aneurysm of the internal carotid artery has frequently caused paralysis of this nerve, and an associated involvement of the trochlear, abducent, and ophthalmic division of the trigeminal would point to the localization of the lesion in the cavernous sinus. Finally, it has been suggested that cases of recurrent oculomotor paralysis may be explained by vascular spasm or paralysis of the vaso-motor nerves of these vessels, causing their distension and consequent compression of the nerve.

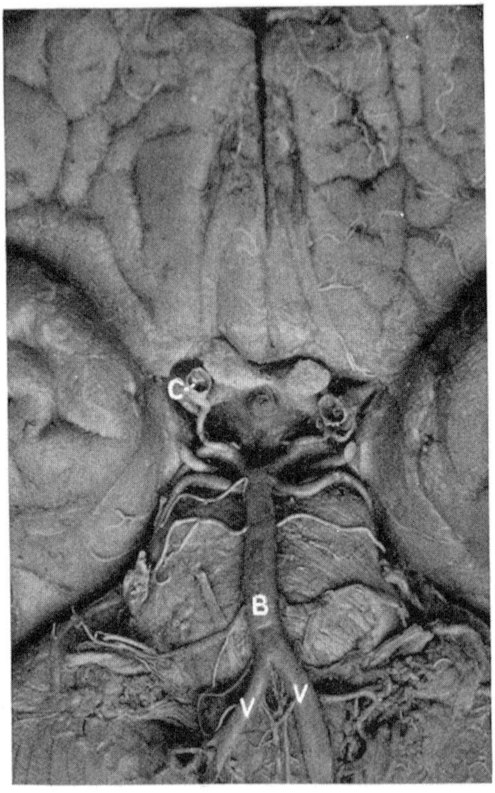

FIG. 182.—The base of the Brain, showing the **relations of the arteries to the nerves.** Natural size.

V, V = the vertebral arteries. B = the basilar; this gives off on each side (opposite B) an anterior inferior cerebellar branch which is seen crossing the sixth nerve, and, in front, the large superior cerebellar artery; immediately afterwards it divides into the two posterior cerebral arteries, each of which is joined to the internal carotid artery (C) of the same side by the posterior communicating artery. The relation of the third nerve to the superior cerebellar and posterior cerebral arteries is well displayed, and the fourth nerve on the reader's left can be seen to present a similar relation to these vessels.

As regards its relation to the base of the skull, the oculomotor nerve is in close contact with the bone at the posterior clinoid process, as seen in Figs. 177 and 178, a position which Charamis (1928) considers exposes it to compression by a hematoma in a fractured base; it is related to the bone also in its passage through the superior orbital fissure, a relation it shares with the abducent and the naso-ciliary nerves.

A part only of the nerve may be affected, but if complete, such a lesion of the oculomotor nerve will cause:

(i) Ptosis, from paralysis of the levator palpebrae superioris. The skin of the eyelid appears abnormally smooth.

(ii) Lateral strabismus, from non-opposition of the lateral rectus and superior oblique muscles. The eye is almost motionless, and cannot be turned upwards, directly downwards, or inwards, though it can be moved down and outwards.

(iii) Diplopia, the chief symptom complained of, is caused by deviation of the affected eye.

(iv) Dilatation of the pupil, with no reaction to light, from paralysis of the sphincter pupillae.

(v) Loss of accommodation from paralysis of the ciliary muscle of the lens.

Variations of the Oculomotor Nerve.—These generally consist of abnormal branches replacing other orbital nerves; for example, the oculomotor may supply the lateral rectus, either in addition to, or even instead of, the sixth nerve (Generali, Harvey), or send a filament to the superior oblique; Martyn (1919) reports a clinical case in which the symptoms suggest an anomalous innervation of the levator by the same inferior branch of the third nerve which supplies the medial rectus; it has been seen to furnish a twig to the sphenopalatine ganglion (Poirier). It may anastomose with the sixth nerve, or the superior division may anastomose with the naso-ciliary. The branch to the inferior oblique may send a double supply to the inferior rectus, and has also been found traversing the latter, and even perforating the ciliary ganglion. It has been seen perforated near its origin by the posterior cerebral artery (Soemmering).

The **Trochlear** or fourth nerve (*nervus trochlearis*), the smallest of the cerebral nerves, supplies only the superior oblique muscle of the globe.

It emerges by separate rootlets at its superficial origin on the dorsal aspect of the mid-brain, just below the inferior colliculus at the side of the frenulum veli; it is the only motor nerve of the cerebro-spinal axis which arises from the dorsal aspect.

Its intra-cranial course is the longest of any of the nerves (40 mm.), and its total length from its superficial origin to the muscle is 85 mm.; it crosses the brachium conjunctivum cerebelli and winds round the cerebral peduncle immediately above the pons, as is seen on pp. 259 and 348. Towards the end of this part of its course it lies in the sub-arachnoid space between and parallel to the posterior cerebral and superior cerebellar arteries, and appears on the base of the brain in the interval between the temporal lobe of the brain laterally and the edge of the pons medially (Fig. 175). It then pierces the dura just beneath the free margin of the tentorium cerebelli, behind the third nerve and the posterior clinoid process, as shown on p. 331, and enters the cavernous sinus. Here it lies in the lateral wall, below the oculomotor nerve and above the ophthalmic nerve, but towards the fore end of the space it crosses and is bound to the third nerve on its lateral side, as seen in Figs. 178 and 185, then comes to lie above this nerve and on the same horizontal plane as the branches of the ophthalmic nerve; in this part of its course it is considered to receive communications from the cavernous plexus of the sympathetic nervous system, and it is joined by a sensory twig from the ophthalmic division of the trigeminal. It enters the orbit through the lateral part of the superior orbital fissure, being the only motor nerve of the eye muscles that enters outside the annulus of Zinn and the cone of muscles.

Its position and short course within the orbit is well displayed in the dissection photographed in Fig. 188, p. 353, where it is seen on the medial side of the frontal nerve, crossing the superior rectus obliquely forwards and medialwards; it lies immediately beneath the periorbita of the orbital roof.

The trochlear nerve consists of about 2,000 coarse fibres, distributed to about the same number of muscle fibres (Macalister, 1889); it shows close to its origin a degenerated ganglion (Gaskell, 1889, but not found by Nicholson). Further details of its constitution have been given in connexion with the oculomotor nerve on p. 334 above. Paralysis of the nerve, the causes of which have likewise been noted above in connexion with the oculomotor, results in loss of function of the superior oblique muscle of the globe, and the patient cannot turn the affected eye down and lateralwards, and diplopia occurs. Little change, however, may be noticed, as the loss of function of the superior oblique may be vicariously performed. The nerve is rarely affected alone. Parsons (1921) describes the comparative anatomy and embryology of the nerve.

As **variations** are recorded a branch to the orbicularis oculi, and communications with the supra-trochlear, infra-trochlear, naso-ciliary, lacrimal, and frontal nerves, which are probably aberrant sensory branches of the trigeminal; it has been seen to pierce the levator on its way to the superior oblique (Shane).

The **Abducent** or sixth nerve (*nervus abducens*) supplies the lateral rectus muscle of the eyeball.

It emerges by numerous and sometimes separate rootlets at its superficial origin on the anterior surface of the hind-brain in the groove between the lower edge of the pons and the upper end of the medulla oblongata, just lateral to the pyramid (Fig. 175, p. 329).

In the first part of its intra-cranial course the nerve is flat and closely applied to the surface of the pons for a distance of about 15 mm., passing beneath and being bound down by the anterior inferior cerebellar artery (Fig. 182), and being contained in the cisterna pontis of the sub-arachnoid space; it is invested by the arachnoid membrane at 15 mm. from its origin, becomes rounded, and pierces the dura opposite the dorsum sellae of the sphenoidal bone, medial to and slightly behind the opening for the fifth nerve, and on the medial side of the inferior petrosal sinus (Fig. 177). It then bends slightly lateralwards through the commencement of the sinus to reach its lateral side, passing forwards through the notch between the apex of the petrous bone and the posterior clinoid process, where it lies under the petro-sphenoidal ligament of Grüber (Fig. 184), and enters the cavernous sinus on the lateral side of the internal carotid artery; here it is closely related to the inferior petrosal sinus, the two passing together into the cavernous sinus through an opening termed 'Dorello's canal', a description of which was given originally by

Gradenigo (1907), and more recently by Vail (1922). Wolff (1928) draws attention to the fact that as the nerve passes up from the posterior cranial fossa it makes a sharp bend forward over the apex of the petrous bone to enter the cavernous sinus (Fig. 183). In the sinus its position relative to the

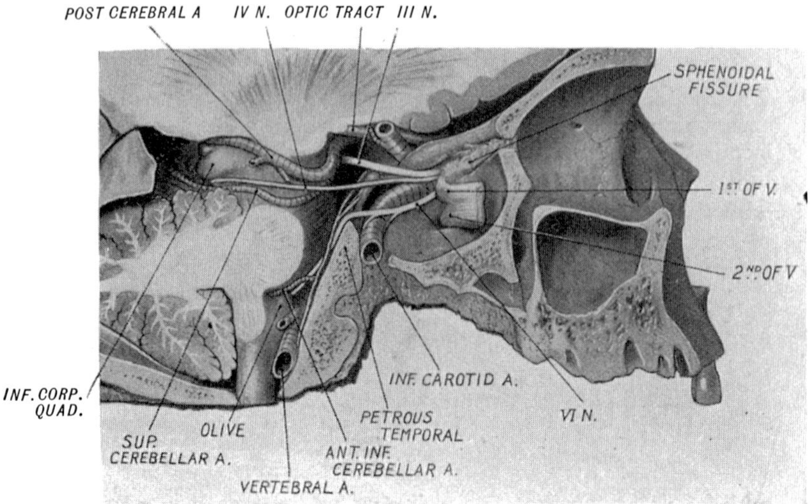

FIG. 183.—Drawing to illustrate the bend in the **Sixth Nerve** as it passes up over the apex of the petrous bone into the cavernous sinus. (From Wolff, 1928.)

artery and the other nerves has already been noted (p. 323, and Figs. 170 and 172); towards the end of its course within the sinus it lies medial to, and is intimately connected with, the ophthalmic nerve. Valentin described an anastomosis between the two nerves in this situation, a fact generally accepted by most authors, and confirmed by Stibbe (p. 334); the nerve is also joined here by filaments from the internal carotid plexus of the sympathetic nervous system; some of these are supposed to supply the dilatator pupillae muscle, since a lesion of the nerve often entails a certain amount of contraction of the pupil (Gray); a branch ascending from the spheno-palatine ganglion to the sixth nerve whilst in the cavernous sinus has been described by Boch and Valentin (Quain), and confirmed by Vitali (1929).

The sixth nerve enters the orbit through the oculomotor foramen, within the cone of muscles, and between the optic nerve and the origin of the lateral rectus muscle (Figs. 142, p. 262, and 147, p. 267); it enters the muscle on its ocular aspect at about the junction of its posterior third and anterior two-thirds, after a total course from its superficial origin of about 55 mm. in length.

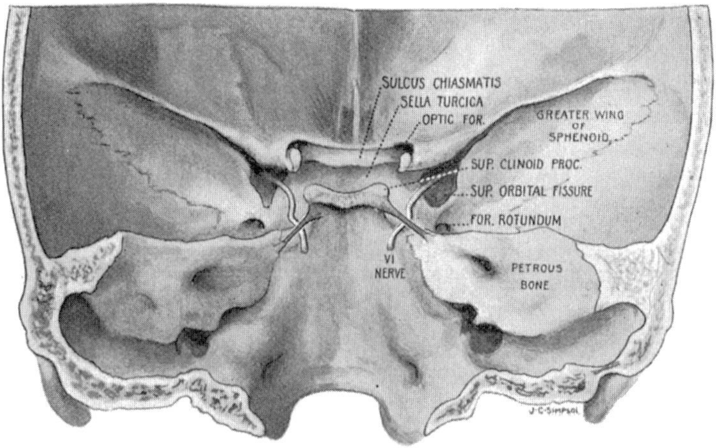

FIG. 184.—The **petro-sphenoidal ligament.** Sketch of the interior of the base of the skull viewed from behind to illustrate a dissection of the abducent or sixth nerve passing beneath the ligament before entering the cavernous sinus. The ligament, which is formed by the deeper or osteo genetic layer of the dura mater, presents differences in its extent and attachments, as is illustrated on the two sides.

Its fibres, mostly large, have been estimated at from 2,000 to 3,600 in number, and are distributed to a muscle containing about 5,000 fibres (Macalister, 1889); the character of the fibres has been discussed above (p. 334).

Paralysis of the abducent nerve is particularly associated with fractures of the base of the skull, owing to its comparatively long course beneath the dura covering the dorsum sellae and its close confinement beneath the petro-sphenoidal ligament. Panas, indeed, first recognized that a fractured base might be manifested by paralysis of the sixth nerve alone. As regards such injuries, Stephenson (1919) states that Hewett, in an analysis of sixty-four cases of fracture of the base, found that in upwards of 82 per cent. the fissure passed through the middle cranial fossa, and that in about one-third of sixty-eight cases examined after death the orbital roof was also found broken; Liebrecht (1906) has specially studied the effect of such fractures on the eyes. Stopford (1917) notes the possible compression of the nerve by the anterior inferior cerebellar artery against the bone before it pierces the dura, and Cushing (1910) its strangulation by lateral branches of the basilar artery in cases of brain tumour; but Wolff thinks that a tumour is more likely to affect the nerve by dragging it against the sharp apex of the petrous bone where it makes the bend described above. The nerve may be affected by infection from the middle ear extending along the carotid tract (Syme, 1924). Accounts of the causes of 'Gradenigo's syndrome' (1907), (abducent paralysis, pain in trigeminal area, with acute suppurative otitis media), together with full descriptions of its anatomical aspects, are given by Wheeler (1918) and Meltzer (1931). In its passage through the cavernous sinus it will be subject to the same affections as noted in the case of the third nerve (p. 339). The results of a paralysis are convergent strabismus and diplopia, and the cornea cannot be moved beyond the centre of the palpebral fissure; the oblique muscles may to some extent counteract the loss of abduction. It is to be noted that a paralysis of the lateral rectus on the one side, accompanied by that of the medial rectus on the other (conjugate deviation), would point to a lesion in the nucleus of origin of the sixth nerve (p. 411). The possibility of dilatator pupillae fibres running in this nerve has already been referred to on p. 344.

As **variations** there are recorded: a case of absence on one side, its place being taken by the third nerve (Generali); Cruveilhier (1886) has seen the origin to be from the olive, and has also described two separate roots of origin which may not unite until they reach the cavernous sinus and between which the artery may pass; this last anomaly has also been seen by Cushing, and is explained by the presence of an aberrant root of the nerve (Stopford, 1916). Krause, on the other hand, has seen the origin placed at 8 mm. above the inferior border of the pons; Bremer (1908, 1921) has written on the aberrant roots and recurrent branches of this nerve in the human embryo. Valentin states that it may receive a branch from the spheno-palatine ganglion or even from the otic instead of from the ophthalmic nerve; it may enter the orbit outside the annulus of Zinn (Poirier). It has been seen to furnish a branch to the superior rectus in addition to the usual supply from the third nerve (Magath, 1919), or it may appear to give off the naso-ciliary nerve (Krause), or a root to the ciliary ganglion.

The **Trigeminal or Fifth Nerve** (*nervus trigeminus*) will be considered briefly before describing those of its branches which enter the orbit. It is the great sensory nerve of the head, collecting impulses from the front of the scalp, face, mouth, nose and anterior two-thirds of the tongue, and it is joined by the motor nerve of the muscles of mastication. It is attached to the side of the pons by a large sensory and a small motor root, as seen in Figs. 175 and 185. Tracing these forwards from the brain stem they are seen in the latter picture to leave the posterior cranial fossa through an opening in the dura just beneath the attached border of the tentorium cerebelli, where the sensory fibres reach the **semilunar** or **Gasserian ganglion**, from the cells of which they actually arise. The ganglion lies in a space, the cave of Meckel (**cavum meckelii**), formed by a separation of the dura, and situated on an impression near the apex of the petrous part of the temporal bone on the floor of the middle cranial fossa (Figs. 177, 185); it lies $2\frac{1}{2}$ inches medial to, but a little higher than, the articular tubercle at the root of the zygoma (Figs. 139, 197). Burr and Robinson (1925) have examined the nature and extent of the cave and find round the root of the nerve and proximal two-thirds of the ganglion an extensive subarachnoid space which communicates freely with the cisterna pontis, whence they consider injections into the ganglion to be fraught with danger. Lockhart (1925, 1927) also studied the dural relations of the ganglion, and described Meckel's cave as being formed by an evagination of the dura of the posterior cranial fossa under that of the middle one, extending forwards into the cavernous sinus to surround and separate the ganglion and the three divisions of the nerve from the other contents of that space; its lateral wall is fused with but can be separated from that of the sinus (Fig. 171, p. 322).

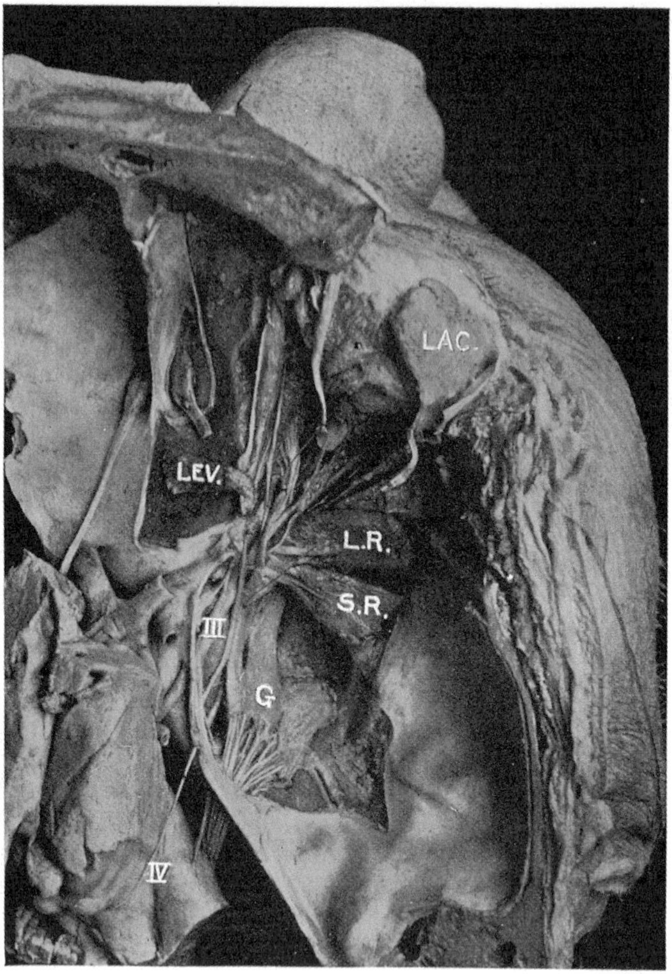

FIG. 185.—Dissection of Orbit, **middle cranial Fossa** and Mid-Brain on right side, to show the **nerves.** The preparation is viewed from above, and has been tilted to the left. Natural size.

G = semilunar or Gasserian ganglion with the fifth nerve entering it behind, and dividing into its three divisions in front; the letter G is in line with the maxillary division and has the mandibular division to its right; on the lateral side of this and marked by a black line is the great superficial petrosal nerve. The lateral wall of the cavernous sinus has been removed to show the nerves. III = oculomotor nerve. The whole course of the trochlear or fourth nerve (IV) is displayed, but the tilting of the specimen makes it appear continuous with what in the cavernous sinus is really the sixth nerve.

From the fore edge of the ganglion three large nerve trunks arise as the three divisions of the fifth nerve; the uppermost is the ophthalmic, the middle one is the maxillary (or superior maxillary), and the lowest and most lateral is the mandibular (or inferior maxillary) nerve. The motor root of the fifth nerve enters the cave of Meckel beneath the main nerve, and passes beneath the ganglion to become a component of the mandibular nerve, as is seen in Fig. 172, p. 323. The arterial supply of the ganglion is derived from the internal carotid and the middle and accessory meningeal vessels; the greater superficial petrosal nerve runs beneath the ganglion to enter the Vidian canal, as seen in Fig. 185. Its morphology has been studied by Frazier and Whitehead (1925) on early human embryos; they find the ophthalmic division more precocious in development than the other two.

A feature of practical importance in the **distribution of the trigeminal nerve** arises from a study of the development of the face. As described on p. 187, this is early formed by five processes, a median frontal-nasal composed of symmetrical right and left halves, with a maxillary and mandibular process on both sides below them; thus three main processes are formed on each side. The three great divisions of the trigeminal nerve grow into these processes as the ophthalmic, maxillary, and mandibular nerves respectively. One would naturally expect fairly definite cutaneous fields of distribution to be formed when the processes have coalesced to form the adult face (Figs. 186, 187), but it is not always realized that each division of the nerve also supplies with sensation deeper structures, such as the mucous membrane lining of cavities and the teeth, which have developed in its own process. Thus behind the frontal cutaneous field are the eyeball, conjunctiva, lacrimal gland and sac, the frontal, ethmoidal, and sphenoidal air sinuses, and the forepart of the nasal cavity, all of which are supplied by the ophthalmic nerve; behind the maxillary field lie the upper teeth and gums, the antrum, the nasal septum and conchae, supplied by deep branches of the corresponding nerve; the mandibular field covers the lower teeth and gums and the tongue, and its nerve supplies them, as well as extending on to the ear and temporal region by an auriculo-temporal branch. It is from foci of disease in such deeper lying regions that pain may be 'referred' to the corresponding cutaneous areas of the face, and an explanation be afforded of, for example, supraorbital neuralgia from glaucoma or frontal sinus disease, or of pain in the ear from a diseased tooth. The urgency of the pain impulse, apart from its own special pathways, apparently over-

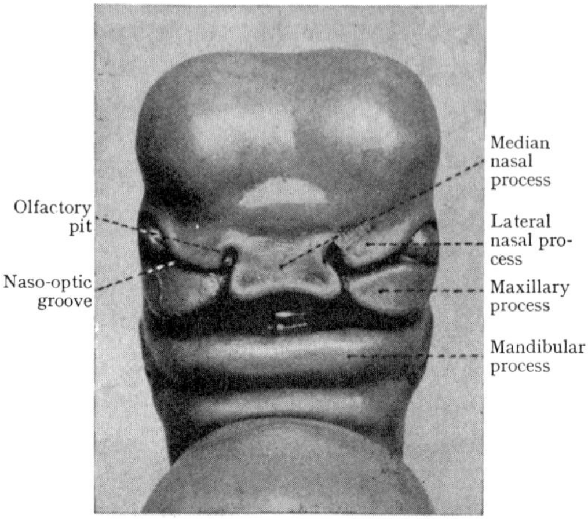

FIG. 186.

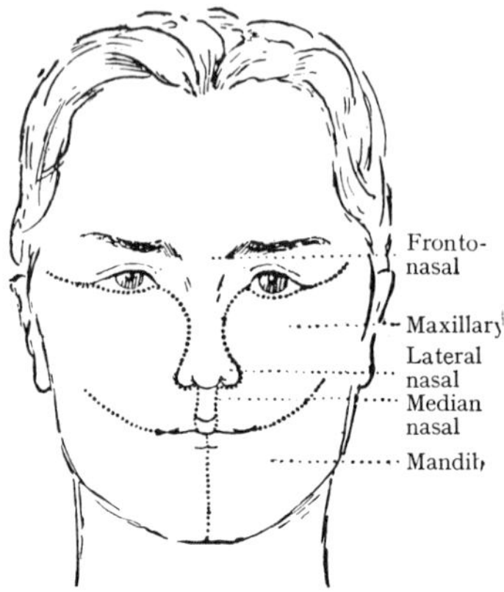

FIG. 187.

FIG. 186.—Model of the **Face of an Embryo** (reconstruction by His) to show the processes which form it. The two mandibular processes have fused at the stage taken.

FIG. 187.—Showing the **corresponding parts of the Adult Face.**

flows at the first synapse or relay-station of its subservient nerve-fibres into a closely related tract accustomed to transmit more ordinary sensations to the sensorium.

The **Ophthalmic Nerve** (*nervus ophthalmicus*) is the smallest of the three divisions of the trigeminal and is purely sensory. It supplies the eyeball and conjunctiva, the lacrimal gland, caruncle, and sac, the forepart of the nasal mucous membrane, and the frontal air-sinus, the upper eyelid, forehead, and front part of the scalp, and the root and front part of the nose; the cutaneous area is shown in Fig. 94, p. 179.

It arises from the fore and medial part of the semilunar ganglion and immediately enters the cavernous sinus; it then passes upwards and forwards, embedded in the lateral wall of the sinus below the trochlear nerve and above the maxillary nerve, as is shown in Fig. 185. Whilst in the sinus the nerve first gives off a recurrent branch (*nervus tentorii*, or recurrent nerve of Arnold), which crosses and is adherent to the trochlear nerve and is distributed to the tentorium cerebelli and possibly the dura over the posterior pole of the brain; it is then generally considered to give off three slender filaments, which join the oculomotor, trochlear, and abducent nerves and so provide sensory fibres for the muscles they supply. These fibres are regarded as proprioceptive in function by Woollard, Wilkinson, and Hines, and the first named traces their origin from the mesencephalic root of the trigeminal (p. 336). The ophthalmic nerve receives filaments from the cavernous plexus.

After a total length from the ganglion of 25 mm., or an inch, the nerve breaks up in the forepart of the sinus into three terminal branches, lacrimal, frontal, and naso-ciliary, though often the last alone is here separated off. All three enter the orbit through the superior orbital fissure, but the naso-ciliary nerve alone traverses the oculomotor foramen, inside the annulus of Zinn, and between the two divisions of the third nerve; the lacrimal and frontal nerves enter through the lateral part of the fissure and outside the cone of muscles (Figs. 142, 145). These nerves run forwards through the orbit, each accompanied in the greater part of its course by an artery, and all three end by finally supplying the skin in regions beyond the orbital margin. The branches of the ophthalmic, like the other sensory parts of the fifth nerve, anastomose on the face with branches of the facial nerve.

A complete lesion of the ophthalmic nerve occasions anaesthesia of the forehead, the upper but not the lower eyelid, the root and tip of the nose, the conjunctiva, cornea, and eyeball, and there is no reflex winking on irritation of the conjunctiva. The area of cutaneous anaesthesia is much less than might be expected from the distribution of the branches of the nerves, owing to overlapping by other adjacent branches of the trigeminal (Fig. 94, p. 179). Taste and smell are said to be impaired. The whole nerve is illustrated on p. 355.

(1) The **lacrimal nerve** (*nervus lacrimalis*), the smallest of the three, is closely bound to the fourth nerve at the fore end of the cavernous sinus, but leaves it to pass obliquely through the lateral extremity of the fissure, so intimately enveloped by the dura that it is difficult to dissect out. It then runs along the lateral orbital wall, embedded in fat in company with the lacrimal artery and above the lateral rectus, toward the lacrimal gland. Just behind, or sometimes within its substance, the nerve divides into two branches. The uppermost one traverses the gland, to which it gives off most of its fibres, and after piercing the septum orbitale ends in the conjunctiva and skin of the upper eyelid in the lateral region (Fig. 110, p. 209); the inferior branch passes downward on the orbital wall behind the gland, to which it also sends fibres, and anastomoses with the zygomatico-temporal branch of the zygomatic nerve, which is derived from the maxillary division of the trigeminal (Fig. 77, p. 154). The signification of this connexion and the constitution of the fibres of this nerve have been referred to on p. 218; the cutaneous distribution is noted on p. 180, and the conjunctival supply on p. 206.

As **variations** the lacrimal nerve may give off fibres which run for part of their course with the other orbital nerves; such junctions are recorded with the fourth nerve, with the frontal, and with the naso-ciliary (Laffay). It has been seen to give off a ciliary nerve, and to receive a filament from the long root of the ciliary ganglion or from the ganglion itself. A case has been observed where the lacrimal instead of the zygomatic (temporo-malar or orbital) nerve provided the usual zygomatico-temporal branch of the latter (Thane), and, on the other hand, the lacrimal has been found absent, its place being taken by the zygomatic nerve (Turner). The anastomosis between the two nerves has been found wanting (Laffay). The writer has found two well-developed lacrimal nerves, both derived from the ophthalmic nerve; a similar instance is recorded by Delbet, though in his case one of the nerves was derived from the maxillary nerve, and so was apparently equivalent to an aberrant zygomatic branch.

(2) The **frontal nerve** (*nervus frontalis*), the largest branch of the ophthalmic nerve, after entering the orbit above the muscles, runs forwards immediately beneath the periorbita of the roof, resting first upon the origin of the superior rectus, then upon the levator of the upper eyelid. At a variable

point, but usually beyond the middle of the orbit, it gives off a supra-trochlear nerve, beyond which point it is continued forwards under the name of supra-orbital nerve (Figs. 188, 189). The supra-trochlear nerve of Arnold (*nervus supra-trochlearis*), as its name implies, passes inwards and forwards close above and sometimes bound to the pulley of the

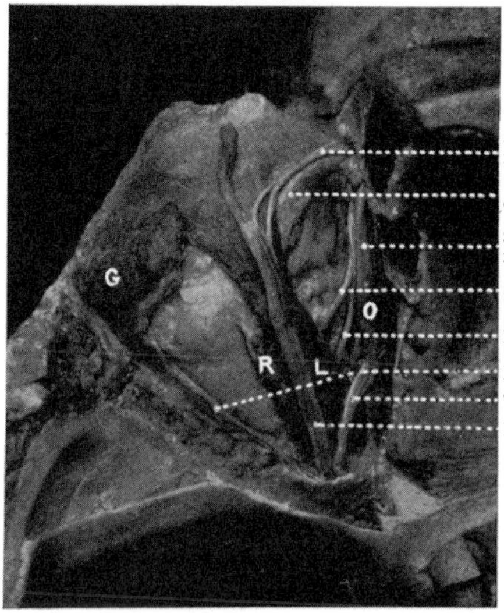

FIG. 188.—Dissection of Left Orbit from above, to show the *levator palpebrae superioris* (L.) and superior oblique (O.) muscles, and the **nerves** in relation. Natural size.

R = superior rectus muscle; G = lacrimal gland.

superior oblique muscle, pierces the septum orbitale, and runs up over the orbital margin through the frontalis and orbicularis muscles in company with the frontal artery, a finger's breadth from the mid-line, to supply by several branches the skin of the forehead above the head of the eyebrow (Figs. 9, p. 21, and 59, p. 127); near the pulley a branch joins the infra-trochlear branch of the naso-ciliary nerve either before or after leaving the orbit, from which superior palpebral twigs supply the skin and conjunctiva of the upper eyelid and side of the nose. The supra-orbital nerve (*nervus supra-orbitalis*) is the direct continuation of the frontal nerve under a different name after it has given off the supra-trochlear nerve. It leaves the orbit by passing through the supra-orbital notch or foramen of the upper margin, and runs up over the forehead, sometimes grooving the bone deeply, in company with the artery of the same name as far as the occipital bone. It divides into a smaller medial and a larger lateral main stem, the former being sometimes called the internal frontal (which has been confused with the supra-trochlear) and the latter the external frontal or supra-orbital nerve proper; the division may occur within the orbit, and the former branch then grooves the orbital margin in a frontal notch. Twigs are given to the eyelids, and, whilst traversing the supra-orbital notch, to the diploë and frontal sinus. The nerve is illustrated in Fig. 179, p. 333, and on page opposite.

From a clinical point of view, it is worth noting that a neuralgia may recur after section of the supra-orbital nerve owing to the escape of the supra-trochlear, and it was only after resection of the frontal nerve itself that Koenig was able to effect a cure in such a case.

As **variations** of the frontal nerve, the writer has seen it provide a very fine infra-trochlear branch instead of the naso-ciliary; also a well-developed twig was found passing directly to the superior oblique muscle, which, when traced backwards, was found to be derived from the ophthalmic nerve, and so represented the sensory filaments usually accompanying the fourth nerve to the muscle. The supra-trochlear may abnormally arise as a branch of the supra-orbital. The writer has found it duplicated.

(3) The **naso-ciliary** or nasal nerve (*nervus nasociliaris*) has a long and complicated course, since it is found in the orbit, in the cranial cavity, in the nasal cavity, and at the tip of the nose; its parts and branches do not always receive the same names in the literature. After entering the orbit through the annulus of Zinn between the two divisions of the oculomotor nerve, it courses obliquely beneath the superior rectus muscle, to the tendon of which it may adhere, and over the optic nerve along with the ophthalmic artery towards the medial wall of the orbit, where it lies between the superior oblique and medial rectus, still in company with the artery; it leaves the orbit by traversing the anterior ethmoidal foramen (hence the name of 'anterior ethmoidal nerve' or *ramus*

ethmoidalis anterius sometimes applied to it here) and enters the cavity of the cranium, where it crosses the forepart of the

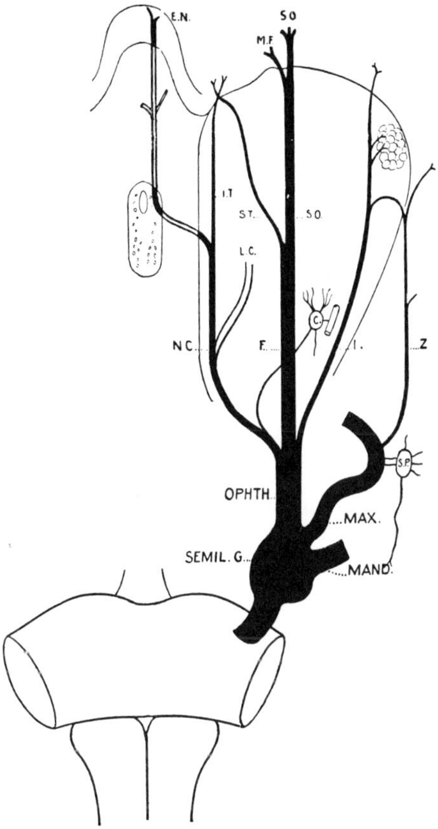

FIG. 189.—Scheme to show the distribution of the **Ophthalmic Division** of the Trigeminal Nerve.

F. = frontal nerve, giving off the supra-trochlear branch (S.T.) and afterwards being known as the supra-orbital (S.O.), which may divide before leaving the orbit into a medial frontal branch (M.F.) and a lateral frontal or supra-orbital proper.

N.C. = the naso-ciliary nerve; it is shown to give off the sensory root to the ciliary ganglion (C.), the two long ciliary nerves (L.C.) to the eyeball, and, just before leaving the orbit through the anterior ethmoidal canal, the infra-trochlear nerve (I.T.); the rest of its course up to its termination on the nose as the external nasal nerve (E.N.) illustrates the description given in the text.

L. = the lacrimal nerve, the anastomosis of which with the zygomatico-temporal branch of the zygomatic nerve (Z.) is represented.

S.P. = the spheno-palatine ganglion. (S.E.W. From Cunningham's Textbook of Anatomy, 1931.)

cribriform plate of the ethmoidal bone beneath the dura of the anterior cranial fossa, running along the side of the olfactory lobe (Fig. 191); it then passes through a slit, the nasal fissure or anterior nasal canal, at the side of the crista galli, enters the nasal cavity (being here related to the medial side of the infundibulum of the frontal sinus), and breaks up into three terminal branches. Two of these, the medial and lateral internal nasal branches (*rami nasales mediales et laterales*), are distributed to the forepart of the nasal cavity, but the third is continued forwards as the external nasal nerve to the end of the nose.

The branches of the naso-ciliary nerve are:

(*a*) The long or sensory root (*radix longa*) of the ciliary ganglion (p. 365), which leaves the nerve soon after or even before its entrance into the orbit, and lies lateral to the optic nerve (Fig. 191); it frequently anastomoses with the long ciliary nerves and with the short root of the ganglion (Valentin).

(*b*) The long ciliary nerves (*nervi ciliares longi*), two, or sometimes three in number, which lie also on the lateral side of the optic nerve, supplying its sheath; after joining some of the short ciliary nerves, they enter the eyeball round the point of attachment of the optic nerve and supply the iris, ciliary muscle, and cornea with sensory filaments, but in addition contain sympathetic fibres from the cavernous plexus for the dilatator pupillae muscle (see pp. 371, 344). Braunstein (1893) concluded from his experiments that the two long ciliary nerves supply different regions of the iris, lateral and medial. It is noteworthy that through these nerves, together with the sensory root of the ciliary ganglion, the naso-ciliary receives all the sensory impulses from the eyeball, and there is an intimate physiological relation between the sensitive forepart of that organ and the skin and mucous membrane areas to which the rest of this nerve is distributed.

(*c*) The infra-trochlear nerve (*nervus infratrochlearis*, sometimes referred to as the external nasal branch in contradistinction to the main trunk, which after this point of separation is then known as the internal nasal nerve) appears to be the direct though more slender continuation of the main nerve forward from the point where the latter turns to enter the anterior ethmoidal foramen; it runs beneath the superior oblique muscle and trochlea, forms a junction with the supra-trochlear nerve, and dividing into several branches, usually before piercing the septum orbitale, is distributed to the parts of the medial canthus of the eyelid, namely to the skin and conjunctiva mainly of the upper eyelid and to a smaller extent of the lower, the root of the nose, lacrimal sac, canaliculi, and caruncle. Ledouble describes a filament passing through the lacrimal bone at the point dacryon into the nose.

(*d*) The three terminal branches in the nose are: (*a*) the medial internal nasal or septal branch (*ramus nasalis medialis*), which supplies the mucous membrane of the anterior part of the septum; (*b*) the lateral internal nasal branch (*ramus nasalis lateralis*, but sometimes confusedly called the external nasal branch, or again, sometimes described as a branch of the latter), which supplies the forepart of the middle and inferior conchae and lateral nasal wall; and (*c*) the external nasal branch (anterior nasal, *ramus nasalis externus* or *nasalis extremus*), the continuation of the main nerve, which passes along a groove on the under surface of the nasal bone, emerges between its lower edge and the upper lateral nasal cartilage, and supplies the skin over the cartilaginous end of the nose.

As an inconstant branch within the orbit, there may be found in two out of six subjects (Delbet) the spheno-ethmoidal nerve of Luschka (who also described such twigs as arising from the spheno-palatine ganglion), or posterior ethmoidal branch of Krause (*ramus ethmoidalis posterior*), which supplies the mucous membrane of the sphenoidal air-sinus and posterior ethmoidal cells. Twigs are also said to be supplied by the naso-ciliary nerve to the frontal and anterior ethmoidal cells as it passes through the anterior ethmoidal canal (Merkel).

As **variations**, the naso-ciliary nerve has been seen to send branches to the superior and medial recti and levator palpebrae superioris (Krause), doubtless representing the sensory twigs usually received by the motor nerves to these muscles in the cavernous sinus. An anastomosis with the lacrimal nerve has been found (Laffay). Instances are recorded of absence of the infra-trochlear nerve, the deficiency being provided for by the supra-trochlear (Testut). The internal nasal branches may be absent (Merkel), or they may enter the nose through the posterior ethmoidal canal. Other minor variations will be found recorded by Cruveilhier (1886).

The **Maxillary Nerve** (superior maxillary, *nervus maxillaris*) or second division of the trigeminal is, like the ophthalmic division, purely sensory. It supplies the skin of the cheek, front of the temporal region, lower eyelid, side of nose, and upper lip (Fig. 94, p. 179); the teeth of the upper jaw; and the mucous membrane of the nose, naso-pharynx, maxillary sinus, posterior ethmoidal cells, soft palate, tonsil, and roof of the mouth.

It arises from the middle of the semilunar ganglion and passes forwards in the lower part of the lateral wall of the cavernous sinus, the position being illustrated in Figs. 170, 172, 178, 185.

It leaves the middle cranial fossa through the foramen rotundum, bends downwards and crosses the upper part of the pterygo-palatine or spheno-maxillary fossa, and curving laterally round the orbital process of the palatine bone as it forms the apex of the floor of the orbit, enters that space through the inferior orbital fissure at about its mid-point. It runs forwards along the orbital floor parallel to the plane of the medial wall, beneath the periorbita in the infra-orbital groove and canal, and emerges on the face through the infra-orbital foramen. In the orbital part of its course it is called the infra-orbital nerve.

Branches are given off from it in the cranium, in the pterygo-palatine fossa, in the infra-orbital canal, and on the face.

(i) The cranial branch is the recurrent or middle meningeal nerve to the dura mater.

(ii) The branches given off in the fossa are the zygomatic, the spheno-palatine, and the posterior superior alveolar.

The **zygomatic** nerve (temporo-malar or orbital nerve; *nervus zygomaticus*; *nervus subcutaneus malae*) enters the orbit through the inferior orbital fissure, and running along the lateral wall divides into two branches, temporal and malar. The temporal branch (*nervus zygomatico-temporalis*) sometimes grooves the orbital wall as it runs forwards in close contact with it; it forms an important communication with the lacrimal nerve, as illustrated in Fig. 77, p. 154, the significance of which has been noted on p. 217, and passing either through a canal in the zygomatic bone or through the spheno-zygomatic suture, enters the temporal fossa. After

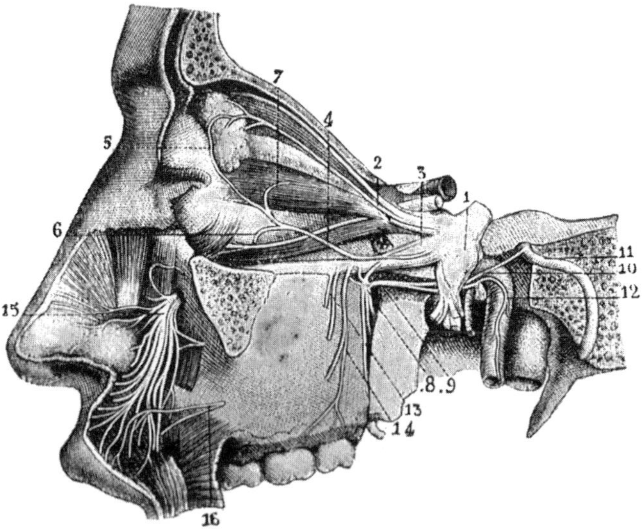

FIG. 190.—The **Maxillary Division** of the Trigeminal Nerve, its branches and their connexions. (From Sappey, after Hirschfeld and Leveillé.)

1 = semilunar or Gasserian ganglion; 3 = the maxillary nerve; 4 = the zygomatic branch with its anastomosis (5) with the lacrimal nerve (2); 11 = the facial nerve; 10 is the greater and 12 the deep petrosal nerve which join to form the Vidian nerve (9) and enter the spheno-palatine ganglion (8). The other branches can be followed from the text.

ascending and piercing the temporal fascia about 25 mm. or an inch above the zygoma, it supplies the skin of the forepart of the forehead up to the lateral side of the orbit. The malar branch (*nervus zygomatico-facialis*) passes along the lower lateral angle of the orbit, and after traversing a canal of the same name in the zygomatic bone, perforates the orbicularis oculi muscle to supply the skin of the cheek; its emergence on the face is seen in Figs. 58, p. 126, and 95, p. 180. Both branches of the nerve inosculate with twigs of the facial nerve.

As **variations**, the whole nerve may pass through the bone before dividing, both branches may traverse separate canals, or the temporal branch may pass out by the fore end of the inferior orbital fissure; either branch may be absent, its fellow supplying the deficiency; the infra-orbital nerve may replace the distribution of the malar, or a twig from the lacrimal may be substituted for the temporal, or *vice versa,* as previously noted on p. 218.

The spheno-palatine branches (*nervi sphenopalatini*) are two or three short twigs which pass directly downwards into the **spheno-palatine or Meckel's ganglion**; only a small part of their fibres actually traverse the ganglion, the majority passing lateral to or in front of it to be continued directly on as its branches, but whilst in neither case are these trigeminal fibres interrupted in the ganglion, in both instances they receive sympathetic nerve fibres from it which accompany them to their destination (Piersol, 1919). The ganglion lies in the upper part of the pterygo-palatine fossa, close to the spheno-palatine foramen; it receives three kinds of fibres, sensory by its branches from the maxillary nerve, motor (and probably also sensory) by the great superficial petrosal nerve from the geniculate ganglion of the facial nerve, and sympathetic by the great deep petrosal nerve from the carotid plexus, these two petrosal nerves running together in the latter part of their course to enter the ganglion as the Vidian nerve or nerve of the pterygoid canal. Among the numerous branches of the ganglion, which chiefly supply the palate and the nasal cavity, there are to be noted here as of immediate interest the ascending or orbital branches (*rami orbitales*), two or three filaments which enter the orbit through the inferior fissure and supply the periorbita, including the involuntary orbitalis muscle, and also traverse the posterior ethmoidal canal or a minute special aperture, to be distributed to the sphenoidal sinus and posterior ethmoidal air-cells (Luschka); it is stated (Cunningham) that minute filaments also supply the lacrimal gland (see p. 218), and others have been traced to the sixth nerve, ciliary ganglion, and optic nerve sheath (Quain). Vitali (1929), in a microscopical study of the ganglion in thirty-five orbits, invariably found numerous filaments supplying the orbitalis muscle, with minute ganglia not only on them but in the fibro-muscular tissue itself; similar bodies were also seen on a communication to the sixth nerve; twigs were found passing to the cavernous plexus and the sheath of the optic nerve but not to the ciliary ganglion (as described by Tiedemann, Arnold, and Valentin), nor could he find the spheno-ethmoidal branches of Luschka. Ruskin (1925) reviews these communications.

The posterior superior alveolar or dental branches (*rami alveolares superiores posteriores*), usually two in number, arise just before the maxillary nerve enters the inferior orbital fissure; they descend on the zygomatic surface of the maxilla and enter minute canals to supply the molar teeth, gums, and mucous membrane of the maxillary sinus.

(iii) The branches of the maxillary nerve given off in the infra-orbital canal are the middle and anterior superior alveolar or dental nerves (*ramus alveolaris superior medius et anterior*). The middle arises in the posterior part of the canal and runs downwards in the lateral wall of the antrum to supply the two premolar teeth. The anterior (usually double) leaves the lateral side of the main nerve in the forepart of the canal, curves medially beneath the infra-orbital foramen in the anterior wall of the antrum, and then bends downwards to supply the incisor and canine teeth; its course is indicated in Fig. 5, p. 8, and further reference has been made to it on p. 16.

(iv) On the face, after emerging by the infra-orbital foramen, buried in fat about 10 mm. from the skin surface, and under cover of the orbicularis oculi and the caput infra-orbitale of the quadratus labii superioris muscles, the maxillary nerve terminates by dividing into three groups of branches: the inferior palpebral (*rami palpebrales inferiores*), distributed to the skin and conjunctiva of the lower eyelid; the lateral or external nasal (*rami nasales externi*), which supply the skin of the side of the nose; and the superior labial (*rami labiales superiores*), which are the largest branches, and are distributed to the skin of the cheek and the mucous membrane and skin of the upper lip. All these branches inosculate freely with the zygomatic branch of the facial nerve, forming the important infra-orbital sensori-motor plexus.

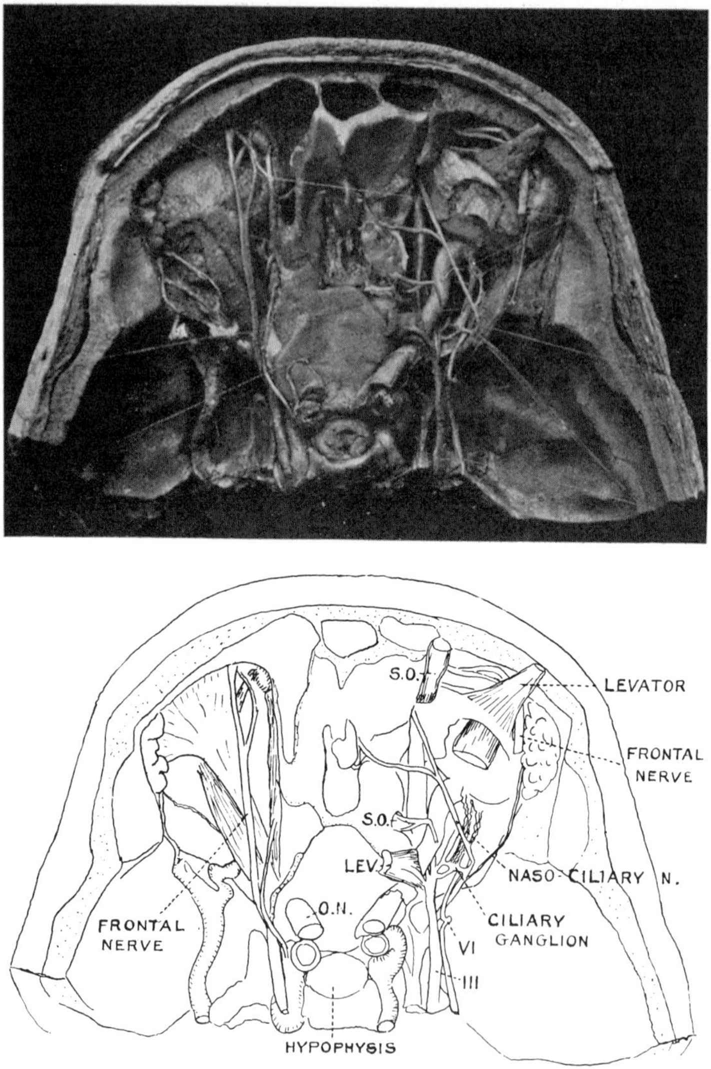

FIG. 191.—Dissection of both Orbits from above; superficial on the left and deep on the right, to show the **nerves.** S.E.W.; Oxford Anatomical Museum. Note the abnormal origin of the ophthalmic artery from the middle meningeal seen in the left orbit: the usual vessel was absent on each side.

S.O. = superior oblique muscle; LEV. = *levator palpebrae superioris*; O.N. = optic nerve.

The **Ciliary Ganglion and Sympathetic Nerve Supply of the Eyeball.**—The ciliary ganglion (*ganglion ciliare*; lenticular or ophthalmic ganglion), which has been referred to above in connexion with the oculomotor and naso-ciliary nerves, is a small mass of cells lying in the apex of the orbit. Like the spheno-palatine, otic, sublingual, and submaxillary ganglia connected with the other two divisions of the trigeminal nerve, it corresponds to the terminal ganglia of the sympathetic system, being chiefly composed of multipolar elements of various types, closely resembling stellate cells of undoubted sympathetic nature found elsewhere, though it contains also the bipolar cells of a cerebro-spinal ganglion. Developmentally, in the human embryo its cells migrated from the semilunar ganglion along the ophthalmic nerve, and relatively few are contributed *via* the oculomotor (Kuntz, 1920). It is essentially a relay station in which certain fibres, contributed by the various nerves which form its so-called roots, end in arborisations round the ganglion cells, the axons of which pass forwards as the short ciliary nerves to the globe; in other words, it serves as a centre for the supply of motor, sensory, and sympathetic nerves to the eyeball. It is noteworthy, however, that since the ciliary nerves which leave the ganglion contain many more fibres than do the roots, probably each of the entering fibres is connected with several cells, and it has been found to contain association neurons and reflex within itself. Pines (1927), in an exhaustive study of the ganglion in man, states that the majority of its cells are doubtless multipolar parasympathetic neurons whose functions are essentially motor; other neurons resembling spinal ganglion cells he regards as receptive in function, and a third kind, whose axons do not extend beyond the ganglion but end on the bodies of neighbouring cells, may serve to complete reflex arcs which do not traverse the central nervous system. According to this view the ciliary ganglion is not purely motor in function, but constitutes an independent reflex centre (Kuntz, 1929). Beauvieux and Dupas (1926) made a special topographical, anatomical and histological study in man (80 subjects) and other animals. Previous papers worthy of reference are by Langley and Anderson, 1892; Retzius,

1894; Marinesco, 1908; Nicholls, 1915. The earlier literature is given by Van Gehuchten, 1906.

It is a small pink mass about the size of a pin's head, 2 mm. long, quadrilateral in outline and bilaterally compressed, though the shape is obscured by fat and connective tissue. It lies in the orbital fat near the apex of the orbit, between the lateral rectus muscle and the optic nerve, and close to the ophthalmic artery (as is especially well seen in the section on p. 263, Fig. 143), a centimetre in front of the annulus of Zinn and about 15 mm. behind the eyeball; its distance from the infero-lateral angle of the orbital margin was found to be from 35 to 42 mm. in twenty-two measurements made by Beauvieux; it is most easily found by following backwards the nerve to the inferior oblique muscle, just above which it lies and to which it is connected by a short stout nerve twig, its motor root. This root, along with two others, the sensory and the sympathetic (which may run together from the naso-ciliary nerve), enters its superior border, whilst from the corners of the anterior margin there arise two bundles of very delicate fibrils, at first sight not easily distinguishable from the connective tissue strands of the orbital fatty mass; these are the short ciliary nerves, and they can be traced forwards to the eyeball, which they enter round the optic nerve.

The **roots** convey to the ganglion:

(i) Motor fibres (parasympathetic) from the oculomotor nerve;

(ii) Sensory fibres from the naso-ciliary branch of the trigeminal nerve;

(iii) Sympathetic fibres from the cavernous plexus on the internal carotid artery (which may travel by the naso-ciliary);

The motor or short root, or parasympathetic root (*radix brevis*), is the thickest, and is derived from the branch of the oculomotor nerve which supplies the inferior oblique muscle. It enters the postero-inferior part of the ganglion and carries two orders of fibres:

(*a*) Contractile to the iris (the sphincter or *constrictor pupillae* fibres); these originate from cells situated in the anterior part of the nucleus of the third nerve in the mid-brain, run along the main trunk of the nerve, and end in the ciliary ganglion, whence the impulses are carried by fresh fibres through the short ciliary nerves to the globe; degeneration does not occur in the motor fibre constituents of these nerves after section of the oculomotor proximal to the ganglion (Van Gehuchten); and the path is therefore interrupted in the ganglion. Contraction of the pupil has been experimentally produced in animals by excitation of the root of the third nerve, and by stimulation of the ciliary ganglion or of the short ciliary nerves (Starling, 1920).

(*b*) Contractile to the ciliary muscle, the *accommodation fibres*; these were for many years regarded as arising from the Edinger-Westphal part of the oculomotor nucleus, but such origin has been negatived recently by many investigators, and it seems to have been proved beyond doubt that cells of this group give rise to no fibres which pass direct to the ciliary ganglion or the eye itself (see p. 408). Wilkinson (1927) has suggested that the contraction of the pupil which normally always occurs during accommodation is really associated with the accompanying convergence by a reflex initiated through the proprioceptive nerve endings in the extrinsic muscles of the globe. Thus the sensory limbs of the arcs concerned in reflex contractions of the pupil on accommodation and to light travel by entirely different afferent pathways (the accommodation reflex along the oculomotor nerve to its nucleus, the light reflex along the optic nerve and tract to the superior colliculus), and this forms the neurological basis of the explanation of the Argyll-Robertson pupil. He states that it is now generally agreed that in such phenomenon the light reflex path is interrupted in the mesencephalon in the region of the aqueduct of Sylvius; the afferent pathways involved in the reaction of accommodation, if the above association be true, naturally escape (see also p. 371).

The sensory, nasal, or long root (*radix longa*), arises from the naso-ciliary branch of the ophthalmic division of the trigeminal nerve; it leaves it in the oculomotor foramen, and runs along the lateral aspect of the optic nerve as a long slender filament which enters the upper posterior angle of the ganglion. Its fibres are chiefly sensory or centripetal from the whole eyeball, but the most important are those from the cornea, and they are not interrupted in the ganglion; this root is considered by some authors to contain also sympathetic fibres effecting dilatation of the pupil (*vide infra*), and other fibres which cause relaxation of the ciliary muscle (Poirier).

The sympathetic root (*radix media vel sympathica*), commonly multiple, is a minute filament which arises from the cavernous plexus and runs forwards to enter the upper hinder part of the ciliary ganglion. It conveys vaso-constrictor fibres to the whole eyeball, vaso-dilator to its anterior segment, and, it is sometimes stated, also pupillo-dilator to the iris (p. 371). The fibres are not interrupted in the ganglion.

Variations in the number of roots are common, trivial, and easily explainable. The motor root may be absent, double or multiple, or may bifurcate before entering the ganglion; it has been found arising from the trunk of the third nerve or from the sixth (Lagrange). The sensory root is commonly absent, the impulses passing backwards through the long ciliary nerves instead of partly through the ganglion; it is frequently fused with the sympathetic root; on the other hand, it may be multiple; it may arise abnormally either from the semilunar ganglion, from the trunk of the ophthalmic nerve, or from the supra-orbital or lacrimal nerve. The sympathetic root frequently fuses with the sensory before entering the ganglion and so appears wanting; it is so often multiple that some authors consider this to be the normal condition; some of these fibrils may accompany the third nerve. Additional roots of the ganglion may be provided by the superior division of the third, fourth, sixth, or lacrimal nerves, or by the spheno-palatine ganglion. As regards the ganglion itself, it may be very small or even absent, possibly due to the scattering of its constituent neurones along the nerves, or there may be two or more accessory ganglia; it has been seen perforated by a ciliary artery. Further details are given by Beauvieux (1926).

The **short ciliary nerves**, or efferent branches (*nervi ciliares breves*), though very variable up to about a dozen in number, usually consist of five or six fine filaments, which leave the anterior margin of the ganglion most often in three groups; they subdivide to form from twelve to twenty fibres, and pass forwards between the lobules of the orbital fat above and below the optic nerve, the lower group being the more numerous; they mix with, and a few are united to, the long ciliary nerves. They pierce the eyeball around the optic nerve entrance along with the short posterior ciliary arteries, run forwards between the sclera and the chorioid, to which they give filaments, and end in a rich plexus round the ciliary body (Fig. 165, p. 307). This plexus supplies the cornea, ciliary muscle, and iris with sensory, motor, and sympathetic fibres respectively, the origin and course of which are illustrated below in Fig. 192.

All the short ciliary nerves are mixed, and all, save two or three, contain pupillo-constrictor fibres. The motor fibres supply the intrinsic muscles of the globe, that is, the ciliary muscle or muscle of accommodation of the lens, and the sphincter or constrictor pupillae muscle of the iris; the fibres are the axons of cells situated in the ciliary ganglion, and transmit the impulses received from the oculomotor nerve, which are consequently interrupted here. The sensory fibres of the ganglion receive impulses from all parts of the globe, but chiefly from the anterior epithelium of the cornea; they are uninterrupted in the ganglion, and pass backwards by its sensory root into the naso-ciliary nerve. The sympathetic fibres end in the walls of the blood-vessels of the eyeball, but also provide pupillo-dilator fibres to the iris.

Besides these short ciliary nerves of the globe, the ganglion also supplies a few filaments to the sheath of the optic nerve, and others appear to be lost in the orbital fat near the globe. Tiedemann described a sympathetic branch which enters the optic nerve along with the central artery of the retina and bears his name.

The ciliary ganglion has been supposed to play a direct role in the formation of the aqueous humour, and to become of importance in glaucoma, but this view is now proved untenable. Landolt found that it contained no excito-glandular filaments, and it neither plays any part in the production, nor does its ablation prevent the formation of the aqueous; Chimanowski showed also that the ganglion bears no relation to the intra-ocular tension. It may be added that de Schweinitz (1903) states that experimental lesions of the ganglion do not cause neuro-paralytic keratitis, and its extirpation does not form trophic lesions, contrary to the results of resection of the cervical sympathetic. Finally, it is to be noted that the nerve plexus in the iris appears to be able to act independently of the ciliary ganglion (see p. 371).

The probable origin and course of the **sympathetic** nerve supply to the eyeball is illustrated in Fig. 192. Briefly, it is first to be noted that excitation of certain cortical regions on the brain surface of animals (for example, in the occipital region or in the sensori-motor area) cause dilatation of the pupil, accompanied by all the symptoms of excitation of the sympathetic cord in the neck, and Schäfer has found that constriction of the pupil can be obtained by stimulation of the quadrate lobe in monkeys, though these results may be associated effects, and do not necessarily imply cortical 'centres' for pupillary movements (Parsons, 1904). Then, the pupillo-dilator fibres have been shown by Karplus and Kreidl (1912) to arise from a definite centre situated in the hypothalamus; and recently Beattie, Brow, and Long (1930) have adduced experimental evidence indicating the presence in cats of a higher sympathetic centre also in the hypothalamus. Finally, Dupuy has shown that puncture of the restiform body (inferior cerebellar peduncle) in the medulla produces exactly opposite effects to destruction of the sympathetic cord in the neck. There would therefore appear to be evidence of cerebral centres which may influence the sympathetic nerve supply to the eye.

The next place, however, from which the course of the sympathetic fibres can more definitely be traced is the spinal cord at the root of the neck, there being strong clinical and experimental evidence for the assumption of a spinal centre (the *centrum ciliospinale inferius* of Budge) situated between the levels of exit of the sixth cervical and fourth thoracic nerves. The *pupillo-dilator* fibres have been traced in animals to leave the spinal cord by the ventral roots of the first three thoracic nerves (Langley), though in man only the first is certain, and to pass from them by the white rami communicantes into the first three thoracic ganglia (or first thoracic only) of the sympathetic cord. The fibres then pass up through the inferior and middle cervical ganglia to reach the superior ganglion, round the cells of which they terminate; this is their first interruption since leaving the spinal cord,

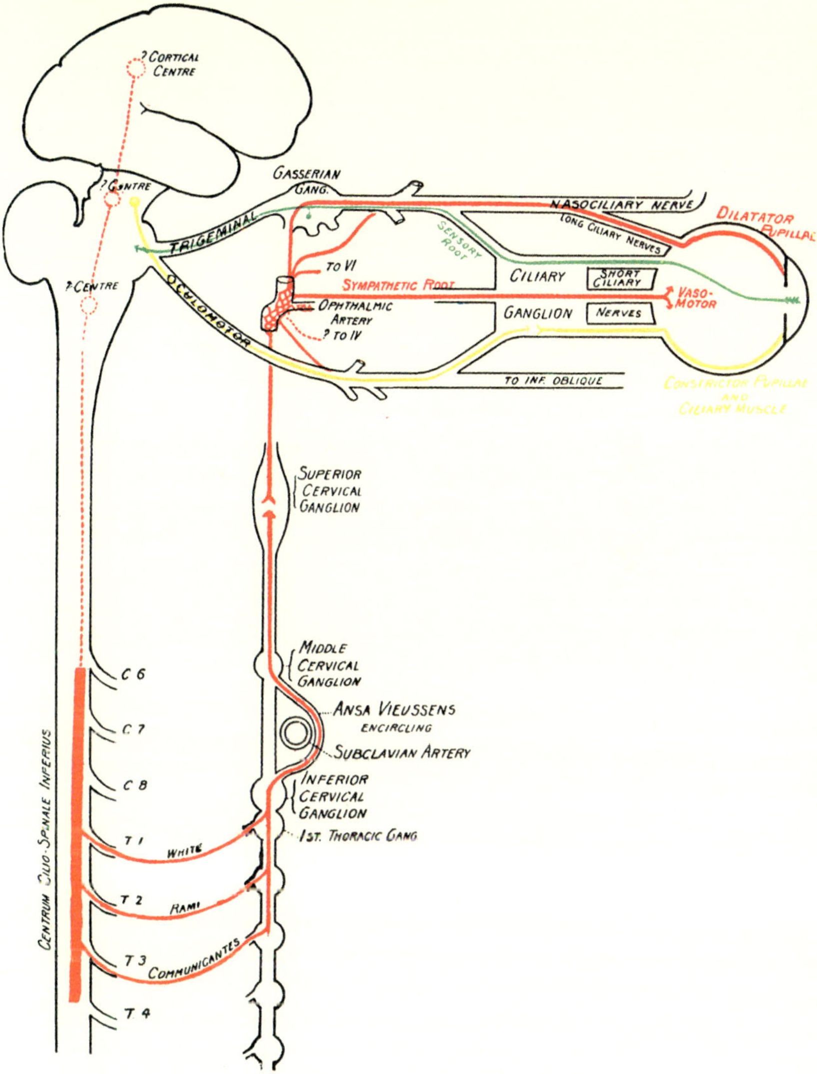

FIG. 192.—Diagram of the **Ciliary Ganglion** with its connexions, and the origin and course of the pupillo-dilator **sympathetic nerve** fibres (in red). For simplification, sensory fibres, though present, are not drawn in the long ciliary nerves. For description see text.

and the only one in their passage to the eyeball. From the cells of the superior cervical ganglion fresh fibres carry the impulses by its superior, ascending, or carotid branch (*nervus caroticus internus*) which accompanies the artery through the carotid canal into the skull. Here the nerve breaks up into a plexus of fine nerve fibres closely adherent to the vessel, and subdivided into a carotid plexus (*plexus caroticus internus*) below, which lies on the lateral side of the artery near the apex of the petrous bone, and a cavernous plexus (*plexus cavernosus*) above, lying medial and inferior to the artery as it traverses the cavernous sinus.

Numerous fine branches arise from these sympathetic plexuses, and their terminal filaments, continued along the branches of the carotid artery, are connected around the anterior communicating artery of the circle of Willis with the corresponding system on the opposite side. The branches which enter the orbit nearly all arise from the cavernous plexus, and proceed as follows:

(i) To the third nerve, at its point of division (recognizable).
(ii) To the fourth nerve (not established).
(iii) To the sixth nerve as it crosses the internal carotid artery in the cavernous sinus (from the carotid plexus).
(iv) To the semilunar ganglion and also to the ophthalmic division of the fifth nerve (recognizable on its under surface).
(v) To the ciliary ganglion, as its sympathetic root from the fore-part of the cavernous plexus, entering the orbit through the superior orbital fissure.
(vi) Vascular branches as a very fine plexus are continued along the ophthalmic artery and each of its twigs (derived from both plexuses).

These sympathetic fibres control in the orbit the dilatator pupillae muscle fibres of the iris, the vaso-constrictor mechanism of the blood-vessels of the orbit and eyeball, and the palpebral involuntary muscles, and provide trophic fibres to all the structures; as regards the vaso-motor actions it may be concluded from the latest experiments that the intra-ocular tension responds passively to all variations in the general blood-pressure. The lacrimal gland, though affected by experimental excitation of the sympathetic, which causes secretion of tears, probably also receives secretory fibres from the seventh nerve (see p. 217). Accommodation of the lens is probably unaffected by the action of the sympathetic, and it has not been shown that the latter can provoke a negative accommodation.

As regards the **pupillo-dilator** nerve fibres (the cerebral connections of which have been described on p. 368), it is first to be noted that the dilatation of the pupil is accompanied by contraction of the radial muscle fibres (the existence of which is now definitely proved) on the one hand, and by inhibition of the sphincter fibres on the other, the reverse of what happens in contraction of the pupil, and its movements are independent of the state of its blood-vessels. Evidence of an independent nerve plexus in the iris, like those of Auerbach and Meissner in the intestine, is adduced by Pollock (1913), who found it to function after complete dissociation from the central nervous system by removal of the ciliary ganglion and superior central sympathetic ganglion. Recently, however, although Jirman (1929) found ganglion cells of all three types in the iris of albino rabbits and observed the isolated iris of the cat to have autonomic spontaneous movements, yet Balado (1927) found no nerve cells present by either Golgi's or Ehrlich's staining method. If present, an explanation would be afforded of the contraction of the pupil in the excised eyeball on exposure to light (Brown-Séquard), and its dilatation by atropin (de Rupter). Ingalls (1922) contrasts the two mechanisms of the iris, and emphasizes how intimately the extremely sensitive one of dilatation (of which Redslob, 1928, maintains the existence) is associated with other general sympathetic reactions (cutaneous stimuli, psychical disturbances, &c.). Magitot (1921) considers the nerve control of the iridial musculature to be a function of the sympathetic system; Langley and Anderson (1892) observed that a general injection of nicotine paralysed both actions of the iris.

There is a difference of opinion as to whether the dilator fibres reach their destination by passing (*a*) along the sympathetic root of the ciliary ganglion and into the short ciliary nerves (Merkel, Poirier, Piersol, Pegler), or (*b*) along the sympathetic branch to the semilunar ganglion, and thence into the naso-ciliary branch of the ophthalmic nerve, to reach the globe by the long ciliary nerves, a path which avoids the ciliary ganglion (Testut, Parsons, de Schweinitz, Gray, Weeks, Halliburton, Wilbrand, and Saenger). Since excitation of either the semilunar ganglion itself or the long ciliary nerves causes a dilatation of the pupil, the latter path appears the chief one, a point upon which Testut is emphatic. Starling (1920) states that they travel by both routes; the short ciliary nerves contain both constrictor (oculomotor) and dilator (sympathetic) fibres, and when they are stimulated electrically both the sphincter and the radial muscles of the iris contract, but the sphincter, being the more powerful, will overcome the other, and therefore the pupil will contract. Cutting the cervical sympathetic causes contraction, and section of the oculomotor nerve produces dilatation, of the pupil (see also p. 344). The mechanism of the pupillary 'light-reflex' is described on p. 402.

The **Optic Nerve** (*nervus opticus*), the second cranial nerve, and the special nerve of sight, extends from the eyeball to the optic chiasma, and is a part only of the visual tract. Nearly all of its fibres originate as the axis cylinder processes of the ganglion cells which comprise the third or inner fundamental layer of the retina. It is interesting to note that more than one-third of the fibres in the optic nerve come from the macular area, although this forms only a twentieth of the whole retina (Elliot Smith). Passing backwards from the eyeball through the orbit and optic canal to the optic chiasma (commissure), where part of them decussate, these fibres are carried on without interruption under the name of the optic tract to certain ganglionic masses of the mid-brain known as the primary or lower visual centres, namely, the lateral geniculate body, the pulvinar of the optic thalamus (see p. 401), and the superior colliculus or quadrigeminal body. Here the great majority of the retinal fibres terminate, but fresh fibres of cells in these bodies form a tract, the optic radiation, which transmits the visual impressions to the cortex of the occipital lobe of the brain, where is situated the higher visual centre. The first half of the pathway lies outside the brain, the second part is deeply buried within its substance. These several parts of the whole visual tract from retina to cortex are represented diagrammatically in the figure below.

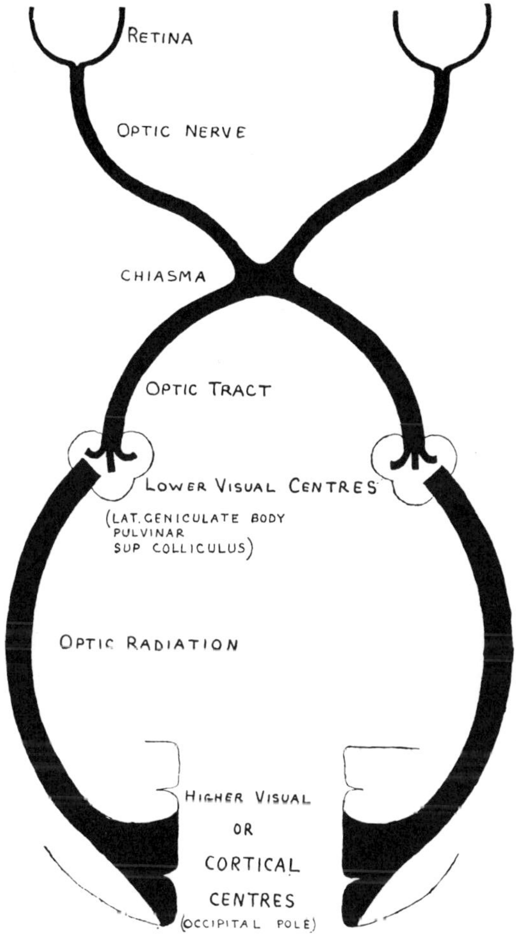

FIG. 193.—Scheme to represent the main parts of the **visual pathways.** Compare with the actual preparation on p. 387, Fig. 201. Note that the connexion with the pulvinar could not be established by Zeeman and Brouwer (see p. 401).

The so-called optic nerve, therefore, is only a part of a continuous tract of fibres extending from the eyeball to the mid-brain. Moreover, since the retinal cells from which these fibres arise are not the primary but the tertiary elements of the visual tract (the visual impressions having previously traversed two sets of neurones, those of the rods and cones and those of the bipolar cell layer), and as is shown further by its development as an outgrowth from the brain, and by its structure and enveloping membranes, the optic nerve is not homologous with the other cerebro-spinal nerves, but is really a part of the central nervous system, and corresponds to an outlying tract of brain fibres.

It is first formed by the development of the optic vesicle, which is an outgrowth from the ventral surface of the fore-brain; the vesicle becomes collapsed by the invagination of its distal wall and forms the optic cup, into which the lens, an ectodermal derivative, is received, whilst the surrounding mesoderm gives rise to the coats of the eyeball. The narrow tube connecting the vesicle to the brain is called the optic stalk, which later becomes solid. The outer layer of the cup forms the hexagonal pigment cells of the retina, and the inner layer develops into the retina proper. The central processes of the neuroblasts of the latter extend backwards along the optic stalk and form the optic nerve, chiasma, and tract. The invagination is on the infero-medial side of the vesicle rather than in front, and the infolding also affects the adjacent part of the stalk, so that the whole formation may be likened to a spoon with a deep bowl and a groove extending down the handle. In this groove anteriorly the central vessels of the retina become enclosed, and as a developmental error the optic disc, or the nerve and its sheath, may present a cleft or coloboma at the side of the infolding. The development of the nerve has been specially studied by Seefelder (1908–9–10), and good accounts are also given by Druault (1912), Keibel and Mall (1912), and by Nussbaum (1912). The growth has also been specially studied in 71 foetuses by Scammon and Armstrong (1925), who found that the increase in length and

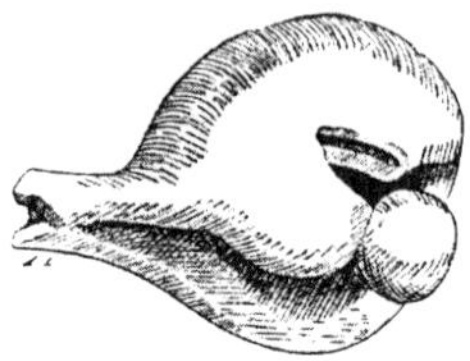

FIG. 194.—Model to illustrate the formation of the **Secondary Optic Vesicle** and the fissure in the Optic Stalk. (From Druault, 1912.)

diameter during the foetal period is directly proportionate to the growth in total body length; the calculated length of the nerve at birth is 24·4 mm., and the diameter 2·7 mm. The development of the fibre medullary sheaths is given by Sättler (1915), who confirms Bernheimer's observation that they develop from the chiasma towards the eyeball, nearly reaching it at birth. Myelination in the tract is described by Keene (1931).

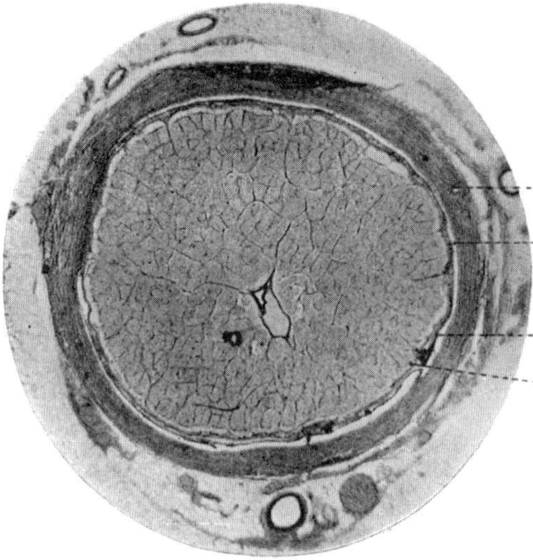

FIG. 195.—Cross-section through the **Optic Nerve and its Sheaths.** The central vein is seen to the right and above the central artery of the retina. × 10.

Structurally, the optic nerve is constituted by 800 to 1,200 fasciculi of myelinated fibres of which there are nearly half a million, divisible into two classes, fine and coarse, the latter being much more numerous (Salzer, Gudden); they have been found to increase in number in growing animals as age advances (Boughton, 1906). The fibres, unlike those of the cerebro-spinal nerves, and like those of the brain tracts, are denude of a sheath of Schwann, and also present similarity to the latter in being supported by neuroglia, the disposition of which has been specially studied by Jacoby (1905); also by Schindler (1926) in eight optic nerves, who considers the septa as ectodermal and not mesodermal in origin, and finds a difference between the architecture of the intra-orbital and intra-cranial parts of the nerve, the septa being less marked in the latter; Haden (1925) also described the ectodermal development of the glial system in human foetuses. It is also not uncommon to find corpora amylacea present (Quain). The fasciculi into which the fibres are collected are separated by connective tissue containing many elastic fibres, and by lymph spaces.

The nerve is pierced on its lower and medial aspect some 5 to 15 mm. behind the eyeball by the central vessels of the retina (p. 306) accompanied by a fine sympathetic plexus, sometimes called the nerve of Tiedemann; the vein usually runs lateral to the artery, and they are enveloped by separate pial sheaths; their position as they run in the centre of the nerve is seen in the above figure, and is specially described on p. 314. The central artery supplies both its anterior and posterior parts whilst in the orbit, where blood is also conveyed to it by ciliary twigs, and in the cranium it receives arterioles from the anterior cerebral artery. Its venules are tributaries of the central vein of the retina and the ciliary veins, and of the cranial veins posteriorly.

The nerve is enveloped by sheaths from each of the three meninges of the brain; the pia mater intimately clothes it throughout its whole extent (though separated from the nerve fibres at intervals by a layer of neuroglia), and sends prolongations into its substance which are continuous with the septa between the fasciculi; the arachnoid membrane comes into contact with it about half-way along its intra-cranial course as it emerges from the cisterna basilis; the dura mater, which is continued as a lining to the optic canal, splits, as previously described, at the orbital opening of this channel to become continuous with the periorbita on the one hand and to form a stout and somewhat loose sheath for the nerve in the orbital part of its course on the other (Figs. 144, 178, 198); the sheath is said to be thicker on the temporal side near the globe (Sanna). All three membranes are continued along the nerve up to the eyeball, where they fuse with the sclera, though part of the pia becomes continuous with the chorioid coat of the globe.

Similarly, the sub-dural and sub-arachnoid spaces of the brain between the membranes are continued along the nerve, and injections in them can be pushed up to the sclera, where they end, as seen in the next figure; they are lined by endothelium. It is stated that with pressure a sub-arachnoid injection will pass between the fasciculi and enter their lymph spaces, indicating the probable path of elimination of the lymph or tissue fluid in the reverse direction; the injection can even be made to enter the peri-chorioidal space of the eyeball (Poirier), but, on the other hand, Behr (1912) found no elimination of fluids from the globe into the nerve. It is to be noted that since the arachnoidea round the nerve is very thin (and, indeed, said to be incomplete) and more closely united to the pia than elsewhere, the sub-arachnoid space is very much reduced and not easy to identify. The sub-dural space, on the contrary, is obvious in any section of the nerve with its sheaths (Figs. 195, 196, 198); it differs from that round the brain in being traversed at intervals by connective-tissue strands. The development of the nerve-sheath has been studied by Lo Cascio (1923) in 30 embryos, and he finds that the differentiation of pial and arachnoid membranes from a primitive exclusively mesoblastic common sheath begins about the third foetal month. The senile changes which occur in the nerve are described by Fuchs (1922).

The nerve is divided, according to its anatomical relations, into four parts, since it traverses successively the sclera, the orbit, the optic canal, and the cranial cavity. The total length, seldom equal on the two sides, is about 45 to 50 mm., or 2 inches, made up by a scleral portion of nearly 1 mm., an orbital part of 20 to 30 mm., an intra-osseous length of 4 to 9 mm., and an intra-cranial part of 10 mm. (varying from 3 to 16 mm.), or, roughly, 1, 25, 7, 10 mm. respectively; Schaeffer (1921), from measurements on twelve cadavera, finds a total length varying from 35 to 55 mm., with an average of 42 mm., the sectional figures not differing materially from those given above; the greatest variations occur in the length of the intra-cranial part. Its diameter in the orbit is 3 or 4 mm., and within the skull varies from 4 to 7 mm., but is commonly 4·5 mm. in cross diameter.

(i) The SCLERAL or intra-bulbar part is composed of the nerve fibres as they collect together at the optic disc and pierce the chorioid and the lamina cribrosa of the sclera; the nerve here is only 1·5 mm. in diameter, since the fibres do not become myelinated or supported by neuroglia until they have passed out of the globe (Fig. 196). For details of the *lamina cribrosa* the reader is referred to the valuable account by

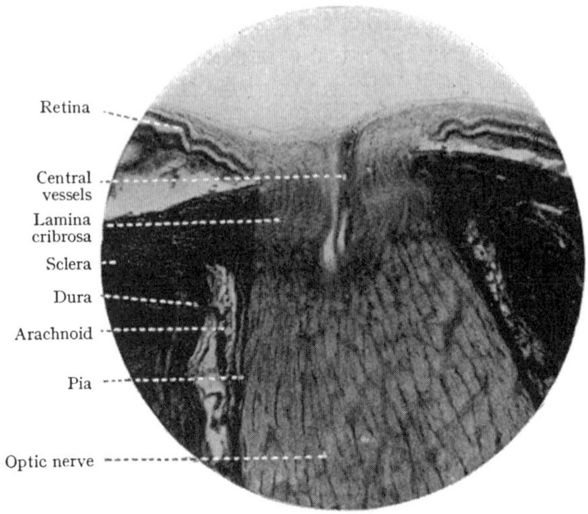

FIG. 196.—Longitudinal section through the **Optic Nerve** at its point of emergence from the Eyeball, and passing through the **Optic Disc.** × 13.

Fuchs (1916), of which résumés are given by Henderson (1918) and Elliott (1921); the development is described by Maggiori (1923), who states that the lamina represents a sharp limit between the non-myelinated and myelinated parts of the optic nerve, apparently acting as an effective barrier to the distal spread of the fibre sheaths. Haden (1925), in a study of nine human foetuses, points out how it serves to anchor the nerve and prevent tension on the retina. Ogawa (1905–6) considers that the pigment cells found in the nerve in the region of the lamina cribrosa and sometimes seen with the ophthalmoscope on the optic papilla are hereditary anomalies, being correlated with the tint of the skin.

(ii) The INTRA-ORBITAL part of the nerve extends from the

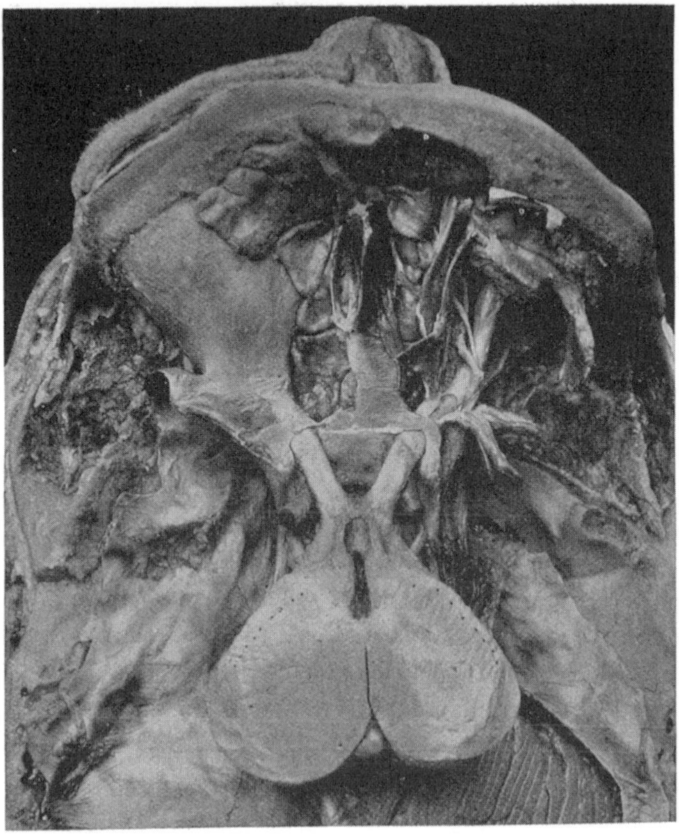

FIG. 197.—Dissection of Head of Adult Male Subject to show the **Orbits from above.** The bone has been removed from above the left orbit, leaving the periorbita intact and showing its connexion with the dura mater lining the middle cranial fossa; the muco-periosteal lining of the frontal and ethmoidal air-cells has also been exposed. On the right side the orbit has been dissected to show the optic nerve crossed by the naso-ciliary nerve and ophthalmic artery. Note the position of the chiasma relative to the sphenoidal bone and *sella turcica.* The course of the optic tracts, indicated by the dots, can be traced round the cut *crura cerebri* to the grey masses of the lateral geniculate body and *pulvinar.* The distance between the optic nerve at its entrance into the eyeball and the lateral orbital margin is 30 mm.; between the optic foramen and zygoma (the edge of which is bared on the left) is 55 mm.; and between the centre of the chiasma and the zygoma is 70 mm.

There is an abnormal double superior oblique muscle in this specimen. The same preparation is viewed from the side in Fig. 139, p. 259. × $\frac{4}{5}$.

eyeball to the anterior optic foramen, and emerges from the former 1 mm. below and 3 mm. to the nasal side of its posterior pole (Fig. 138, p. 257); whilst the course of the nerve roughly corresponds to the long axis of the orbit, it is not straight, but describes two variable curves, usually bending downwards in its forepart and laterally in its hinder half, as seen in Fig. 197; the nerve is therefore a little longer than the distance between the eyeball and the optic foramen, and Weiss in sixty subjects found its length to average 24 mm. (ranging between 20 and 30 mm.), whilst the distance between the globe and the foramen was 18 mm. (14 to 24); this surplus or slack of 5 or 6 mm. is associated with the mobility of the globe, being, like the curves, greater in animals with very mobile eyes, such as the chameleon, and less in exophthalmos. In extreme movements of the eyeball the nerve becomes nearly straight, and indeed it can be stretched enough in certain conditions to affect the optic disc and form a 'conus'.

The relations of the nerve in the orbit are as follows. It is surrounded by the fascia bulbi as it leaves the globe, though this is the weakest and most ill-defined part of the membrane. The posterior ciliary arteries and nerves surround its anterior half and pierce the globe round it (Fig. 138, p. 257). The central vessels of the retina enter on its infero-medial aspect about half-way along its course. It is crossed superficially by the naso-ciliary nerve and also by the ophthalmic artery (Fig. 197), though in 15 per cent. of cases the vessel passes beneath it; farther back the artery lies on its lateral aspect. On this side also is found the inferior division of the third nerve and the ciliary ganglion. At the apex of the orbit it is closely surrounded by the recti muscles at their origin from the annulus of Zinn; the superior and medial recti arise partly from the sheath of the nerve, a fact that probably accounts for painful movements of these muscles in inflammation of the nerve, but the lateral rectus is separated from it by the third, naso-ciliary, and the sixth nerves. Its relative position in regard to the muscles is well displayed in Fig. 147, p. 267, and also in Fig. 198, from which it will also be realized how completely it is surrounded by the orbital fat; the latter, it may be noted, extends farthest into the apex of the orbit below the nerve.

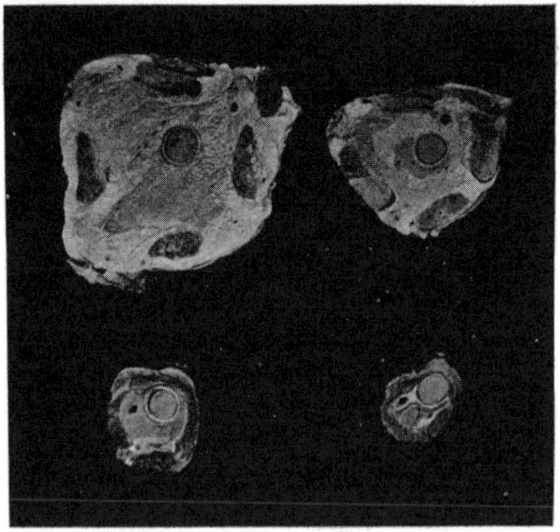

Fig. 198.—Sections through the right orbital contents, seen from in front, after removal intact within the periorbita, to show the **relations of the muscles to the optic nerve.** The ophthalmic artery is also seen presenting different relations in its course forwards through the orbit. The sheath of the nerve is well shown.

The sections were made 5, 15, 20, and 25 mm. behind the eyeball. Natural size.

Lagrange (1903), quoting Lange, gives the following distances in millimetres which separate the nerve from the orbital walls at various points:

	Roof.	Floor.	Lateral wall.	Medial wall.
At 27 mm. from optic foramen (near globe)	11	8–9	10	9
,, 20 ,, ,,	12	6	7	8
,, 14 ,, ,,	15	6	5·5	8·5
,, 10 ,, ,,	6	5	1·1	6·1
,, 7 ,, ,,	4·7	4·3	4·5	4·8
,, 5 ,, ,,	4·5	4·4	4·3	4·4
,, 4 ,, ,,	1·5	2	2	1·2
,, 0·5 ,, ,,	1·5	1·7	2·5	0·6

(iii) The Intra-osseous or canalicular part of the nerve, as it traverses the optic canal, has the ophthalmic artery embedded in its sheath first to the lateral side and then below it; by this sheath the nerve is firmly fixed to the walls of the canal, especially superiorly. The important relations of the canal to the sphenoidal sinus have been described above on pp. 28, 29.

(iv) The INTRA-CRANIAL part of the nerve, from the posterior aperture of the optic canal to the chiasma, is flattened instead of being, as previously, round, measuring about 5 mm. by 3 mm.; it is surrounded by pia mater and arachnoid, and above it lies a falciform fold of the dura mater prolonged a short distance behind the upper border of the canal. Superiorly, it is separated from the medial root of the olfactory tract and lateral part of the anterior perforated substance by the anterior cerebral artery; laterally lies the internal carotid artery (Figs. 182, 185), and infero-laterally the origin of the ophthalmic artery, the varying relations of which to the nerve have been mentioned above on p. 304; the sheath of the nerve is firmly connected to the internal carotid artery at the point of origin of the ophthalmic vessel, since the latter becomes embedded in the sheath (Fig. 172). Stopford (1917) notes the possible compression of the nerve by the internal carotid artery in disease. Beneath the nerve are the diaphragma sellae and the sphenoidal sinus in the body of the sphenoidal bone. The nerve does not lie exactly in the optic groove or sulcus chiasmatis on the upper aspect of this bone, but is placed lateral to it (*vide infra*).

The **Optic Chiasma** or commissure (*chiasma opticum*), formed by the junction of the two optic nerves, is transversely oblong in shape, but varies in size in different subjects. The transverse width in fifteen specimens has been found to average 12·8 mm. (Gérard, 1904), and in thirty-six specimens a mean of 13·29 mm. was found, with extremes of from 10 to 19 mm.; the antero-posterior width in thirty-two cases varied from 4 to 13 mm., with an average of 8 mm. (Zander, 1896); it is from 3 to 5 mm. thick. The dimensions in the specimen illustrated in Fig. 197 are: breadth, 13 mm.; antero-posterior width, 11 mm.; thickness, 3 mm.; in Fig. 175 it is 13 mm. broad also. It is intimately clothed by the pia mater, and lies in the cisterna basalis of the sub-arachnoid space on the body of the sphenoidal bone. As regards the overlying brain, above, the chiasma is related to the third ventricle, in the anterior part of the floor of which it forms a prominence; behind it is the tuber cinereum, from which springs the infundibulum or stalk of the hypophysis cerebri (pituitary body); laterally is the anterior perforated substance (through which there pass the basal branches of the middle and anterior cerebral arteries), and, more anteriorly, the olfactory tubercle at the base of the olfactory tract. A more important relation is the internal carotid artery, which, ascending to the brain after piercing the roof of the cavernous sinus, is in close contact with the lateral side of the chiasma just before dividing into the anterior and middle cerebral vessels (Fig. 182, p. 340); the two anterior cerebral arteries, united by the anterior communicating vessel, lie likewise in contact with the front of the chiasma, and aneurysms of these vessels could compress it against the cranial base (Mitchell and Bramwell).

As regards its relations to the base of the skull, it is to be noted that the chiasma does not lie in the *sulcus chiasmatis* of the body of the sphenoidal bone, as is usually stated, but over the tuberculum sellae (olivary eminence) behind this groove, as in Fig. 177, p. 331, or even farther back still and over the diaphragma sellae, covering the *hypophysis cerebri*, as in Figs. 197, 199; the farther back the chiasma lies the more acute will be the angle formed by the two optic nerves as they enter it. The exact position has been studied by Zander (1896), Lawrence (1894), who describes the development of the bony parts in this region, and Wallis (1917) from the clinical point of view. The last-named author examined eleven subjects, and in only one did the chiasma lie in the generally described position; he found it usually to lie over the diaphragma and partly on the dorsum sellae, as is seen in the figures last referred to; he concludes, in agreement with Zander and Lawrence, that whilst the chiasma does occasionally rest on the sulcus chiasmatis, it is nearly always completely posterior to this groove. Traquair (1916) considers it to lie above the posterior half of the fossa hypophyseos (sella turcica), and its posterior edge is on an average 1·58 mm. behind the dorsum sellae, and only exceptionally in front of this point. The most recent and valuable clinical work is by de Schweinitz (1923), who described 100 cases of pituitary

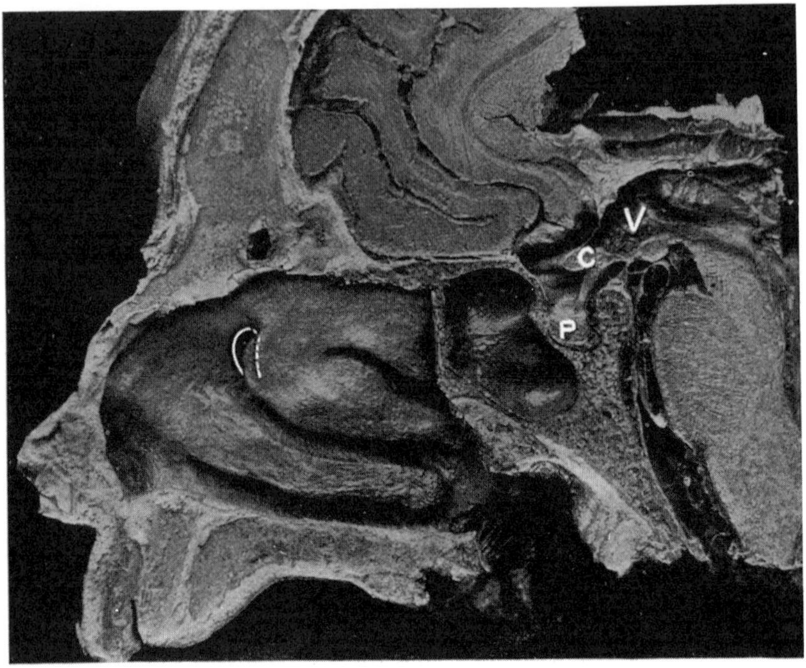

FIG. 199.—Half section of Head, adult male, showing right lateral nasal wall; the relative position of the fossa for the lacrimal sac is indicated by the white outline. The position of the **optic chiasma** (C), lying well above the *hypophysis* or **pituitary body** (P) and behind the *sulcus chiasmatis*, is seen, and the third ventricle (V) is shown in its relation to both. × ¾.

disorder, in conjunction with Schaeffer's co-operative anatomical observations on 125 subjects (1924). The latter author rightly considers the term 'chiasmatic sulcus' to be a misnomer, since he found the chiasma to be related actually to the groove only in five specimens, and in the rest lay wholly or partly over the *diaphragma sellae* covering the *hypophysis cerebri* or pituitary body. The various positions and frequency of occurrence are shown in Fig. 200. Changes in the position of the sella turcica, it may be noted, have been found associated with hereditary optic atrophy (Fisher, 1916; Taylor, 1919). It is to be noted further that the chiasma does not lie in close contact with the diaphragma and the hypophysis, but is separated from them by a space, the cisterna

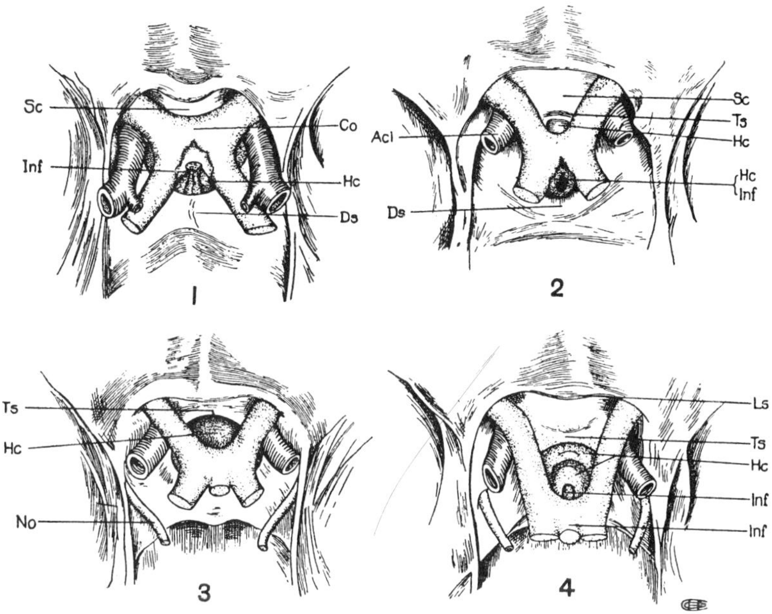

FIG. 200.—Variations in the **position of the Optic Chiasma** relative to the *sulcus chiasmatis* and *hypophysis cerebri*, according to the observations of Parsons Schaeffer (1924) on 125 subjects. Cf. Fig. 32, p. 66.

Position 1—the chiasma (Co) resting partly on the sulcus (Sc), mostly on the hypophysis (Hc) (5 per cent. of cases).
Position 2—resting wholly over the diaphragma (12 per cent.).
Position 3—almost so, but in part projecting backwards onto the dorsum sellae (Ds) (**79 per cent.**).
Position 4—lying on and behind the dorsum sellae (in only 4 per cent.).
Aci = internal Carotid Artery; Ts = tuberculum sellae; No = III nerve; Ls = limbus sphenoidalis.

basalis of the sub-arachnoidea; the interval is from 4 mm. (Traquair) to 7·5 mm. (Hirsch), or even 10 mm. (Cope, 1916); Schaeffer found it to vary from 0 to 10 mm. In the specimen illustrated in Fig. 199 the distance is 8 mm.; in Fig. 139, p. 259, where the interval between the chiasma and the hypophysis is clearly seen, it measures 5 mm.

The relation of the sphenoidal air-sinuses to the chiasma

will naturally depend upon their degree of backward extension (see p. 65). They may extend only as far back as the posterior lip of the olivary eminence, and lie altogether in front of the fossa hypophyseos and the normally situated chiasma, but Cope (1917) found them more often to undermine the whole fossa, and they are often separated from it by only a thin plate of bone, as in Fig. 128, p. 236. Loeb (1909) found that the chiasma is nearly always related in front to both sphenoidal sinuses and never to the posterior ethmoidal cells. The relations of the chiasma to the third ventricle of the brain and to the pituitary body, of the optic nerve to the sulcus chiasmatis, and of the sphenoidal air sinus to all these structures are well shown in Fig. 199 above, in which preparation, it is to be noted, but little shrinkage and displacement of the parts have occurred.

The **optic tract** (*tractus opticus*), after leaving the chiasma, diverges from its fellow of the opposite side in front of the interpeduncular space, and sweeps laterally and backwards between the tuber cinereum and the anterior perforated substance round the lateral surface of the crus cerebri, to which it is intimately applied, with the posterior cerebral artery, which supplies it, just below it; it becomes flatter as it proceeds, and, reaching the postero-lateral aspect of the optic thalamus, becomes divided by a shallow furrow into two so-called roots. The medial root is the smaller, and appears to end in the medial geniculate body; as can be seen in Fig. 201, its fibres, though apparently forming part of the optic chiasma and tract, do not come from the optic nerve, but sweep across the back of the chiasma to the opposite side, forming Gudden's commissure (p. 400). The lateral root, containing all the fibres of retinal origin, can be traced to the lateral geniculate body, the pulvinar (see p. 401), and the superior colliculus. The lateral geniculate body receives the largest number of fibres from the tract (80 per cent., according to Monakow, 90 per cent., according to Brouwer); they partly sink into its interior, where most of them end in the grey matter between its lamellae, but some may traverse it to enter the pulvinar. Most of the fibres to the pulvinar,

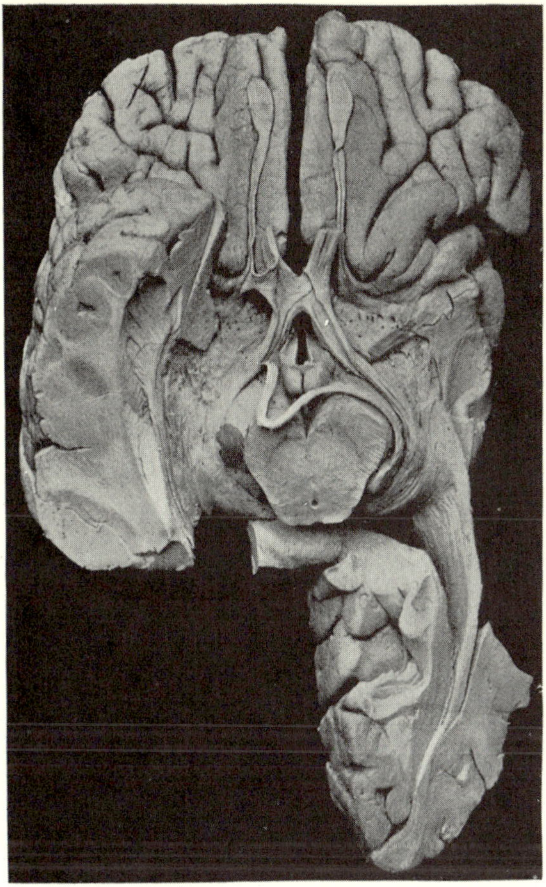

FIG. 201.—Dissection of Brain from inferior aspect to show the **Optic Chiasma** and **Tracts,** the **Lower Visual Centres,** and **the Optic Radiation.** A flap has been turned aside from the chiasma, and the crossed nasal fibres of the left optic nerve indicated. The commissure of Gudden has been dissected away from the back of the chiasma and the continuity of its fibres both here and in the optic tract demonstrated, as far as the medial geniculate body. Lateral and slightly anterior to this body is a swelling, the lateral geniculate body, into which the rest of the fibres of the tract are traceable, and beyond it, to the pulvinar. From these centres the optic radiation is seen sweeping backwards to the occipital pole. × $\frac{3}{4}$.

therefore, have passed either through or over the lateral geniculate body on their way to it. The fibres to the superior colliculus reach it directly through the superior brachium, which lies, however, in close contact with the lower aspect of the lateral geniculate body.

These bodies, which are known as the 'lower visual centres,' are situated on the postero-lateral aspect of the mid-brain (Figs. 201, 203). The lateral geniculate body is an oval mass on the lateral side of the thalamus, and is connected to the superior colliculus by a slender band, the superior brachium. The superior colliculus (anterior or superior quadrigeminal body) lies on the dorsal aspect of the mid-brain above the inferior colliculus and below the thalamus and the pineal body. The pulvinar (posterior tubercle of the thalamus) is the prominence on the posterior end of the thalamus. The blood supply is provided by the posterior cerebral artery (postero-lateral set of basal branches).

The retinal fibres of the optic nerve, chiasma and tracts have thus been traced from their origin in the retina to their termination in the lower visual centres of the mid-brain. In the appendix to this work which follows, the second part of the path taken by the visual impressions will be traced to its ultimate destination in the occipital cortex, and an analysis of the fibres of the whole tract given. It will add to the interest of this section on the nerves of the orbit if a brief account is then given of their nuclei of origin and their connexions, but no attempt will be made here to cover this large and complicated field in detail; the bibliographies given in such papers as are referred to will enable the reader to find the literature bearing upon the subject.

PART IV.—APPENDIX

THE CEREBRAL CONNEXIONS OF THE NERVES

1. The Optic Radiation.
2. The Cortical Visual Centre.
3. Analysis of Fibre Tracts.
4. The Nuclei of Origin.
5. Reflex Paths.

1. THE **Optic Radiation** (*radiatio occipito-thalamica of Gratiolet,* geniculo-calcarine path) is the second or intra-cerebral link in the pathway of the visual impressions from the retina to the perceptive centre in the cortex, and it extends from the lower visual centres of the mid-brain, runs through the central white matter of the posterior part of the cerebral hemisphere, and ends in the surface of the occipital pole. It consists of a fairly narrow (3 mm.) but deep band of fibres, which arise as processes of cells mainly of the lateral geniculate body, but some also, according to most authors, from the pulvinar of the thalamus; they emerge from the lateral side of these bodies, and passing through the retro-lenticular or posterior part of the internal capsule, bend backwards to run along the lateral wall and round the posterior horn of the lateral ventricle, separated from it by the tapetum of the corpus callosum, to reach the cortex of the occipital pole. The dorsal fibres of the radiation pass directly backwards into the cortex, but the ventral ones make a loop forward into the temporal pole and sweep round the anterior horn of the lateral ventricle before turning back to reach their destination, as described by Flesching, Henschen, and confirmed by Meyer (1907); a series of 33 cases where a tumour of the temporal lobe produced defect of vision by implication of this loop is given by Harvey Cushing (1922). All the fibres ultimately end in arborizations round the cells of the fifth or granule layer of the cortex in an area which constitutes the 'higher visual centre'.

2. The **Cortical Visual Centre** (higher visual 'centre,' or visuo-sensory area) is comprised in an extensive area of the

cortex on the medial surface and pole of the occipital lobe, chiefly in the region of the cuneus (the wedge-shaped area of cortex lying between the parieto-occipital and calcarine fissures). The extent of this area, as deduced from clinical observations of cases of disease or injury and from experimental investigation, shows much variation and is difficult to

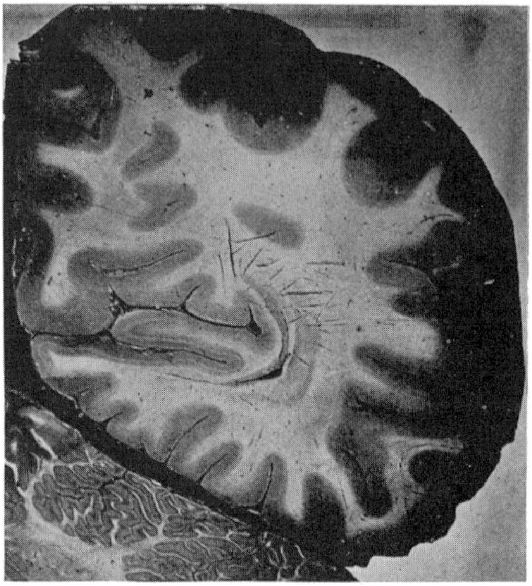

FIG. 202.—Transverse section through the Right Occipital Lobe of the Brain, seen from behind, to show the **white line of Gennari** in the cortex of the lower lip of the calcarine fissure (marked C in the figure opposite).

define with exactness. The cortex in this region is characterized, however, by the presence of a white line in the grey matter, visible to the naked eye and first noticed by Gennari in 1776, and the region is consequently named the *area striata*. The appearance, which is illustrated in the figure above, is caused by a thick granule layer of cells with a dense plexus of medullated fibres situated in the fifth cortical layer, where the fibres of the optic radiation have been traced to end. The exact limits of the striate area have been mapped out by microscopical examination of this cell layer (Mott, Bolton), by the character of the nerve fibres (Campbell), by the period of myelinization of the fibres (Flechsig), and by the naked eye differences in the cortex as seen in sections of fresh brains (Elliot Smith, 1907). These various delimitations practically coincide, and the account of the region which follows is taken from the description of the author last named.

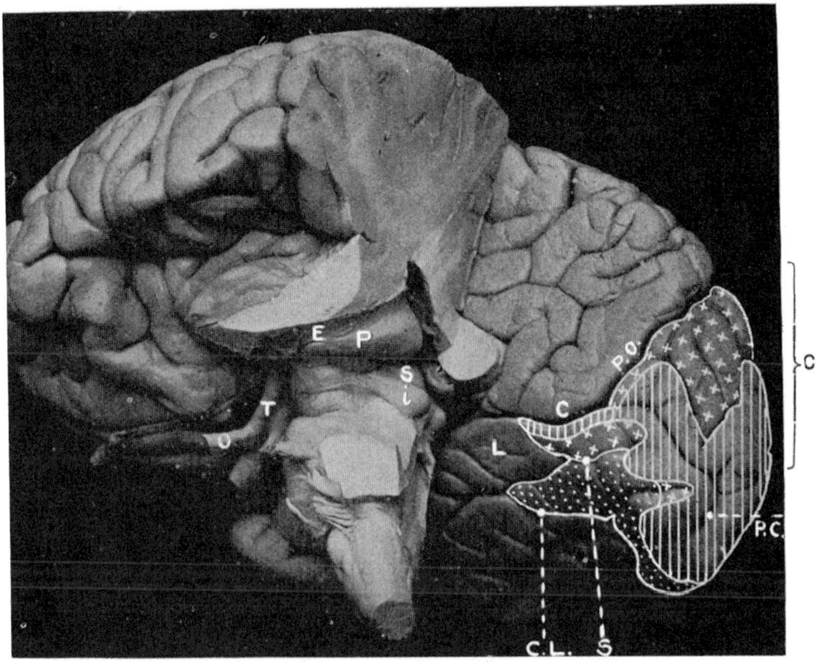

FIG. 203.—Dissection of Brain, viewed from left side and tilted over to show the **cortical visual areas,** according to the findings of Elliot Smith. Half of the left hemisphere has been cut away to expose the medial surface of the right occipital lobe. Compare with posterior pole of next figure. × about $\frac{2}{3}$.

The striate area is marked by vertical lines, the para-striate area by crosses, and the peri-striate area by dots. C. = the calcarine sulcus; P.C. = the terminal part of the posterior calcarine sulcus; P.O. = the parieto-occipital sulcus; C.L. = the collateral sulcus; S. = the sagittal sulcus of the gyrus lingualis (L.). The lower visual centres are also displayed: O. = the optic nerve; T. = the optic tract, traceable into a small elevation just below E., the lateral geniculate body, and P., the *pulvinar*. Below P. is the more conspicuous medial geniculate body. Below the hinder end of the pulvinar are seen the colliculi, superior (S.) and inferior (*i.*); CN = Cuneus.

If this area of cortex be excised and spread out flat, it will be found to present an elongated ovoid form, which covers about 3,000 sq. mm. (varying from 2,700 to 4,000 in different brains). The narrow end of the ovoid lies just behind and below the splenium of the corpus callosum, and the rest of it extends backwards from this point to the occipital pole, and beyond it on to its lateral aspect. In the course of development this area becomes folded along its axis during the sixth month, and so forms the calcarine fissure or sulcus (of Huxley). By a fold which develops in the occipital lobe above it (the parieto-occipital sulcus), it is divided into two parts: an anterior, the calcarine fissure proper (*sulcus calcarinus*), and a posterior (*sulcus calcarinus posterior*); the area overlapping the lateral surface of the pole may also become folded into a *sulcus calcarinus lateralis*. The anterior part of the fissure is much deeper, more constant in form and position, and more precocious in development than the posterior part, and phylogenetically it is much older. There is also a fundamental structural difference between them, for the stria of Gennari is found only in the lower wall and lip of the anterior part of the fissure or sulcus calcarinus proper, as is seen in Fig. 202, whereas it extends throughout both walls of the posterior calcarine sulcus, and in most cases beyond its lip on to the surface of the cuneus and the gyrus lingualis, that is, over both lips of the fissure. The upper and lower boundaries of the area are often demarked from the adjacent cortex by shallow limiting sulci. Cohn and Papez (1930), in a comparative study of the two areas in human brains, find the right one to be generally longer (8·8 cm. against 7·9 cm.) and larger (7·9 sq. cm. against 7·1 sq. cm.) than the left.

The area striata is held to be the visuo-sensory area or the perceptive 'centre' for the retinal impressions, since its destruction entails blindness of parts of each eye. It should, however, be realized that no cortical area can properly be described as an exclusive centre of a particular function, but such 'centres' are merely nodal points in an exceedingly complex system of neurones which must act as a whole to perform any function whatsoever (Herrick, 1922). The projection of the retinal fields in this area is given on p. 405.

This area is surrounded by two peripheral concentric bands of cortex, which again present structural differences, though not so well marked: an inner, called the *area para-striata*, and an outer, the *area peri-striata*, which extend considerably on to the lateral aspect of the occipital lobe. These areas are doubtless visuo-psychic or visual memory 'centres', since injury to this part of the cortex entails 'mind blindness': the objects are seen, but convey nothing to the patient's mind.

The whole visual area in the occipital lobe appears divisible, therefore, into two parts: an inner, the striate area, closely following the calcarine fissure and serving for the reception of primary visual impulses, and an outer, the para- and peri-striate areas, destined for the further elaboration and storing of the visual sensory impressions. Further subdivision of these adjacent regions into areas specially concerned with appreciation of colour, form, places, names, space, &c., has been suggested, largely on evidence derived from clinico-pathological cases, for a consideration of which, together with that of other 'centres', the reader may consult the article by Mills in Posey and Spiller's work (1906).

The cortex of the angular gyrus (A. in Fig. 204) is considered by some authors to be an important, if not the most important, visual area; it is said to be concerned with central direct vision of the opposite eye (and possibly with the pulvinar, p. 401), whilst the occipital pole area is connected only with the peripheral parts of the retina, a view that has not met with general acceptance. Injury of this part of the cortex more certainly entails 'word-blindness', the patient being unable to understand printed or written matter, though spoken words are understood; it is therefore termed the visual word memory 'centre', and is said to be on the left side of the brain in right-handed subjects. This area is also stated to be directly concerned with winking movements of the eyelids.

The motor centre for eye movements is placed on the frontal lobe, on the posterior part of the middle frontal convolution in front of the prae-central sulcus (F.M. in Fig. 204). Stimulation of this area on the right side causes both eyes to turn to the left ('conjugate deviation'), as stated by Riley (1930), Carleton, Gordon Holmes, Duane (1924), but the angular gyrus is also stated to be a centre for this movement (Ferrier, Grasset). It is found, moreover, that stimulation of any part of the occipital lobe produces eye movements; thus, stimulation of its upper surface on the right side causes both eyes to turn downwards and to the left; excitation of its posterior part causes both eyes to turn upwards and to the left; and stimulation of the intermediate zone, specially on the mesial surface, causes lateral deviation to the left. Such

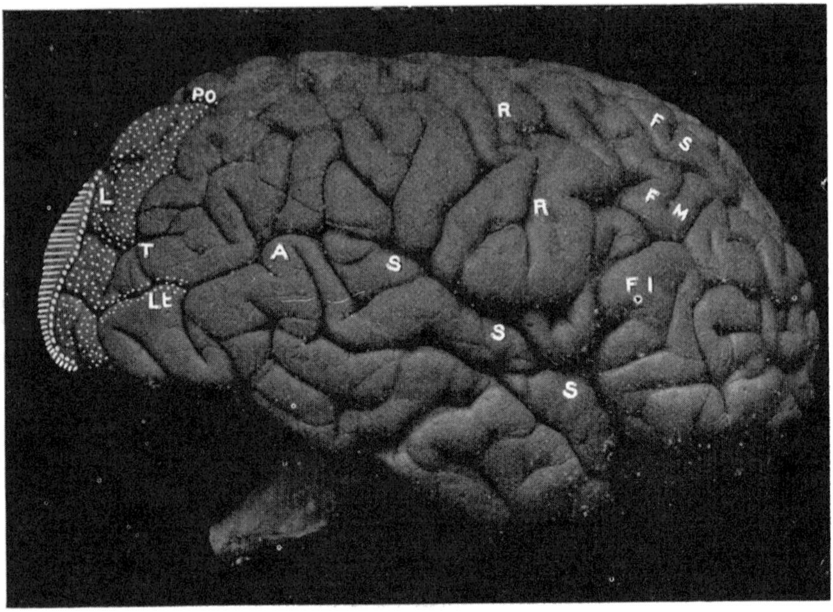

FIG. 204.—Lateral view of Right Hemisphere, to illustrate the approximate position of the '**centres' connected with vision** (see text).

On the occipital lobe, P.O. = parieto-occipital, L. = lunate, T. = transverse occipital, Lt. = lateral sulcus; the striate area is indicated by cross-lines, the para-striate and peri-striate areas by large and small dots respectively. A. is placed on the angular gyrus, which surrounds the upturned end of the superior temporal sulcus. S., S., S. is the lateral cerebral fissure (fissure of Sylvius). R., R. is the central sulcus (fissure of Rolando). F.S., F.M., F.I. are placed on the posterior parts of the superior, middle, and inferior frontal convolutions respectively; between them and the central sulcus are the prae-central sulcus and convolution.

impulses, no doubt, pass through the descending or corticifugal fibres of the optic radiation. Miceli (1923), in a series of papers on the centres for ocular movements (in the dog), states that no cortical centre for convergence is known, but centres for other eye movements can be established in the pons. The centre correlated with movements of the eyelids is apparently situated in the prae-central convolution, behind the centre for the ocular movements and just above the centre for the face and mouth, though, as already noted, the angular gyrus is said to be directly concerned with winking movements.

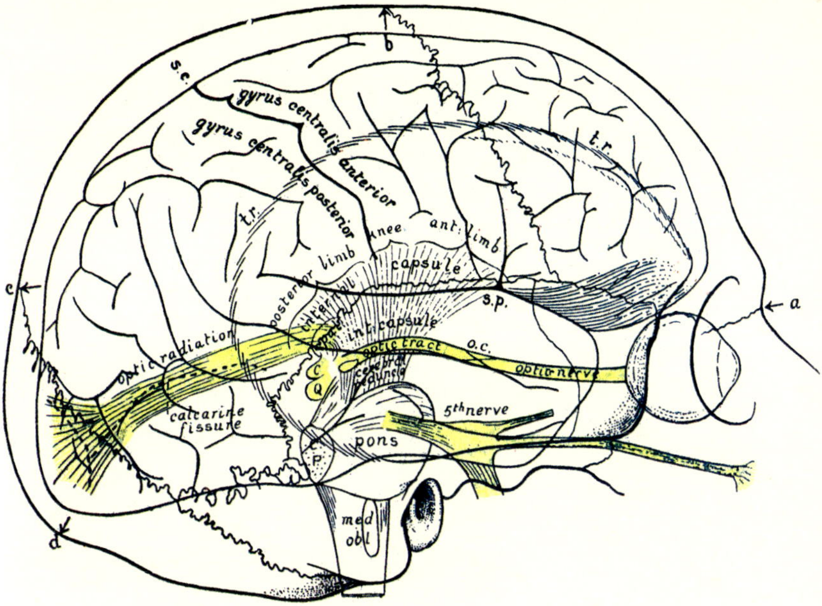

FIG. 205.—Right lateral aspect of the Skull and Cerebral Hemisphere outlined in black, with orthogonal projection of the medulla oblongata, pons, mid-brain, internal capsule, **visual tracts,** and fifth cerebral nerve. Man aged fifty-six years. Cephalic index, 70. (J. Symington.) (From Quain, 1908.)

a = nasion; *b* = bregma; *c* = lambda; *d* = inion; *t.r.* = temporal ridge; *s.c.* = sulcus centralis; *s.p.* = Sylvian point; *C*, *Q* = the colliculi.

The visual cortical area is connected with the corresponding area of the opposite hemisphere by commissural fibres which traverse the splenium of the corpus callosum; with the visual memory, visual word, auditory and speech centres by association fibres; with the lower visual centres, and possibly also with the lower motor nuclei, by descending or corticifugal fibres of the optic radiation.

The arterial supply for almost the whole of the striate area is derived from the calcarine branches of the posterior cerebral artery, the anterior border being often supplied by the middle cerebral (Shellshear, 1927).

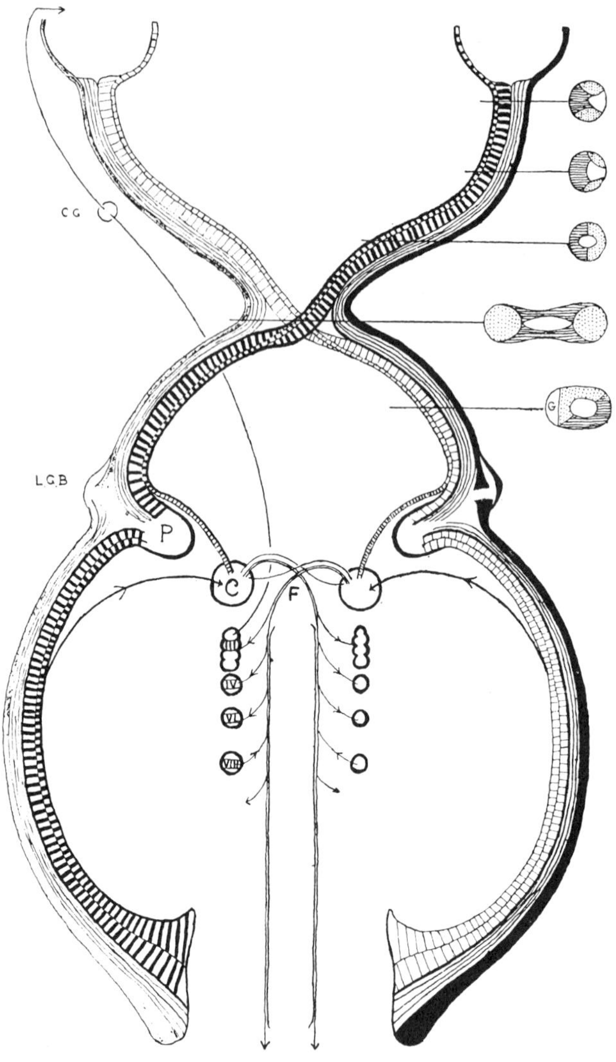

FIG. 206.—Scheme to illustrate the **visual paths** from the retinae to the occipital poles, those from the right eye being emphasized. The nasal or crossed and the temporal or direct fibres are shown passing into the lower visual centres, but the diagram is not intended to convey the idea that any particular band of fibres of the nerve or tract ends in any particular one of these bodies. Note that the connexion with the Pulvinar (P.) could not be established by Zeeman and Brouwer (see p. 401).

L.G.B. = the lateral geniculate body, and P = the *pulvinar*. The relative positions of the bundles in the nerve and tract (from Testut, 1911, after Hensen and Vialet) are shown in the cross-sections, where the crossed fibres are indicated by horizontal lines, the direct fibres by dots, and the macular fibres by the blank area. G = the position of the commissure of Gudden. The connexions of C, the superior colliculus, with the nuclei of the nerves to the muscles of the eyeball by means of the fountain decussation of Meynert (F), and also the fibres it receives from the higher or cortical visual centre, are shown. The path of the pupillary reflex from the oculomotor nucleus (III) through the ciliary ganglion (C.G.) to the *sphincter iridis* muscle fibres of the globe is also indicated.

As regards the influence of the **cerebellum** on the eye movements, Risien Russell (1894) concludes from experimental investigations that this part of the brain exercises a direct and perfectly independent action on the ocular muscles; the influence of the cerebrum, however, in correction of the phenomena of cerebellar deficiency is to be noted, since it has been found that the severity of the symptoms (atonia, asthenia and astasia in the muscles of the same side) occasioned by destruction of one half of the cerebellum of the dog gradually lessen (Black, 1916). A co-ordinating centre for ocular movements is described as lying in the fore end of the vermis close to the mid-line (Peter).

The relation of the various parts of the visual pathway to the skull is shown in Fig. 205. The external occipital protuberance, that is, the point inion, lies on a level with the lower margin of the occipital pole, and therefore just below the cortical visual centre.

3. **Analysis of the Fibres** in the visual paths and their destination.—In the **optic nerve** the fibres can be divided according to their function into:

(i) Visual fibres conveying visual impressions to the brain;
(ii) Pupillary fibres connecting the retina with the pupillo-motor centres;
(iii) Retino-motor fibres passing from the brain to the retina;
(iv) Inter-retinal fibres, connecting the two retinae;
(v) Possibly trophic fibres;

though our knowledge regarding the course of the paths taken by these groups becomes less certain in passing from the first to the last.

(i) The visual fibres are grouped in the nerve in three bundles according to the retinal fields from which they originate:

(*a*) From the temporal side of the retina; these temporal or 'direct' fibres pass through the chiasma into the optic tract and lower visual centres of the same side.

(*b*) From the nasal side of the retina; these nasal or 'crossed' fibres, more voluminous than the last since they come from about the medial two-thirds of the retinal field, pass across in the chiasma to the opposite tract and lower visual centres.

(*c*) From the cells of the fovea centralis of the macula lutea area of the retina; these central or 'macular fibres' comprise one-third of the total number, and coming from the area of direct and most distinct vision are the most important. In the nerve they lie at first lateral to and a little below the others, but later come to occupy a central position; at the chiasma they are generally considered to undergo partial decussation, half of the fibres passing across to the opposite side, though this is denied by some authors; in the tract they lie centrally, and they terminate like the other visual fibres in the lower visual centres. The relative positions of these bundles in the nerve and tract have been studied experimentally by Zeeman and Brouwer (1926) on monkeys; in the nerve they found fibres from the temporal side of the retina to lie laterally and the nasal ones medially; the upper retinal fibres lay above in both nerve and tract and the lower ones below; the macula fibres, large in number, were scattered but chiefly passed from a lateral position to a central one in both nerve and tract. With slight differences, these points are illustrated in Fig. 206: Gradle (1923), in describing a case, discusses the intra-neural position of the circumpapillary retinal fibres, and favours the theory of Leber and Fuchs that the peripheral ones pass into the outside and not into the centre of the nerve (as Wilbrand and Saenger on the contrary assume). The cortical representation is discussed on p. 406.

(ii) The 'pupillary fibres' of the optic nerve are of finer calibre, and are concerned with contraction of the iris on exposure to light; it is known that they run on the surface of the tract and superior colliculus (Ingvar 1927, 1928), and they probably enter only this lower visual centre; the whole path of this light-reflex is referred to later on p. 402.

(iii) The retino-motor group consists of exceedingly fine fibres, some of which form direct connexion between the cortical centre and the retina, whilst others appear to arise in the lower visual centres and pass thence to the retina; they may be vaso-constrictor or trophic in function, or concerned with certain movements of the retinal elements.

(iv) The inter-retinal fibres are considered to pass from one retina to the other, though the path is not definitely known; their presence may afford an explanation of sympathetic inflammation of the eyes, or of binocular contrast and after-images.

In the **optic chiasma**, the two orders of fibres, crossed and indirect, are not clearly separated (though an artificial demonstration can be prepared, as in Fig. 201); the lateral parts of the chiasma, however, are largely occupied by direct fibres and the median part by crossed fibres, the latter taking a somewhat curved course and some of them mounting in the root of the opposite nerve to a height of as much as 3 mm. (Gallemaerts, 1900). A similar arrangement was found by Wilbrand (1926) in a case of optic nerve atrophy, the medial fibres forming a loop in the opposite nerve.

In the **optic tract** of one side, in consequence of the partial decussation of the optic nerves, there are found four bundles of fibres of retinal origin:

Direct or temporal fibres from the eye of the same side,
Crossed or nasal fibres from the opposite eye,
Direct macular, and
Crossed macular fibres, of similar origins.

Each tract therefore contains all the visual fibres from the corresponding sides of both retinae; the right tract, for example, containing the fibres from the right halves of both retinae, including half the macular or central fibres from each eye, and it carries all the impulses originating from stimuli occurring in the left field of vision. The relative positions of the bundles, though by no means sharply defined, are seen in the last figure, from which it appears that the fibres from the different parts of the retina preserve the relative position of their origins in the optic nerve and tract. Zeeman and Brouwer (1926) found definitely in the tract of monkeys that fibres from the periphery of the upper retinal quadrant lie dorso-medially and those from the lower quadrant lie ventro-laterally, the two areas being largely separated by macular fibres forming a wedge-shaped mass to the upper lateral aspect. It may also be said that the different regions of the retina correspond point by point to similar regions of a transverse section of the optic nerve of the same side and to similar regions of a transverse section of the optic tract of the opposite side.

The tract, besides these visual fibres and the other groups noted above as constituents of the optic nerve, contains additional fibres, all of which, however, appear to bear no certain relation to vision; they are as follows:

The commissure of Gudden (*commissura inferior Guddeni*) is a well-marked bundle of fibres passing from the medial geniculate body of one side forwards along the optic tract as its medial root, round the back of the chiasma, and down the opposite tract, to end eventually in the inferior colliculus of the opposite side; the bundle can be dissected away completely from the rest of the tract and chiasma, as shown in Fig. 201; it has no connexion with vision, but is part of the auditory cerebral mechanism.

The commissure of Meynert (*commissura superior Meynerti*) is a much smaller tract of fibres lying on the mesial side of Gudden's commissure and pursuing a similar course; its fibres arise from a small basal optic ganglion in the grey matter between the third ventricle and the chiasma on the side of the tuber cinereum, and end in the sub-thalamic region; they do not appear to have any direct connexion with the optic nerve, though they have been called its basal root, and are described as passing to the retina.

The *tractus peduncularis transversus,* a tract of fibres occurring constantly in many animals, but only occasionally in man (30 per cent.), may be found just in front of the superior colliculus; it passes round the ventral aspect of the cerebral peduncle and enters the tegmentum near the exit of the oculomotor nerve, where it ends in a small nucleus (Gray). Since it has been found to undergo atrophy after enucleation of the eye or after removal of the cortical visual area it is supposed to be related to the secondary visual paths. It may be an aberrant part of the optic tract.

Finally, Darkschewitsch described some fibres of the optic tract as passing to the ganglion habenulae and the pineal peduncle, and thence through the posterior commissure to end in the oculomotor nucleus; he considered them to subserve the reflex changes in the pupil.

Of the **lower visual centres,** the LATERAL GENICULATE BODY is stated to receive 80 per cent. (even, according to Brouwer, 90 per cent.) of the retinal fibres of the optic tract. As in the nerve and tract, there appears to be some relative representation of the retinal fields in this body, since investigations suggest that the fibres from the upper and lower parts of the retina end respectively in its upper and lower halves, with the macular fibres lying chiefly between. For further details the reader should consult the important work of Zeeman and Brouwer (1926), who have made an exhaustive and well illustrated study of the projection of the retina in the primary optic neurons in monkeys, and the admirable survey of the subject given in a series of articles by Putnam (1926). Minkowski (1922) describes the degenerative effects on particular layers of the lateral geniculate body after destruction of one eye, and secondary degenerations produced by lesions of the striate area of the cortex. The comparative anatomy in certain monkeys is given by Woollard and Beattie (1927).

As regards the PULVINAR, whilst, as Henschen (1890) always insisted, it cannot be regarded as a primary visual centre, yet it may be concerned with vision; it is connected with the lateral geniculate body and with the angular gyrus, and possibly is concerned with such higher visual functions as stereognosis; its influences on pupillary reactions, which Henschen favoured, is uncertain. It is noteworthy that Zeeman and Brouwer were unable to trace degenerated fibres in monkeys by the Marchi method passing from the tract to the pulvinar, though Minkowski holds that a few such fibres do end there; it must be remembered, however, that negative Marchi results are not in themselves sufficient evidence to deny the existence of such fibres, although it is now generally agreed that the lateral geniculate body alone is directly concerned with vision.

The SUPERIOR COLLICULUS (or anterior quadrigeminal body) receives a much smaller proportion of retinal fibres from the optic tract; fibres also enter it from the lemniscus (whereby it is connected with the sensory tracts of the medulla and spinal cord), and from the cerebral cortex, the latter, together with the retinal fibres, entering it through the superior brachium; but, unlike the lateral geniculate body and the pulvinar, it sends no fibres outwards to the cortex. It is joined to its fellow of the opposite side by decussating fibres. Its most important efferent connexion is by means of fibres which cross the median plane, forming the 'fountain decussation of Meynert' (*decussatio tegmenti dorsalis*), to accompany the medial longitudinal bundle as the tectobulbar tract, whereby it is connected with the nuclei of origin of the oculomotor, trochlear, and abducent nerves. Other fibres from the colliculus descend as the tectospinal tract to the spinal medulla and cord (Fig. 206). These connexions afford pathways through which the reflex and possibly also the voluntary optic and ocular mechanisms may be conducted, and the superior colliculus thereby becomes of great importance. It receives impressions from the eye through the optic nerve and tract on the one hand, and impulses from the cerebral cortex through the optic radiation on the other, and it can link up either path with the nuclei of the motor nerves or with the tracts in the medulla and cord by means of the fountain decussation and the tectal tracts. Thus the superior colliculus may be regarded as a ganglionic centre for the co-ordination of visual impressions with those of other regions of the body in influencing its movements. A localization of the retina in these bodies is merely suggested by Zeeman and Brouwer's experiments; Montalti (1928) confirms the fact that the optic nerve fibres enter by two distinct bundles.

The student of morphology should read the works of Kappers (1929), and Edinger (1899), who display in lower animals the interesting origin of the superior colliculus as the primitive 'sight brain'; and the papers of Elliot Smith (1924, 28, 30), who traces in higher forms the evolution of visual impressions of infinitely greater importance than the mere appreciation of light and their transference to the cerebral cortex. It is interesting to note that it is the primitive function that is affected in the Argyll-Robertson pupil phenomenon.

The fibres concerned with the **pupillary light-reflex** most probably pass through these paths, though the exact course is not known for certain. The reflex contraction of the pupil may be provoked by light falling upon any part of the retina, though the macular region is the most important, and a bilateral effect upon the pupils (consensual reflex) is obtained by throwing light upon either eye. The fibres traverse successively the optic nerve, the chiasma (where they are generally considered partly to decussate), the optic tract and superior brachium to enter the superior colliculus. In traversing the brachium they present a large number of non-medullated fibres and lie superficially, a point emphasized by Ingvar (1928) in a paper on the parthenogenesis of the Argyll-Robertson pupil. In the colliculus, according to most authorities, they end, and fresh fibres convey the impulses through the fountain decussation to either the sub-nucleus or the lateral principal nucleus of the oculomotor nucleus (see p. 406), though this last connexion has not actually been traced. A third set of fibres convey them along the oculomotor nerve and into the ciliary ganglion by its motor root; and after a final interruption the path to the constrictor iridis muscle is provided by the short ciliary nerves. Such course has been indicated in Fig. 206; it is also presented in the scheme devised by Behr (1924), which is shown with clinical bearings in Fig. 207. Adler (1926) gives a plan which differs from that of others in that the afferent pupillary fibres decussate not in the chiasma but partially so between the superior colliculi and the oculomotor nuclei. Parsons (1924), in a dissertation on the physiology and pathology of the pupillary reactions, gives a list of the requirements to be met practically by any such scheme; he states that each macula sends fibres to both constrictor centres. On the other hand, some authors consider that the first axons from the retinal cells pass independently of the colliculus and directly to the oculomotor nucleus; and yet another path is that noted on p. 400 as given by Darkschewitsch. (See also pp. 364, 372, 344.)

The **optic radiation,** or geniculo-calcarine path, is composed mainly of ascending fibres arising from cells of the lateral geniculate body and partly from the pulvinar; none comes from the superior colliculus; it also contains descending fibres passing from the visual cortical area to these lower visual centres, together with commissural and association fibres. It resembles the optic tract of the same side in that it receives impressions from the corresponding lateral halves of both retinae, and probably from the macular field of each. As regards the relative position of its constituent fibres, they are considered to represent the retinal areas in the same manner as do the fibres of the optic tract described above on p. 399 (Zeeman, Brouwer, Putnam).

In the **higher visual centre** round the calcarine fissure,

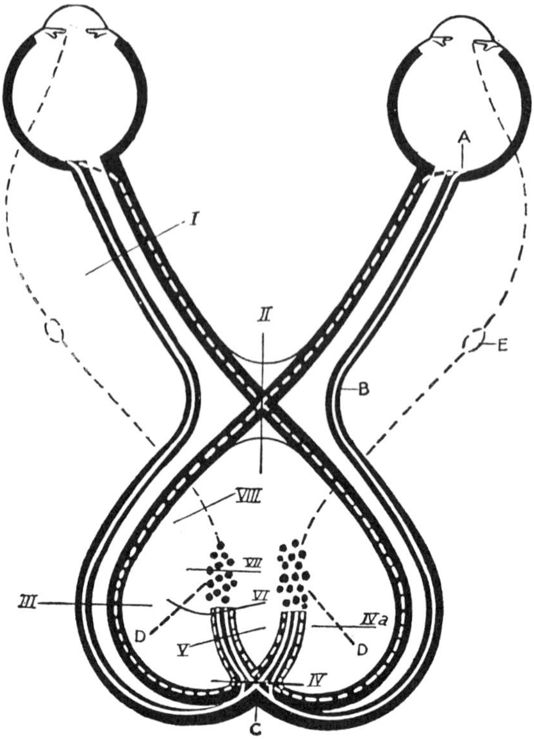

FIG. 207.—Scheme to illustrate the nerve-fibre paths concerned in the **Pupillary Reflex to Light.**

The dotted and continuous lines represent the crossed and uncrossed macular fibres. A, macula; B, chiasma; C, mesencephalic decussation; D, path of convergence reaction; E, ciliary ganglion. The sites of the lesions giving rise to various types of pupillary anomalies are indicated by numbers. The numbered lesions denote sections which give rise to the following symptoms:

I. Optic nerve: *unilateral amaurotic paralysis* (abolition of the direct reaction on ipso-lateral and the indirect on the contra-lateral side; retention of the indirect on the ipso-lateral and direct on the contra-lateral side).

II. Optic chiasma: *bitemporal hemianopic paralysis.*

III. Optic tract: *contra-lateral hemianopic paralysis.*

IV. Central decussation: *bilateral reflex paralysis* (bilateral abolition of the light reflex, retention of convergence reaction and lid-reaction).

IVa. Isolated abolition of convergence: retention of light reflex.

V. Between central decussation and mesencephalic centre; *ipso-lateral reflex paralysis* (ipso-lateral abolition of the direct and indirect light reaction, with retention of both contra-laterally; normal convergence and lid-reaction bilaterally).

VI. Supra-nuclear complete paralysis of light, convergence, and lid reflex on ipso-lateral side; contra-lateral side normal.

VII. Ipso-lateral pupillary paralysis.

VIII. Absolute paralysis to light, direct and indirect and to convergence and the lid-reaction; contra-lateral pupil is normal.

(After Behr from Duke Elder, 1927.)

it is first to be observed that Elliot Smith (1928, 1930) identifies clearly a macular part of the striate area which extends round the occipital pole as far as the lunate sulcus and is as extensive in size as the whole peripheral part; he discusses the consequences of the evolution of this region and the eminent advantages which man possesses by its use. As

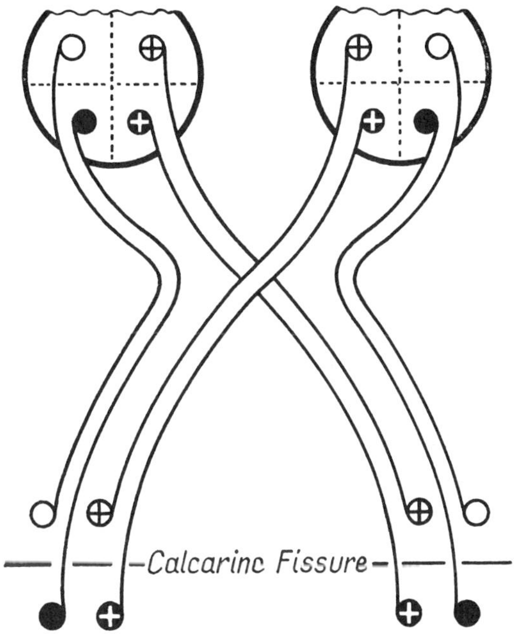

FIG. 208. Scheme to represent the **Projection of Retinal Fields** in the Calcarine Area of the higher **Visual Cortical Centre.**

regards the representation of the retinal fields in these areas, first it appears established that each hemisphere receives visual impulses from the complete ipsi-lateral half of each retina (consequently from the contra-lateral half of the field of vision): in other words, the right halves of both retinae are related to the cortex of the right occipital lobe and the left halves to that of the left lobe. Then the upper half of each retina is represented in the upper part and the lower half of each retina in the lower part of both visual areas, by ipsi-lateral quadrants to each, the line of separation being the calcarine fissure (Fig. 208). Further, the centre for the peripheral part of the retina lies in the anterior part of the area, whilst that for the macular region is regarded as diffusely spread over the whole cortical area by Monakow, but as localized in the posterior part by Henschen and Wilbrand. Lastly, Pfeifer (1925, reviewed by Kronfeld, 1929) described commissural visual fibres traversing the splenium of the corpus callosum, thus supporting Heine and Lenz's theory of a bi-cerebral representation of each macula, contrary to the view of Holmes and Brouwer. These conclusions are drawn from the study of war wounds by Holmes and Lister (1916, 1920), Riddoch (1917), and Monbrun (1918), and generally agree with clinical observations such as those of Putnam (1926), Lenz (1926), and Brouwer (1930).

4. The **Nuclei of Origin** of the orbital nerves and their connexions:—

The **oculomotor nucleus** (*nucleus nervi oculomotorius*) lies in the upper part of the cerebral peduncle, parallel to and just beneath the aquaeductus cerebri (aqueduct of Sylvius), below and coextensive with the superior colliculus; it is from 5 to 6 mm. long, and extends nearly the whole length of the floor of the aqueduct from the posterior part of the wall of the third ventricle in front to the anterior part of the fourth ventricle behind, where it is in close relation with the nucleus of the fourth nerve. The two oculomotor nuclei are separate above but touch below, and their infero-lateral faces are in contact with the medial longitudinal fasciculus of each side. The subdivision of each mass into smaller centres for the supply of the different muscles of the eyeball is now generally recognized, though their limits are not sharply defined, and their identity varies according to the views of different investigators. As generally described, there is a large-celled lateral, principal or 'chief' nucleus on each side, a large-celled mass lying inferior and medial to it and fused with its fellow of the opposite side to form a 'central nucleus' (of Perlia), and a small-celled group on either side in front (the Edinger-Westphal nuclei). Tsuchida (1906) found only the large-celled nuclei anatomically constant, the smaller ones showing marked variations. As regards functional association, the lateral chief nucleus is recognized as supplying the extrinsic muscles of the globe, and the scheme of its subdivision as devised by Bernheimer has the approval of Brouwer (1918), as shown in Fig. 210. The latter authority, however, con-

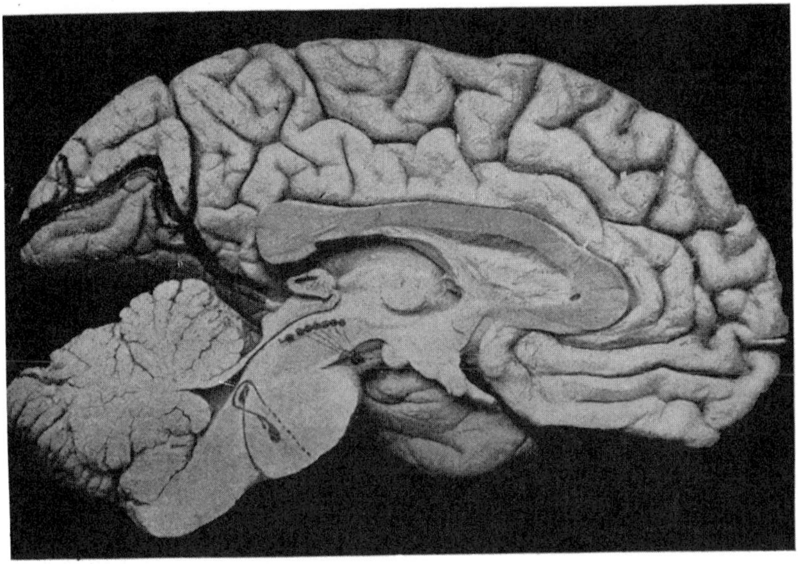

FIG. 209.—Mesial section of Brain, left half viewed from the medial aspect to show, schematically, the nuclei of the third or oculomotor and fourth or trochlear nerves, represented by the row of black dots; the position of the abducent or sixth nerve nucleus is also shown lying within the loop formed by the facial or seventh nerve in the pons; the nucleus of the latter is represented as consisting of a main mass and a secondary, the superior facial nucleus.

The calcarine fissure has been outlined in black. × $\frac{2}{3}$.

siders that the central nucleus of Perlia is formed by fused portions of that part of the lateral chief nucleus which is associated with the medial rectus in subserving its function of convergence in near vision (the remainder of the medial rectus centre being concerned with other movements of the muscle). He states that the nucleus of Perlia is the last portion of the oculomotor nucleus to be differentiated, and its appearance in the primate series accords with the inception of its now generally recognized function of convergence, although Bernheimer considered it to subserve accommodation by the ciliary muscle.

As regards the function of the Edinger-Westphal nucleus opinions differ. It makes its appearance in phylogeny before Perlia's nucleus, and is specially developed in mammals with binocular vision. It is said to receive its fibres from the optic tract direct (Cunningham, 1931). Clinical evidence associating it with pupillary movements has been presented by Lenz (1924), and the assumption that it is the nucleus of accommodation, supplying the ciliary muscle and sphincter iridis, is favoured by Brouwer and most standard works. The idea, however, has been negatived by a number of investigators; indeed Latumeten, according to Le Gros Clark, seems to have proved beyond doubt that the cells of this group give no fibres which pass direct to the ciliary ganglion or the eye itself (proceeding to the ciliary muscle and sphincter iridis), though no alternative suggestion as to the origin of such fibres is forthcoming. Parsons (1924) stated that recent work tended to prove the truth of the theory that the sphincter iridis centre lies near the front inner end of the oculomotor nucleus, the centre for convergence on the middle inner aspect, and that for accommodation at the hinder end of the nucleus. Le Gros Clark (1926), from a morphological study, infers that there is more reason to describe localization of function in the nucleus rather than representation of individual muscles, though it may be pointed out that the fourth and sixth nuclei are segregated. His evidence indicates a division of the large-celled elements

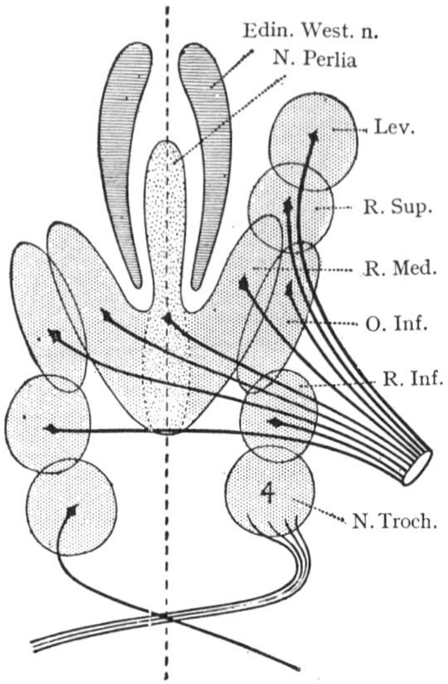

FIG. 210.—Diagram of the **Oculomotor Nucleus** (from Cunningham's Text-book, 1931, after Brouwer).

in a *dorso-ventral* rather than a cephalo-caudal direction; he describes three fundamental groups, formed by typical large pyramidal motor cells: a dorso-lateral (probably concerned with upward movements of the eyes), a ventro-medial (downward movements), and a para-medial or central (inward movements), which agrees with Kapper's divisions. This scheme seems to harmonize with the divergent views of many observers, and is also compatible with the data of comparative anatomy. Le Gros Clark further recognizes the antero-lateral group of small stellate cells (the Edinger-Westphal nucleus) on each side, but adds a small caudal central group; he does not consider that the central position of the nucleus of Perlia necessarily implies association with convergence or with binocular vision. On the other hand, Paton and Mann (1925) conclude on embryological grounds that the Edinger-Westphal nucleus is concerned with pupillary movement and the central one of Perlia with convergence. Riley (1930) discusses the question from both neurological and clinical aspects, drawing material from Wilbrand and Saenger's comprehensive presentation (1921), and reproduces twelve figures of different observers' findings from Brouwer's article. A discussion on the function of the different parts of the nucleus is given also by Duke-Elder (1927).

It may be added that there is a group of large nerve cells lying dorsal to the upper end of the oculomotor group, called the 'nucleus of Darkschewitsch'; it is connected with the medial longitudinal fasciculus, but is not supposed to be concerned with ocular movements and its function is unknown.

The majority of the nerve fibres pass out from that side of the nucleus on which lie the cells from which they originate, but there is a decussation of fibres principally from the posterior parts of the lateral nuclei, so that the inferior rectus muscle appears to receive crossed fibres only, the superior rectus and levator direct fibres only, but the medial rectus and inferior oblique muscles both crossed and direct fibres; the intrinsic muscles of the globe receive their fibres direct.

The fasciculus so formed by fibres from the nucleus of the same side, but also by some crossed fibres from the opposite nucleus, passes forwards through the medial longitudinal bundle, the tegmentum, the red nucleus, and the substantia nigra, and emerges on the medial side of the cerebral peduncle at the superficial origin of the nerve.

The connexions of the oculomotor nucleus are:

(i) With the superior colliculus, through which impulses reach it from the visual system, as described above.

(ii) With the cortical motor centre for eye movements.

(iii) With the cerebellum through the superior cerebellar peduncle.

(iv) With the trochlear, abducent, and vestibular, and possibly other nuclei by means of the tectobulbar tract and medial longitudinal fasciculus.

As regards these last communications, it has been supposed that some of the fibres from the oculomotor nucleus pass down the fasciculus into the facial nucleus and then run along with that nerve to supply the orbicularis oculi, corrugator supercilii, and frontalis muscles, though this has been questioned (see p. 187).

Another interchange of fibres through the same path is suggested as explanatory of the harmonious action of the lateral rectus (sixth nerve) with the medial rectus (third nerve) in producing simultaneous movements of the two eyeballs to the right or left (conjugate deviation), for which several explanations have been given. The sixth nucleus sends no fibres directly into the oculomotor nerve, but it may send fibres, or be connected with fibres, which ascend in the medial longitudinal fasciculus, and establish connexions with that part of the oculomotor nucleus from which the crossed fibres to the medial rectus of the opposite side arise (Fig. 211). Marquez (1930) gives an exhaustive review of the subject, as does Adler (1930) in a consideration of three clinical cases.

The **trochlear nucleus** (*nucleus nervi trochlearis*) lies beneath the floor of the Sylvian aqueduct opposite the upper part of the inferior colliculus; it appears to be the detached end of the column of grey matter which forms the oculomotor nucleus (Figs. 209 and 210). The fibres from this origin run lateralwards and downwards through the tegmentum, then turn back and inwards into the upper part of the anterior medullary velum; here they decussate, though not completely (Kidd, 1922), with the corresponding nerve of the opposite side, and emerge from the surface of the brain at the side of the fraenulum veli just below the inferior colliculus, as the superficial origin (Fig. 203). Parsons (1921), in an article on the morphology of the nerve, finds no satisfactory explanation of the decussation. Its central and cortical relations are similar to those of the oculomotor nucleus.

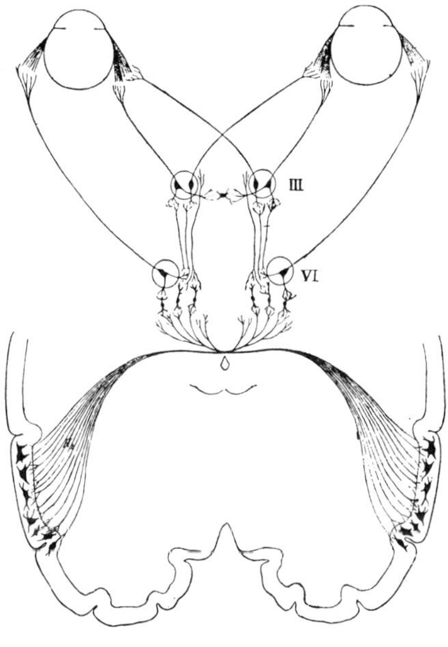

FIG. 211.—Scheme of innervation of conjugate and **convergent movements** of the eyes. (From Bernheimer, 1904.)

III = oculomotor nucleus; VI = abducent nucleus.

The **abducent nucleus** (*nucleus nervi abducentis*) lies beneath the colliculus (eminentia teres) of the floor of the fourth ventricle close to the mid-line and immediately above the striae medullares. It lies on the ventral side within the loop formed by the facial nerve, but has no connexion with it (Fig. 209). The fibres from these cells pass forwards and downwards on the medial side of the superior olivary nucleus, and farther on to the lateral side of the pyramidal tract, and after thus traversing the whole thickness of the pons, emerge at the superficial origin at the lower border of this body; they are direct and do not decussate. In lower animals other fibres arise from an 'accessory' or ventral sixth nucleus supplying the third eyelid and retractor bulbi (Terni, 1922.)

The central connexions are, like those of the oculomotor nucleus, with the visual and motor cortical areas, with the superior colliculus, and with the nuclei of the third, fourth, and vestibular nerves by means of the medial longitudinal fasciculus and also with the superior olive. By the medial longitudinal fasciculus a connexion is formed between eye movements and equilibration; amongst its constituents are fibres which arise in the lateral vestibular (Deiter's) nucleus of the auditory system and ascend, either on the same or opposite side, to form connexions with the nuclei of the third, fourth, and sixth nerves. A reflex arc is formed which is concerned with movements of the head and eyes in response to the vestibular stimulation of the nerve-endings in the semicircular canals of the ears, a stimulation arising from every movement and change of poise of the head. The superior olivary nucleus and its pedicle similarly afford a connexion between the cochlear or true auditory nerve and the abducent and oculomotor nuclei, explanatory of the simple auditory reflex of turning the eyes in the direction whence a sound comes.

It has been found that strong stimulation of either the semicircular canals, the vestibular nerve, or the nucleus causes an abnormal oscillatory side-to-side movement of the eyes known as *nystagmus*, a consideration of the mechanism of which is given by Wilson and Pike (1915), and Myers (1925). It may also be of interest to note that Fox and Holmes (1926), after a description of 'railway nystagmus', consider the value of the sign in localization of cerebral tumours; their numerous clinical observations support Stenver's hypothesis that the reflex centres concerned lie in the occipital lobe (centripetal) and second frontal convolution (the motor centre for eye-movements), connected by a path through the white matter of the hemisphere interruption of which, or of one of the centres, accounts for the phenomenon.

The **trigeminal nucleus.**—Three nuclear masses are associated with the trigeminal nerve in the brain-stem:—

(1) the MOTOR NUCLEUS, concerned with the muscles of mastication together with the mylohyoid and anterior belly of the digastric; it receives fibres from the pyramidal tract of the opposite side and from the rubro-spinal tract of the same side, and its emerging fibres, forming the 'motor root' of the trigeminal, join the mandibular division.

(2) The MESENCEPHALIC NUCLEUS, receiving fibres (the 'mesencephalic root' of the trigeminal) from the mandibular division, which are believed by Kappers to convey proprioceptive impulses arising in the neuro-muscular and neurotendinous end-organs in the muscles of mastication. Woollard (1931) has adduced evidence that other proprioceptive fibres reach this nucleus incorporated certainly in the oculomotor nerve and probably also in the other nerves of the extrinsic muscles of the eyeball. (See also the paper by Allen, 1925.)

(3) The SENSORY NUCLEUS, a long mass extending from the pons at the level of the nerve entry down to the second cervical segment of the spinal cord. It is divisible into two parts, (i) an upper enlarged mass, the superior or 'chief sensory nucleus' (the *nucleus sensibilisia* of Winkler, 1927), and (ii) a long descending bulbo-spinal or 'nucleus of the spinal tract' (the *nucleus b* and *gelatinosus* of Winkler), which by its descending root is continuous below with the *substantia gelatinosa Rolandi* of the spinal cord. The fibres of the sensory root of the trigeminal (other than the mesencephalic ones) on entering the pons divide into short ascending and long descending branches. The short ascending ones enter the superior or chief nucleus and convey to it epicritic (gnostic) sensibility from the whole trigeminal area, and probably some, if not all, of the proprioceptive impulses (Winkler, Gray); it is connected by fibres joining the mesial fillet of the opposite side with the thalamus, and thence is relayed to the cortex. The descending fibres form the 'spinal tract' of the trigeminal, and enter the bulbo-spinal nucleus, which thereby receives protopathic (vital—touch, pain, and thermal) impulses from the whole trigeminal area. The fibres of the three divisions of the nerve enter this spinal

nucleus in inverse order, the mandibular ones above, the maxillary in the middle, and the ophthalmic below (Gerard, 1923, and Stopford, 1925). It is pointed out by Paton (1926) that since this last region is on the same level as that of the origin of the great occipital nerve an explanation may thereby be afforded of the frequency of sub-occipital headache in eye disease. The spinal nucleus sends fibres into the spino-thalamic tract of the opposite side, and from the thalamus is connected with the cortex.

The secondary central connexions of these sensory nuclei are very extensive, since impulses of great importance are collected by the nerve from so wide and diverse a field. They are linked up with the nuclei of the other cerebral nerves, with the thalamus above and the spinal cord below, and it is possible, as Paton suggests, that such extensive distribution of the nuclei and their numerous relationships may explain why headache so frequently results from many different causes and is so vague in its distribution. (See on this the paper by Halphen, 1929.) A main path of inter-communication is afforded by the medial longitudinal fasciculus, which has been referred to in the preceding accounts of the third and sixth nuclei and is here to be considered further.

The **medial longitudinal fasciculus,** or posterior longitudinal bundle, consists of ascending and descending fibres which extend from an 'interstitial nucleus' (of Cajal) in the lateral wall of the third ventricle, down the pons, and into the spinal cord. It exists in all vertebrates, and its fibres are amongst the very earliest in the body to become myelinated, beginning as early as the 14th foetal week (Keene, 1931). It is accompanied by the tectobulbar and tectospinal tracts from the superior colliculus (p. 401), but its chief contributions are made by fibres from the vestibular nuclei, and doubtless its prime function is to co-ordinate movements of the eyes and head in response to stimulation of the eighth nerve, and to deal with maintenance of posture and equilibrium, but notably it (i) links up the great sensory trigeminal nuclei with the nuclei of the motor cerebral nerves; (ii) forms paths of inter-communication for the latter; and (iii) connects them with motor centres of the upper cervical cord

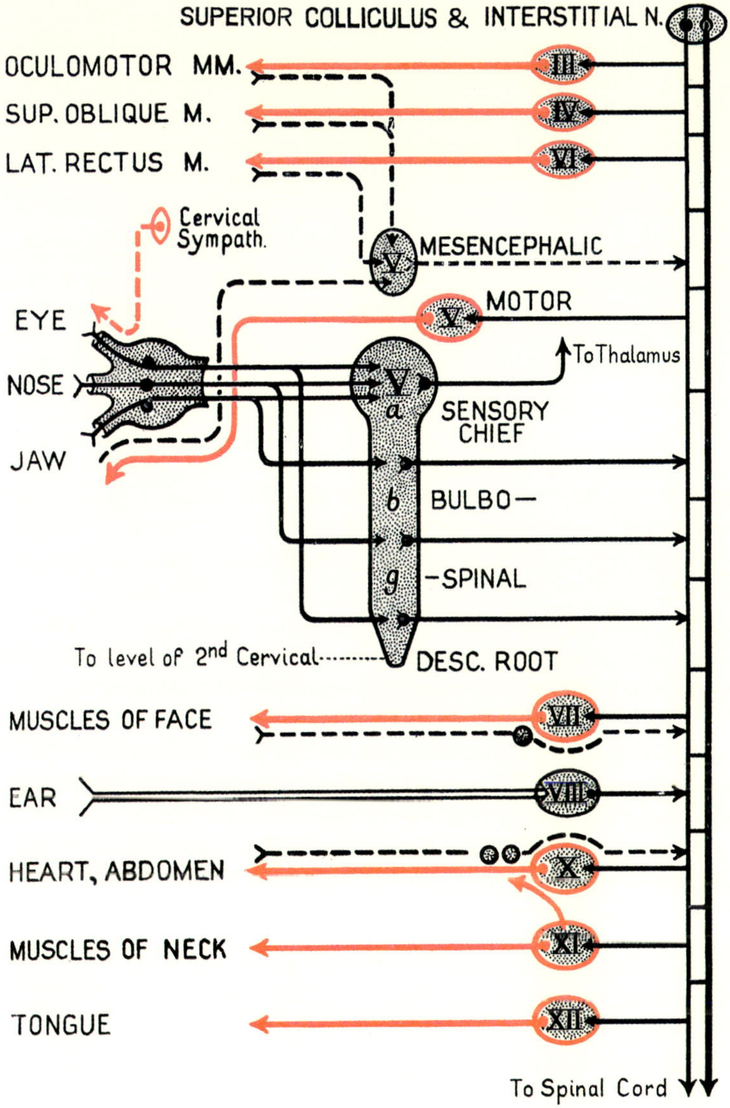

FIG. 212.—Scheme of **Inter-nuclear Connexions of the Cerebral Nerves** by means of the Medial Longitudinal Fasciculus and tectal tracts (arising from cells of the opposite side) to explain various Ocular, Facial, and Auditory Reflexes and Associated Movements. The roman numerals indicate the nuclei and the red lines the efferent paths, the connecting tracts being indicated by parallel lines on the right.

supplying the neck muscles; there is evidence that it is especially related to the nuclei of the ventral horn of the cord in the regions of the cervical and lumbar enlargements. The efferent fibres concerned must obviously come from both sides, but for simplicity in the diagram (Fig. 212) the contralateral connexions are not shown.

Although it is doubtful whether the fasciculus really functions in the general way indicated by the illustrations that follow, and in spite of the differences of opinion still held in regard to the details of its connexions, yet such a scheme as is displayed in Fig. 212 would afford a reasonable explanation of a number of **ocular, auditory and facial associated movements** and reflexes:

i.	Conjugate movements of the eyes	(III nucleus and VI nucleus)
ii.	Ciliary muscle and orbicular group	(III and VII)
iii.	Levator and orbicularis in opening eye	(III and VII)
iv.	Corneal blinking and tear reflex	(V to VII)
v.	Sneezing reflex	(V to X)
vi.	Sucking reflex	(V to VII, XII)
vii.	Various reflexes of mastication	(V to VII, XII, ? IX)
viii.	Turning head to touch	(V to XI, C 1–3)
ix.	Turning eyes to sound	(VIII to III, VI)
x.	Closing eyes to sound	(VIII to VII)
xi.	Turning head to sound	(VIII to XI, C 1–3)
xii.	Co-ordination of lips and tongue	(VII and XII)

The remarkable 'oculo-cardiac' reflex (first described by Aschner in 1909), where the pulse-rate is regularly slowed by pressure on one or both eyeballs, and the vomiting attendant on acute glaucoma, may likewise be explained by such connexion between the Vth sensory nuclei and the visceral motor nuclei of the Xth nerve.

The phenomenon of 'jaw-winking', where there are associated movements of the jaw and eyelids (motor V and VII), is explained by Lutz (1919) as due to alteration of the supranuclear nerve connexions which travel in this bundle to the muscles concerned. Bishop Harman (1903, 1904) traces the morphological evolution of this phenomenon, and suggests that it is an atavistic anomaly, the orbicularis palpebrarum being the descendant of the original spiracle or gill musculature, which in fishes works rhythmically with the deep spiracle muscles now represented by the pterygoids.

As of clinical interest in these matters, the reader may refer to papers by Carmichael and Critchley (1925), who give six types of association between actions of the ocular movement of the eye and other cranial muscles, and likewise suggest that some of them may have a phylogenetic origin; by Molinie (1916), who distinguishes even six types of oculo-auditory reflexes; by Lebensohn (1929), who demonstrated experimentally oculo-visceral reflexes; and by Petzetakis, (1929) who made one hundred and forty observations on a 'supraorbital-cardiac reflex', where pressure on the nerve affected the cardiac rhythm.

BIBLIOGRAPHY

An attempt has been made to give a fairly complete list of the papers and books published since 1900 relating to the anatomy, histology, and embryology of the subject, but other references have been added where deemed of sufficient interest. The list (650 articles in all) comprises those authors referred to in the text whose names are followed by a date, which is a guide to a particular article where several have been published. The original papers have been consulted in each case, except where an abstract only has been obtainable. Abstracts have also been noted where they may possibly suit the reader's convenience of reference.

For the earlier literature the reader should consult the chronological bibliographies given in the *Graefe-Saemisch Handbuch der gesamten Augenheilkunde.*

Ackland, W. and Lang, W. 'Discussion on the Relationship of Dental Sepsis to Diseases of the Eye,' Oxford Ophthalmological Congress, 1923. Trans. Ophth. Soc. United Kingdom, 1923, xliii, p. 417.

Adachi, B. (i) 'Die Orbita und die Hauptmass des Schädels der Japaner,' Zeitschr. f. Morphol. u. Anthropol., 1904, vii, p. 379.

(ii) 'Das Arteriensystem der Japaner,' 1928, Kioto.

Adler, F. H. (i) 'Pupilloscopic Findings in Lesions in Different Parts of the Reflex Arc,' Arch. of Ophth., 1926, lv, p. 262.

(ii) 'Reciprocal Innervation of the Extra-ocular Muscles,' Arch. of Ophth., 1930, n.s. 3, p. 318.

(iii) 'The Local Control of the Ocular Circulation,' Arch. of Ophth., 1924, liii, p. 1.

Adrogue, E. and Gavina Alvarado, E. 'Anomalies of Lid Pigmentation,' Arch. de oftal., Buenos Aires, 1929, iv, p. 736. Also in Prensa med. argentina, Buenos Aires, 1929, xvi, p. 246. Abst. in Amer. Jour. Ophth., 1930, xiii, p. 740.

Ahrens, E. 'Die Cribra orbitalia und die Spina trochlearis, etc.,' Inaug. Diss. Göttingen (1904), abst. in Nagel's Jahresb. f. Ophth., 1905, p. 39.

Allen, W. F. 'Function of the Nerve Cells in the Motor Root of the Nervus Trigeminus in the Cat,' Jour. Comp. Neurol., 1925, xxxviii, p. 349.

Ambialet. 'Orbites et conformations craniennes,' Ann. d'Ocul., 1905, cxxxiv, p. 176.

ARCHAMBAULT, LA SALLE. 'The Inferior Longitudinal Bundle and the Geniculo-calcarine Fasciculus,' Albany Med. Ann., 1909, xxx, p. 118.

ARGAUD and FALLOUEY. 'Sur la structure du tarse palpébral et son indépendance vis-à-vis de la glande de Meibomius,' Compt. Rend. Soc. de Biologie, 1913, lxxiv, p. 1068.

ARLT, F. (i) 'Über den Thränenschlauch,' Graefe's Arch. f. Ophth., 1855, i. 2, p. 135.

(ii) 'Über den Ringmuskeln der Augenlider,' Graefe's Arch. f. Ophth., 1863, ix. 1, p. 64.

ASK, F. (i) 'Anthropometrische Studien über die Grösse und Gestaltung der Orbitalmündung bei den Schweden, etc.,' Zeitschr. f. Augenheilk., 1906, xvi, p. 146.

(ii) 'Über die Entwicklung der Caruncula lacrimalis beim Menschen, etc.,' Anat. Anzeiger, 1907, xxx, p. 197.

(iii) 'Über die Entwicklung der Lidränder, etc.,' Anat. Hefte (Merkel und Bonnet), 1908, xxxvi, p. 189.

(iv) 'Entwicklung des Drüsenapparatus der Bindehaut beim Menschen,' Anat. Hefte (Merkel und Bonnet), 1910, xl. i, p. 489.

ASK, F. and VAN DER HOEVE, J. (v) 'Beiträge zur Kenntnis der Entwickelung der Tränenröhrchen, etc.' Graefe's Arch. f. Ophth., 1921, cv, p. 1157. Abst. in Amer. Jour. Ophth., 1922, v, p. 211.

AUBARET, E. (i) 'Les Replis valvulaires des canalicules et du conduit lacrymo-nasal, etc.,' Arch. d'Opht., 1908, xxviii, p. 211. (Abst. in Ophthalmoscope, 1908, vi, p. 900.)

(ii) 'Des rapports des faisceaux lacrymaux de l'orbiculaire des paupières, etc.,' Compt. Rend. Soc. de Biologie, 1909, lxvii, p. 235.

(iii) 'Morphologie du conduit lacrymo-nasal chez l'homme,' Bibliographie Anatomique, 1910–11, xx. i, p. 97.

AUBARET et BONNEFON. 'Des rapports du conduit lacrymo-nasal avec le méat moyen et la gouttière de l'infundibulum,' Arch. d'Opht., 1910, xxx, p. 469.

AUGIER, M. 'Les Os frontaux accessoires,' Compt. Rend. de l'Assoc. des Anatomistes, 14e Réunion, 1912, p. 22.

AXENFELD, D. (i) 'Bemerkungen zur Histologie und Physiologie der Thränendrüsen', Berichte der 27. Vers. d. Ophth. Gesellsch. zu Heidelberg, 1898 (abst. in Nagel's Jahresb. f. Ophth., 1898).

(ii) Über die feinere Histologie der Thränendrüsen, etc.' (Abst. in Nagel's Jahresb. f. Ophth., 1900, p. 33.)

BACH, L. 'Die Nerven der Augenlider und der Sklera beim Menschen und Kaninchen,' Graefe's Arch. f. Ophth., 1895, xli. 3, p. 50.

BAILEY, J. H. 'Surgical Anatomy of the Lacrimal Sac,' Amer. Jour. Ophth., 1923, vi, p. 665.

BAKER, F. Article on the Eyeball in Norris and Oliver's System of Diseases of the Eye, Philadelphia, 1900, i, p. 109.

BALADO, M. 'El Iris. Estudio Anatomico y Fisiopatologico de su Innervación.' Semana med., Buenos Aires, 1927, xxxiv, p. 1441. Abst. in Jour. Amer. Med. Assoc., 1927, lxxxix, p. 1099.

BARRAGUER, J. 'Anatomie du fond de l'orbite et du sinus sphénoïdal,' Compt. Rend. Société Française de l'Opht., 1896, xiv. Internat. Cong. Méd., Madrid, 1903. (Abst. in Ann. d'Ocul., 1903, cxxx, p. 189.)

BARRATT, J. 'Observations on the Structure of the Third, Fourth, and Sixth Cranial Nerves,' Jour. Anat. and Physiol., 1901, xxxv, p. 214.

BARTELS, M. (i) 'Darstellung der Manz'schen Drüsen und ihre Beziehung zu den Cysten,' Zeitschr. f. Augenh., 1908, xx, p. 193.

(ii) Article 'Das Lymphgefässsystem,' Bardeleben's Handbuch der Anat. des Menschen, 1909, iii. 4, p. 103.

BEATTIE, J. 'The Oculomotor Nerve,' Anat. Anzeiger, Verhandl. d. anat. Gesellsch., 1930, lxxi, p. 168.

BEATTIE, J., BROW, G. and LONG, C., 'Physiological and Anatomical Evidence for the Existence of Nerve Tracts connecting the Hypothalamus with Spinal Sympathetic Centres,' Proc. Roy. Soc., London, B, 1930, cvi, p. 253. (See also under Woollard.)

BEAUVIEUX, J. and DUPAS, J. 'A Topographical, Anatomical and Histological Study of the Ciliary Ganglion in Man and Other Animals,' Arch. d'Opht., 1926, xliii, 641. Abst. in Amer. Jour. Ophth., 1928, xi, p. 490. Also in Brit. Jour. Ophth., 1927, xi, p. 403.

BEAUVIEUX, J. and RISTITCH, K. 'Les vaisseaux centraux du nerf optique,' Arch. d'Opht., 1924, xli, p. 352. Abst. in Brit. Jour. Ophth., 1925, ix, p. 532.

BEHR, C. (i) 'Besteht beim Menschen ein Abfluss aus dem Glaskörper in den Sehnerven?' Graefe's Arch. f. Ophth., 1912, lxxxiii, p. 519.

(ii) 'Die Untersuchungsmethode,' Graefe-Saemisch Handbuch d. g. Augenh., 1924, ed. III, ii, p. 129.

BENJAMINS, C. *See* ROCHAT.

BERGER, E. (i) 'Anatomie normale et pathologique de l'œil,' 1893, Paris.

(ii) 'Variétés de la paroi optico-sphénoïdale,' Arch. d'Opht., 1894, xiv, p. 545.

BERNHEIMER, S. (i) 'Ätiologie und pathologische Anatomie der Augenmuskellähmungen,' and 'Die Wurzelgebiete der Augennerven,' Graefe-Saemisch Handbuch d. g. Augenh., 2nd ed., 1904, Lief. 39 and 16.

(ii) 'Weitere experimentelle Studien zur Kenntnis der Lage des Sphinkter- und Levatorkerns,' Graefe's Arch. f. Ophth., 1909, lxx, p. 539.

BIRCH-HIRSCHFELD. (i) 'Die Krankheiten der Orbita.' Article in Graefe-Saemisch Handbuch d. g. Augenh., 1909, Lief. 147–170, S. 261.

(ii) Graefe-Saemisch Handbuch d. g. Augenh., 1930, ed. II, ix, Cl. xiii, p. 361.

BIRNBAUM, H. 'Exophthalmometrische Masse bei Normalen und deren Beziehungen zur Grösse der Orbitalöffnung,' Arch. f. Ophth., 1915, xc, p. 378.

BISCHOFF, E. 'Mikroskopische Analyse der Anastomosen der Kopfnerven,' 1865, München.

BLACK, D. (i) Cerebellar Localisation in the Light of Recent Research,' Jour. Lab. and Clin. Med. (St. Louis), 1916, i, p. 467.

(ii) 'The Motor Nuclei of the Cerebral Nerves in Phylogeny,' Jour. Comp. Neurol., 1917, xxvii, p. 467; 1917, xxviii, p. 379; 1920, xxxii, p. 61; 1922, xxxiv, p. 233.

BLOCH, A. 'De l'œil mongoloïde,' etc., Bull. et mém. Soc. d'Anthropol. de Paris, 1911, 6e série, 11, 3 and 4, p. 326.

BOCHDALEK. 'Beitrag zu den anomalen Muskeln der Augenhöhle,' Prager Vierteljahrsschrift, 1868, Band iv.

BOCK, E. 'Zur Kenntniss der Gesunden und kranken Thränendrüsen,' Wien, 1896 (quoted by Schirmer and Picou).

BOEKE, J. (i) 'Die doppelte (motorische und sympathische) efferente Innervation der quergestreiften Muskelfasern,' Anat. Anzeiger, 1913, xliv. 15–16, p. 343.

(ii) 'On the Termination of the Efferent Nerves in plain Muscle-cells and its bearing on the Sympathetic (Accessory) Innervation of the Striated Muscle Fibre.' (Abst. in Ophthalmology (Seattle), 1915–16, xii, p. 696.)

(iii) 'Die morphologische Grundlage der sympathischen Innervation der quergestreiften Muskelfasern.' Part 1, Jahrbuch f. Morphol. u. mikros. Anat.; Part 2, Zeitschr. f. mikros.-anat. Forschung, 1927, viii, p. 561.

BORDLEY, J. 'The Optic Nerve in its Relation to Posterior Nasal Sinuses,' Virginia Med. Monthly, 1921, xlviii, p. 144.

BORN, G. (i) 'Über die Nasenhöhlen und den Thränennasengang der Amphibien,' Morphol. Jahrbuch, 1876, ii, p. 577.

(ii) 'Über die Nasenhöhlen und den Thränennasengang der amnioten Wirbelthiere,' Morphol. Jahrbuch, 1883, viii, p. 188.

BOUCHERON. (i) 'Nerfs de l'hémisphère antérieure de l'œil,' Compt. Rend. Soc. de Biologie, 1890, ii. (Mémoires), p. 71.

(ii) 'Nerfs ciliares superficiels,' *ibid.*, 1891, iii (Mémoires), p. 59.

BRAILEY. 'Congenital Distichiasis,' Brit. Med. Jour., July 21, 1906, p. 142.

BRAIN, W. R. and STRAUSS, 'Recent Advances in Neurology,' 1929, Blakiston, London.

BRANCA, A. Article on the Skin and its Derivatives in Poirier's Anatomy, 1912.

BRAUNSTEIN, 'Ein Beitrag zur Kenntnis der Innervation der Irisbewegungen,' 1893, Charkow.

BRAV, A. 'Oxycephaly and Optic Atrophy,' Ann. of Ophth. of St. Louis, 1912, xxi. i, p. 1.

BREMER, J. (i) 'Aberrant Roots and Branches of the Abducent and Hypoglossal Nerves,' Jour. Comp. Neurol. and Psychol., 1908, xviii, p. 619.

(ii) 'Recurrent Branches of the Abducens Nerve in Human Embryos,' Amer. Jour. Anat., 1921, xxviii. 2, p. 371.

BRIGGS, H. 'Hereditary Congenital Ptosis with report of Sixty-four Cases conforming to the Mendelian Rule of Dominance,' Amer. Jour. Ophth., 1919, ii, p. 408.

BROCA, P. 'Recherches sur l'indice orbitaire,' Arch. d'Opht., Paris, 1875. Mémoires d'anthropol., Paris, 1871–77.

BRODMANN, 'Individuelle Variationen der Sehnsphäre und ihre Bedeutung für die Klinik der Hinterhauptschüsse,' Deutsche med. Wochenschrift, 1918, xliv, p. 783.

BROMAN and ASK. Development of Glands of Eyelids in Mammals, Deutsche Südpolen Expedition, 1901–3, Berlin, 1910, xii.

BROUGHTON, T. 'The Increase in the Number and Size of the Medullated Fibres in the Oculo-motor Nerve of the White Rat and of the Cat at Different Ages,' Jour. Comp. Neurol. and Psychol., 1906, xvi, p. 153.

BROUWER, B. (i) 'Über die Projektion der Makula auf die Area Striata des Menschen,' Jour. f. Psychol. u. Neurol., 1930, xl, p. 147.

(ii) 'The Oculomotor Nucleus,' Zeitschr. f. d. g. Neurol. u. Psychiat., 1918, xl, p. 152. Abst. in Amer. Jour. Ophth., 1921, iv, p. 475.

See also ZEEMAN and BROUWER.

BRUCE, A. and PIRIE. 'On the Origin of the Facial Nerve,' Rev. Neurol. and Psychiat., 1908, vi, p. 685.

BUDGE, E. S. 'Anatomical Variations of Accessory Nasal Sinuses,' California and Western Medicine, 1930, xxxiii, p. 576.

BUDGE, J. 'Beschreibung eines neuen Muskels,' etc., Zeitschr. f. rationelle Medizin, 1859, 3, vii, p. 273.

BURKARD, O. 'Über die Periorbita der Wirbeltiere und ihre muskellosen Elemente,' Arch. f. Anat. u. Physiol., Anat. Abt., Suppl. Band, 1902, p. 79. (Abst. in Nagel's Jahresb. d. Ophth., 1902, p. 31.)

BURKITT, A. N. and LIGHTOLLER, G. 'The Facial Musculature of the Australian Aboriginal,' Jour. of Anat., 1926, lxi. 1, p. 14; 1927, lxii, p. 33.

BURR, H. and ROBINSON, G. 'An Anatomical Study of the Gasserian Ganglion, with Particular Reference to the Nature and Extent of Meckel's Cave,' Anat. Record, 1925, xxix. 4, p. 269.

BYERS, W. G. M. 'A Case of Encapsulated Angioma of the Orbit,' Arch. of Ophth., 1924, liii. 3, p. 280.

CAMERON, J. (i) 'The Naso-orbito-alveolar Index,' Amer. Jour. Phys. Anthropol., 1920, iii, No. 1, p. 63.

(ii) 'Contour of Orbital Aperture in representatives of modern and fossil Hominidae,' Amer. Jour. Phys. Anthropol., 1920, iii, No. 4.

CAMPBELL, D., CARTER, J. and DOUB, H. 'Roentgen Ray Studies of the Nasolacrimal Passageways,' Arch. of Ophth., 1922, li, p. 462.

CAMPOS. 'Recherches expérimentales et cliniques sur les nerfs sécréteurs des larmes,' Arch. d'Opht., 1897, xvii, p. 529.

CARLINI, V. 'Il tessuto elastico in rapporto con le glandule di Moll,' Ann. di Ottal., 1907, xxxvi, p. 231.

CARMICHAEL, E., and CRITCHLEY, M. 'The Relations between Eye Movements and Other Cranial Muscles,' Brit. Jour. Ophth., 1925, ix, p. 49.

CHARAMIS, J. 'Paralysies du nerf moteur oculaire commun dans les fractures de la base du crâne,' Arch. d'Opht., 1928, xlv, p. 759. Abst. in Amer. Jour. Ophth., 1929, p. 332.

CHARPY, A. (i) 'Orifices adipeux de la base de l'orbite,' Compt. Rend. de l'Assoc. des Anatomistes, 10e Réunion, 1908.

(ii) 'Le Coussinet adipeux du sourcil,' Bibliog. Anatomique, 1909, xix. i, p. 47.

(iii) 'Capsule de Tenon,' Bibliog. Anatomique, 1908–9, xviii. 4–5, pp. 215, 270.

(iv) 'Plis et sillons des paupières,' Bibliog. Anatomique, 1910, xx, p. 1.

(v) 'Structure topographique des paupières,' Bibliog. Anatomique, 1911, xxi. 2, p. 65.

(vi) Article on the Muscles of the Eyeball in Poirier's Anatomy, 1912, tome v, fasc. 2, p. 539.

CIACCIO, G. (i) 'Über den Bau der Bindehaut des menschlichen Auges,' Moleschott's Untersuch. zur Naturlehre des Menschen u. d. Thieren, 1874, xi, p. 420. (Reference from Virchow.)

(ii) 'Osservazione Intorno alla Struttura della Congiuntiva Umana,' Mem. Accad. d. sc. d. Ist. di Bologna, 1873, 3 s., iv, p. 460.

CLARK, W. E. Le Gros, 'The Mammalian Oculomotor Nucleus,' Jour. of Anat., 1926, lx. 4, p. 426.

CLERMONT. 'Le Muscle releveur de la paupière supérieure et le septum orbitaire,' Compt. Rend. de l'Assoc. des Anatomistes, 1909, xviii. supplément, xi, p. 264.

COFFIN, L. 'The Development of the Accessory Sinuses of the Nose,' Amer. Jour. Med. Sciences, 1905, cxxix, p. 297.

COHN, H. and PAPEZ J. 'A Comparative Study of the Visuosensory or Striate Area in the Two Hemispheres of the Human Brain,' Amer. Jour. Phys. Anthropol., 1930, xiv. 3, p. 405.

COLLINS, E. TREACHER. 'Changes in the Visual Organs correlated with the Adoption of Arboreal Life and with the Assumption of the Erect Posture,' Bowman Lecture, Trans. Ophth. Soc. United Kingdom, 1921, xli.

CONGDON, E. 'The Distribution and Mode of Origin of Septa and Walls of the Sphenoid Sinus,' Anat. Record, 1920, xviii. 2, p. 97.

CONTINO, A. (i) 'Über Bau und Entwicklung des Lidrandes beim Menschen,' Graefe's Arch. f. Ophth., 1907, lxvi, p. 505.

(ii) 'Über die Entwicklung der Plica semilunaris beim Menschen,' Graefe's Arch. f. Ophth., 1909, lxxi, p. 1.

COPE, V. (i) 'The Pituitary Fossa and the Surgical Methods of Approach to it,' Lancet, March 18, 1916, p. 601; also in Brit. Jour. of Surgery, July 1916, p. 107.

(ii) 'The Internal Structure of the Sphenoidal Sinus,' Jour. of Anat., 1917, li, p. 127.

COPPEZ, H. 'Études sur la pigmentation de la conjonctive,' Bull. de l'Acad. Roy. de Méd. de Belgique, 1905, série 4, xix, p. 443.

CORNING. 'Über die vergleichende Anatomie der Augenmuskelatur,' Morphol. Jahrbuch, 1900–2, xxix, p. 94.

COSMETTATOS, G. 'Über einige angeborene Anomalien der Tränenwege,' Arch. f. Augenh., 1906, lv, p. 362.

CREVATIN, F. 'Beitrag zur Kenntniss der epithelialen Geflechte der Hornhaut der Säugetiere,' Anat. Anzeiger, 1903, xxiii, p. 151.

CRITCHLEY, M. *See* CARMICHAEL.

CRUVEILHIER, J. 'Anatomie descriptive,' 1834–36.

CRYER, M. 'Some Variations in the Frontal Sinus,' Jour. Amer. Med. Assoc., 1907, xlviii, p. 284.

CUNNINGHAM. Text-book of Anatomy, 6th ed., 1931, London and New York.

CURNOW, J. 'Notes of some Irregularities in Muscles and Nerves,' Jour. Anat. and Physiol., 1873, vii, p. 304.

CUSHING, HARVEY. (i) 'Strangulation of the Nervi Abducentes by Lateral Branches of the Basilar Arteries in Cases of Brain Tumours,' Brain, 1910–11, xxxiii, p. 204.

(ii) 'The Field Defects Produced by Temporal Lobe Lesions,' Brain, 1922, xliv. 4, p. 341.

CZERMAK, W. 'Die topographischen Beziehungen der Augenhöhle, etc.' Augenärztliche Unterrichtstafeln (Magnus), 1895, ix, Breslau.

DARWIN, C. 'On the Expression of the Emotions in Man and Animals.' 1872, London.

DAVIDA, E. 'Über die Varietäten des Verlaufes der Sutura infraorbitalis sagittalis,' Anat. Anzeiger, 1913, xliv, pt. 10, p. 203.

DAVIDSON BLACK. See BLACK, DAVIDSON.

DAVIS, F. A. 'Tower Skull, Oxycephalus,' Amer. Jour. Ophth., 1925, viii. 7, p. 513.

DAVIS, L. E. 'The Deep Sensibility of the Face,' Arch. Neurol. and Psychiat., 1923, ix, p. 283.

De Bono and Frisco. 'Sulla permeabilità verso i microrganismi delle mucose congiuntivale, etc.,' Archiv. di Ottal., 1901, viii, p. 401. (Abstr. in Arch. of Ophth., 1901, xxx, p. 113.)

Dedekind, F. 'Beiträge zur Entwicklungsgeschichte der Augengefässe des Menschen,' Anat. Hefte (Merkel und Bonnet), 1909, xxxviii, p. 1.

Dewey, K. 'A Contribution to the Study of the Lymphatic System of the Eye,' Anat. Record, 1920, xix, p. 125.

Deyl, J. 'Über den Eintritt der Arteria centralis Retinae in den Sehnerven beim Menschen,' Anat. Anzeiger, 1896, xi, p. 687.

Dieulafé, L. 'Topographie des voies lacrymales: applications chirurgicales,' Bulletin Médical, 1905, p. 179.

Dixon, A. 'On certain Markings due to Nerves and Blood-Vessels upon the Cranial Vault,' Jour. Anat. and Physiol., 1904, xxxviii; New Series, xviii, p. 377.

Dock, G. 'Oxycephaly and Exophthalmos,' The Osler Dedication Volume of Contributions to Medical and Biological Research, 1919. Hoebner, New York.

Dogiel, A. (i) 'Die Nervenendigungen in der Tränendrüse der Säugetiere,' Arch. f. mikros. Anat., 1893, xlii, p. 632.

(ii) 'Die Nervenendigungen im Lidrande, und in der Conjunctiva palpebrarum des Menschen,' Arch. f. Mikros. Anat., 1895, xliv, p. 15.

(iii) 'Die Endigungen der sensiblen Nerven in den Augenmuskeln und deren Sehnen beim Menschen und den Säugetieren,' Arch. f. mikros. Anat., lxviii, 1906, p. 501.

Donders, F. 'Untersuchungen über die Entwicklung und den Wechsel der Cilien,' Graefe's Arch. f. Ophth., 1858, iv. 1, p. 286.

Dor, L. 'La Fatigue oculaire.' 1900, Paris.

Dorfmann, R. 'Über Pathogenese und Therapie des Turmschädels,' Graefe's Arch. f. Ophth., 1908, lxviii, p. 412.

Druault, A. Article 'Organe de la vision' in Poirier's Traité d'anatomie humaine, 1912, tome v.

Duane, A. 'The Associated Movements of the Eyes,' &c. Amer. Jour. Ophth., 1924, vii. 1, p. 16.

Dubreuil, G. (i) 'Les Glandes lacrymales de l'homme et des mammifères,' Rev. gén. d'Opht., 1907, xxvi, p. 339.

(ii) 'Les Glandes lacrymales et les glandes annexes de l'œil des vertébrés,' Rev. gén. d'Histol., 1908, viii, p. 695; also 1908, Masson, Paris.

Del Duca, M. (i) 'Ricerche Anatomiche sul Canale Ottico.' Riv. oto-neuro-oftal., 1929, vi, p. 215. Abst. in Arch. of Ophth., 1930, iii, p. 461.

(ii) 'Considerazioni Radiografiche e Cliniche sul Canale Ottico.' Riv. oto-neuro-oftal., 1929, vi, p. 307. Abst. in Arch. of Ophth., 1930, iii, p. 462.

DUCKWORTH, W. Morphology and Anthropology, 1904, Cambridge.

DUKE-ELDER, W. S. (i) 'The Contraction of the Extrinsic Muscles of the Eye by Choline and Nicotine,' Proc. Roy. Soc., London, B, 1930–31, cvii, p. 332.

(ii) 'Recent Advances in Ophthalmology,' 1927, Churchill, London.

DUVERNEY. 'L'Art de disséquer méthodiquement les muscles, etc.,' 1749, Paris.

DWIGHT, T. (i) Article on the 'Anatomy of the Orbit' in Norris and Oliver's System of Diseases of the Eye.

(ii) Article 'The Skeleton' in Piersol's Human Anatomy.

EDINGER, L. 'The Anatomy of the Central Nervous System of Man and of Vertebrates in General,' (Trans. by Hall). 1899, Davis, Philadelphia.

EGGELING, H. 'Zur Morphologie der Augenlider der Säuger,' Jenaische Zeitschr. f. Naturwissenschaft, 1904, xxxix, p. 1.

ELLIOT, R. H. 'The Yielding of the Optic Nervehead in Glaucoma,' Brit. Jour. Ophth., 1921, v, p. 307.

ELLIOT SMITH, G. *See* SMITH, G. ELLIOT.

EMMERT, E. 'Auge und Schädel.' 1880, Berlin.

ENSLIN, E. 'Die Histologie der Caruncula lacrymalis des Menschen,' Arch. f. Augenh., 1905, li, p. 253.

EVANS, J. HOWELL. (i) 'The Applied Topographical Anatomy of the Sinuses accessory to the Nasal Cavities in their Relations to the Orbit and to its more important Contents,' Ophthalmoscope, 1908, vi, No. 4, p. 259.

(ii) 'Tendo Oculi and Pars Orbitalis in Different Races.' Amer. Jour. Phys. Anthropol., 1925, viii. 4, p. 411.

EVATT, E. 'A Method for determining the Position of the Eye-socket,' Jour. Anat. and Physiol., 1907, xli, p. 304.

FAWCETT, E. (i) 'The Origin and Intracranial Course of the Ophthalmic Arteries and the Relation they bear to the Optic Nerve,' Jour. Anat. and Physiol., 1896, xxx, p. 49.

(ii) 'The Development of the Human Maxilla, etc.' Jour. Anat. and Physiol., 1911, xlv, p. 378.

FEIN, J. 'Über Beziehungen zwischen Kieferhöhle und Tränennasengang,' Arch. f. Laryngol. u. Rhinol., 1912, xxvi, p. 29.

FERRON, M. 'Note sur la constitution des parois du sinus caverneux et les rapports des nerfs, etc.,' Jour. Méd. de Bordeaux, 1913, No. 6, p. 88.

FESTAL, A. 'Recherches anatomiques sur les veines de l'orbite,' 1887, Thèse. Paris.

FILETI, A. 'Embryology and Morphology of the Optic Canal,' Ann. di ottal. e clin. ocul., 1927, lv, p. 493. Abst. in Amer. Jour. Ophth., 1928, xi, p. 583.

FISCHER, E. 'Über Pigment in der menschlichen Conjunctiva,' Anat. Anzeiger, 1905, Verhandl., xxvii, p. 140.

FISHER, J. H. (i) Ophthalmological Anatomy. 1904, Hodder & Stoughton, London.

(ii) 'Leber's Disease (Hereditary Optic Atrophy), a Suggestion as to its Cause,' Ophthalmoscope, 1916, xiv, p. 398.

FLECKER, H. 'Observations upon Cases of Absence of the Lacrimal Bones and of Existence of Perilacrimal Ossicles,' Jour. Anat. and Physiol., 1913, xlviii. 1, p. 52.

FLEISCHER, B. (i) 'Beiträge zur Histologie der Tränendrüse, etc.,' Anat. Hefte (Merkel und Bonnet), 1904, xxvi, p. 101.

(ii) 'Die Entwicklung der Tränenröhrchen bei den Säugetieren,' Graefe's Arch. f. Ophth., 1906, lxii, 379.

(iii) 'Musculus retractor bulbi, etc.,' Anat. Anzeiger, 1907, xxx, p. 465.

FLOWER, T. Catalogue of the Museum of the Royal College of Surgeons (2nd ed.), 1907, London.

FORSTER, A. 'Zur Morphogenese des Epicanthus und der Faltenbildungen der Haut in der Nasenwurzelgegend,' Anat. Anzeiger, 1919–20, lii, p. 49.

FOX, J. C. and G. HOLMES. 'Optic Nystagmus and its Value in the Localisation of Cerebral Lesions.' Brain, 1926, xlix, p. 333.

FRANCIS and GIBSON. 'The Anatomical Relation between the Sphenoidal Sinus and Orbit,' Ophthalmoscope, 1911, ix. 3, p. 172.

FRAZIER, C. H. and WHITEHEAD, E. 'The Morphology of the Gasserian Ganglion,' Brain, 1925, xlviii, p. 458.

FRIEBERG, T. (i) 'Untersuchungen über die Mechanik der Tränenableitung,' Zeitschr. f. Augenh., 1918, xxxvii, pp. 42, 211, 324.

(ii) 'Weitere Untersuchungen über die Mechanik der Tränenableitung,' Zeitschr. f. Augenh., 1918, xxxix, p. 266.

FROHSE, F. 'Die oberflächlichen Nerven des Kopfes.' 1895, Berlin.

FRÜND, H. 'Die glatte Muskulation der Orbita, etc.,' Beitrag zur klinischen Chirurgie, lxxiii, 1911, p. 755. (Abst. in Zeitschr. f. Augenh., 1912, xxviii, p. 164.)

FRY, W. E. 'Variations in the Intraneural Course of the Central Vein of the Retina,' Arch. of Ophth., 1930, iv. 2, p. 180.

FUCHS, E. (i) 'Zur Anatomie der Blut- und Lymphgefässe der Augenlider,' Graefe's Arch. f. Ophth., 1878, xxiv. 3, p. 1.

(ii) 'Beiträge zur normalen Anatomie des Augapfels,' Graefe's Arch. f. Ophth., 1884, xxx. 4, p. 1.

(iii) Text-book of Ophthalmology (Duane's Translation), 1917, London and Philadelphia.

(iv) 'Ueber die Lamina Cribrosa.' Arch. f. Ophth., 1916, xci, p. 435. (See abst. under Henderson, 1918; Elliot, 1921.)

(v) 'Senile Changes of the Optic Nerve,' Amer. Jour. Ophth., 1922, v, p. 215.

(vi) 'Ueber die Ciliargefässe.' Arch. f. Ophth., 1929, cxxii.

p. 219. Abst. in Brit. Jour. Ophth., 1931, xv. 3, p. 182; also in Amer. Jour. Ophth.; 1930, xiii, ser. 3, p. 75.

FUTAMURA, R. 'Über die Entwickelung der Facialismuskulatur des Menschen,' Anat. Hefte (Merkel und Bonnet), 1906 (30 Band, Heft 91), p. 435.

GALLEMAERTS, E. (i) 'Recherches sur la fente sphénoïdale,' Bull. l'Acad. Roy. Méd. de Belgique, Bruxelles, Séance Fév. 27, 1897. Also Policlin. Brux., 1897, vi, p. 365.

(ii) 'Sur l'aponévrose orbito-oculaire,' *ibid.*, Séance Jan. 28, 1899.

(iii) 'Sur la structure du chiasma optique,' *ibid.*, Séance Juin 28, 1900.

GASKELL, W. 'On the Relation between the Structure, Function, Distribution, and Origin of the Cranial Nerves, etc.,' Jour. of Physiol., 1889, x, p. 153.

GAYAT, J. 'Essais de mensuration de l'orbite,' Ann. d'Ocul., 1873, lxx, p. 1 (Volume of the Orbit).

GEGENBAUR, C. (i) 'Über die Pars facialis des Lacrimalis des Menschen,' Morphol. Jahrbuch, 1881, vii, p. 173.

(ii) 'Vergleichende Anatomie der Wirbeltiere,' Morphol. Jahrbuch, 1901, ii.

GEHUCHTEN, VAN. Anatomie du système nerveux de l'homme, 1906, édition iv, Louvain.

GÉRARD, G. (i) 'Les Voies optiques extra-cérébrales,' Jour. de l'Anat. et de la Physiol., 1904, xl. i, p. 22.

(ii) 'Des obstacles naturels capables de compliquer le cathétérisme des voies lacrymales,' Ann. d'Ocul., 1907, cxxxvii, p. 193.

GERARD, M. W. 'Afferent Impulses of the Trigeminal Nerve,' Arch. Neurol. and Psychiat., 1923, ix, p. 306.

GERLACH. Anatomie des Auges, 1880, Leipzig.

GEROTA, D. 'Zur Technik der Lymphgefässinjektion,' Anat. Anzeiger, 1896, xii, p. 216.

GERRISH. Text-book of Anatomy, 1902, Lea Brothers & Co., Philadelphia.

GIACOMINI, C. Several articles on the Anatomy of the Negro, in Le memorie Giorn. della R. Accad. di Med. di Torino, 1882, 1884, 1887, 1892, 1898.

GIFFORD, H. (i) 'Congenital Defects of Abduction and other Ocular Movements and their Relation to Birth Injuries,' Amer. Jour. Ophth., 1926, ix, p. 3.

(ii) 'The "Mongolian Eye",' Amer. Jour. Ophth., 1928, xi, p. 887.

VAN GILSE, P. H. G. 'Investigations on the Development of the Sphenoidal Sinus,' Jour. Laryngol. and Otol., 1926, xli, p. 137.

GOALWIN, H. 'One Thousand Optic Canals. A Clinical, Anatomic, and Roentgenologic Study,' Jour. Amer. Med. Assoc., 1927, lxxxix, p. 1745.

Goldenburg, M. 'Tower Skull with Double Optic Nerve Atrophy,' Amer. Jour. Ophth., 1918, i, p. 760.

Goldnamer, W. 'The Anatomy of the Human Eye and Orbit,' 1923, Professional Press, Chicago.

Goldstein, I. 'The Embryology of Tenon's Capsule,' Arch. of Ophth., 1923, lii, p. 327.

Gosselin, L. 'Über die Ausführungsgänge der Tränendrüsen,' Arch. gén. de la Méd., Paris, 1843. (Reference from Merkel and Kallius.)

Gowers, W. 'The Movements of the Eyelids,' Medico-Chirurgical Trans., 1879, lxii, p. 429.

Göz, A. 'Untersuchung von Tränendrüsen aus verschiedenen Lebensaltern,' Inaugural Dissertation, 1908, Tübingen. (Abst. in Nagel's Jahresb. f. Ophth., 1909, p. 28, and quoted by Picou, 1912.)

Gradenigo, G. 'Ueber die Paralyse des Nervus Abducens bei Otitis,' Arch. f. Ohrenh., 1907, lxxiv, p. 149.

Gradle, H. (i) 'The Intraneural Course of the Optic Nerve Fibres,' Trans. Amer. Acad. Ophth. and Otolaryngol., 1923, xxviii, p. 234.

and S. J. Meyer, (ii) 'The Blind Spot,' Amer. Jour. Ophth., 1929, xii, p. 802.

Graefe-Saemisch. Handbuch der gesamten Augenheilkunde, 2nd ed. 1901– , Engelmann, Leipzig.

Gray. Text-book of Anatomy, Descriptive and Applied, 24th ed., 1930, Longmans, Green & Co., London.

Green, Leedham. 'Über die Bedeutung der Becherzellen der Conjunctiva,' Graefe's Arch. f. Ophth., 1894, xl. i, p. 1.

Greig, D. M. (i) 'Hypertelorism,' Edinburgh Med. Jour., 1924, xxxi, p. 560.

(ii) 'Oxycephaly,' Edinburgh Med. Jour., 1926, xxxiii, pp. 189, 280, 357.

Groyer, F. (i) 'Zur vergleichenden Anatomie des Musculus orbitalis und der Musculi palpebrales (tarsales),' Sitzungsbericht d. k. Akad. d. Wiss. in Wien, 1903, cxii. iii, p. 51; also Wien. klin. Woch., 1903, xvi, p. 959.

(ii) 'Zur Anatomie des Musc. palpebral. sup. des Menschen,' Zeitschr. f. Augenh., 1905, xiv, p. 365.

(iii) 'Über den Zusammenhang der Musculi tarsales (palpebrales) mit den geraden Augenmuskeln, etc.,' Internat. Monatsschr. f. Anat. und Physiol., 1906, xxiii, p. 210.

Grüber, W. Mémoires de l'Académie Impériale de St-Pétersbourg, 1877. Quoted by Ledouble, 1906, p. 148, on the *ossiculum infra-orbitale marginale*.

Grunert, K. 'Die Lymphbahnen der Augenlider,' Arch. f. Augenh., 1901, xliv, p. 189.

GRÜNWALD, L. 'Der Recessus ethmolacrymalis,' Anat. Hefte (Merkel und Bonnet), 1910, xli, p. 373.

GURWITSCH, M. 'Über die Anastomosen zwischen den Gesichts- und Orbitalvene,' Graefe's Arch. f. Ophth., 1883, xxix. 4.

HADEN, H. 'The Development of the Connective Tissue Framework of the Human Optic Nerve with especial Reference to the Lamina Cribrosa,' Amer. Jour. Ophth., 1925, viii. 1, p. 1.

HALBEN, R. 'Beiträge zur Anatomie der Tränenwege,' Graefe's Arch. f. Ophth., 1903, lvii, p. 61.

HALPHEN, MONBRUN, and TOURNAY. 'Headaches in Oto-neuro-ophthalmology (Physiology, Pathology and Treatment).' Rev. d'oto-neuro-opht., 1929, vii, p. 161. Abst. at length in Arch. Neurol. and Psychiat., 1930, xxiv, p. 390.

HAMBURGER, C. 'Die Grundlagen der Prof. Stilling'schen Theorie, etc.,' Zeitschr. f. Augenh., 1904, xii, p. 351.

HANNES, B. 'Über das Vorkommen und die Herkunft von Plasmazellen der menschlichen Tränendrüsen,' Arch. Path. Anat. Physiol., 1911, ccv, p. 410.

HARDY, W. (i) Article on Congenital Anomalies of the Eye, with Bibliography, in Amer. Encyclopedia Ophth., 1914, iv, p. 2776.
(ii) 'Ocular Disease of Dental Origin,' (Abstracts) Amer. Jour. Ophth., 1917, xxxiv. 4, p. 97.

HARMAN, BISHOP. (i) 'The Innervation of the Orbicularis Palpebrarum Muscle,' Trans. Ophth. Soc. United Kingdom, 1903, xxiii, p. 356; (ii) 1904, xxiv, p. 318.

HAYS, H. 'The Posterior Ethmoid Cells and the Sphenoid Sinus in Relation to Affections of the Optic Nerve,' Amer. Jour. Surg., 1926, xl, p. 22.

HEERFORDT, C. 'Über das Emphysema der Orbita,' Graefe's Arch. f. Ophth., 1904, lviii, p. 123.

HEINLEIN, H. 'Zur mikroskopischen Anatomie der Tränenröhrchen,' Graefe's Arch. f. Ophth., 1875, xxi. 3, p. 1.

HENDERSON, E. (i) 'The Ophthalmic Arteries in the Rabbit and Dog,' Roy. London Ophth. Hosp. Rep., 1903, xv, pt. 3, p. 260.
(ii) 'The Lamina Cribrosa', Brit. Jour. Ophth. 1918, ii. p. 50.

HENLE. Handbuch der topographischen Anatomie. 1853, Wien.

HENSCHEN, S. 'Klinische und anatomische Beiträge zur Pathologie des Gehirns,' 1890, Upsala.

HERRICK, C. 'An Introduction to Neurology,' 1922, 3rd ed. Saunders, Philadelphia.

HESSER, C. 'Der Bindegewebsapparat und die glatte Muskulatur der Orbita beim Menschen im normalen Zustande,' Anat. Hefte (Merkel und Bonnet), 1913, xlix. (See Rousseau for abst.)

HINES, MARION. 'Studies on the Innervation of Skeletal Muscle,' Amer. Jour. Anat., 1931, xlvii, p. 1.

HIWATARI, K. 'Histology of the Corneoscleral Margin,' Arch. of Ophth., 1921, l. i, p. 10.

HOCEVAR, M. 'Zur Topographie der Tränendrüse und tubulo-acinöser Drüsen der Augenlider des Menschen,' Wiener med. Wochenschr., 1900, No. 49, p. 2329, and No. 50, p. 2375.

HOEVE, J. VAN DER. (i) 'Optic Nerve and Accessory Sinuses,' Ann. Otol., Rhinol. and Laryngol., 1922, xxxi, p. 297.

(ii) *See also* under ASK.

HOLLOWAY, T. 'Cranial Deformities associated with Ocular Changes,' Trans. Amer. Ophth. Soc., 1914. (Abst. in Arch. of Ophth., 1915, xliv, p. 469.)

HOLMES, C. R. (i) 'The Sphenoidal Cavity and its Relation to the Eye,' Arch. of Ophth., 1896, xxv, p. 460.

(ii) 'Extirpation of the Lacrimal Gland in Epiphora,' Arch. of Ophth., 1919, xlviii. 4, p. 323.

HOLMES, G. (i) 'The Cortical Localisation of Vision,' Brit. Med. Jour., 1919, ii, p. 193.

HOLMES, G. and LISTER. (ii) 'Disturbances of Vision from Cerebral Lesions, with special Reference to the Cortical Representation of the Macula,' Brain, 1916, xxxix, p. 34; and Brit. Jour. Ophth., 1918, ii, pp. 352, 449, 506.

HOPKINS, G. 'The Innervation of the Muscle Retractor Oculi,' Anat. Record, 1916, xi. 5, p. 199.

HORNER, W. 'Description of a Small Muscle at the Internal Commissure of the Eyelids,' Philadelphia Jour. Med. and Phys. Sci., 1824, viii, p. 70.

HOVELACQUE, A. Article on Venous System in third edition of Poirier's Traité d'anatomie humaine, 1920, Paris.

HOWE, L. The Muscles of the Eye, 1907, Putnam, New York.

HUBER, E. 'Evolution of Facial Musculature and Facial Expression,' 1931, Johns Hopkins Press, Baltimore.

HUEBER, A. 'Transverse Orbital Muscles,' Nederl. Tijdschr. v. Geneesk. 1918, i, p. 198. (Abst. in Amer. Jour. Ophth., 1919, 3rd series, vol. ii, pt. 11, p. 248, and 1920, iii. 9, p. 701.)

HYMES, C. 'The Postnatal Growth of the Cornea and Palpebral Fissure and the Projection of the Eyeball in Early Life,' Jour. Comp. Neurol., 1929, xlviii, p. 415.

HYRTL, J. Corrosionsanatomie, 1873, Vienna.

INGALLS, N. W. 'The Dilatator Pupillae and the Sympathetic,' Jour. Comp. Neurol., 1922–23, xxxv, p. 163.

INGVAR, S. (i) 'On the Pathogenesis of the Argyll-Robertson Phenomenon,' Bull. Johns Hopkins Hosp., 1928, xliii, p. 363.

(ii) 'Concerning the Representation of Certain Eye Muscles in the Oculomotor Trunk,' Trans. Oft. Selskab i København, 1928, p. 13. Abst. in Amer. Jour. Ophth., 1930, xiii, 3, p. 170.

(iii) 'Zur Morphogenese der Tabes,' Acta med. Scandinav., 1927, lxv, p. 645.

IWATA, N. 'Beiträge zur Kenntnis der Formverhältnisse der Tränenwege des Menschen mit besonderer Berücksichtigung ihrer Entwickelung,' Folia anat. japonica, 1927, v, p. 51.

JACKSON, C. and CONNOR. 'A Wax Model of the Nasal Cavity and Paranasal Sinuses,' Ann. of Otol., Rhin., and Laryngol., 1917, p. 585.

JACOBY, E. 'Über die Neuroglia des Sehnerven,' Klin. Monatsbl. f. Augenh., 1905, xliii, i, p. 129.

JIRMAN, J. 'Concerning the Existence of Ganglion Cells in the Iris,' Oft. Sbornik, 1929, iii, p. 211. Also in Časop. lék. česk., 1929, lxviii, p. 90. Abst. in Amer. Jour. Ophth., 1930, s. 3, xiii, p. 266.

JOERSS. 'Zur Frage der Tränensackdrüsen,' Klin. Monatsbl. f. Augenh., 1905, xliii. i, p. 392. (Discussion of Werncke's paper.)

JOHNSON, G. 'Contributions to the Comparative Anatomy of the Mammalian Eye, chiefly based on Ophthalmoscopic Examination,' Philosoph. Trans. Roy. Soc. London, Series B, 1901, cxciv, p. 1.

KALT, E. Article on the Lacrimal Gland in the Encyclopédie Française d'Opht., 1903, Paris.

KAPPERS, C. ARIENS. 'The Evolution of the Nervous System in Invertebrates, Vertebrates and Man,' 1929, Bohn, Haarlem.

KARPLUS, J. P. and KREIDL, A. 'Gehirn und Sympathicus,' Arch. f. d. ges. Physiol. (Pflüger), 1912, cxliii, p. 109.

KEENE, LUCAS, and HEWER, E. 'Some Observations on Myelination in the Human Central Nervous System,' Jour. of Anat., 1931, lxvi, p. 1.

KEIBEL and MALL. Manual of Human Embryology, 1910 and 1912, Lippincott, Philadelphia and London.

KEITH, A. Human Embryology and Morphology, 1913, E. Arnold, London.

KIDD, L. J. 'The Fourth Cranial Nerve,' Brit. Jour. Ophth., 1922, vi, p. 49.

KIDD, W. 'Notes on the Eyebrows of Man,' Jour. Anat. and Physiol., 1904, xxxviii, p. iii.

KILLIAN, G. Die Nebenhöhle der Nase, etc. (Jena, 1903), and Accessory Sinuses of the Nose and their Relations to Neighbouring Organs, Patterson's translation, 1904, Chicago

KIRCHSTEIN, F. 'Über die Tränendrüse des Neugeborenen und die Unterschiede derselben von der des Erwachsenen,' Inaugural Dissertation, 1894, Berlin. (Quoted by Picou, Schirmer, Merkel.)

KISS, F. 'Die Ursprungsweise der Augenmuskeln,' Klin. Monatsbl. f. Augenh., 1919, lxiii, p. 653.

KLODT, J. 'Zur vergleichenden Anatomie der Lidmuskulatur,' Arch. f. mikros. Anat., 1893, xli, p. 1.

KNOTT, J. 'On the Cerebral Sinuses and their Variations,' Jour. of Anat., 1882, xvi, p. 27.

KOCH, S. L. 'The Structure of the Third, Fourth, Fifth, Sixth, Ninth, Eleventh, and Twelfth Cranial Nerves,' Jour. Comp. Neurol., 1916, xxvi, p. 541.

KOENIGSTEIN, L. 'Notizen zur Anatomie und Physiologie der Orbita,' Beiträge zur Augenh., 1896, iii, Heft xxv, p. 399.

KOEPPE, L. 'Klinische Beobachtungen mit der Nernstspaltlampe und dem Hornhautmikroskop,' Graefe's Arch. f. Ophth., 1918, vol. xcvii, pp. 1, 34, 198, 346.

KOFLER, K. 'Ueber einige anatomische Details und Varietäten im knöchernen Aufbau der Tränensackgegend,' Zeitschr. f. Augenh., 1929, lxvii, p. 151. Abst. in Brit. Jour. Ophth., 1931, xv, p. 56.

KOGANEI. 'Cribra cranii und cribra orbitalia,' Mitteil. aus d. med. Fak. d. Kaiserlichen Universität zu Tokyo, 1911–12, x, p. 113.

KRAUSE, C. Handbuch der Anatomie des Menschen, 1842, ii, Hannover.

KRAUSE, W. (i) 'Über die Drüsen der Conjunctiva,' Zeitschr. f. rationelle Medizin, 1854, iv, p. 337.

(ii) 'Termination of the Nerves in the Conjunctiva,' Jour. Anat. and Physiol., 1867, i, p. 346.

(iii) Handbuch der menschlichen Anatomie, 1879, ii, p. 29, Hahn, Hannover (on *spinae orbitales*).

KRAUSS, W. (i) 'Beiträge zur Anatomie, Physiologie und Pathologie des orbitalen Venensystems,' Arch. f. Augenh., 1910, lxvi, pp. 163, 285.

(ii) 'Über die Anatomie der glatten Muskulatur der Orbita und der Lider, speziell die Membrana orbitalis musculosa,' Münchener med. Wochenschr., 1911, lviii. 38, p. 1993. (Abst. in Zeitschr. f. Augenh., 1912, xxviii, p. 167.)

(iii) 'Über die Anatomie der glatten Muskulatur der Orbita und der Lider, speziell die Membrana orbitalis musculosa' (in the new-born), Arch. f. Augenh., 1912, lxxi, p. 277.

KREHBIEL, G. Die Muskulatur der Thränenwege und der Augenlider. 1878, Stuttgart.

KRONFELD, P. 'The Central Visual Pathway,' Arch. of Ophth., 1929, ii, p. 709.

KUDO, T. 'The Facial Musculature of the Japanese,' Jour. of Morphol., 1919, xxxii, p. 637.

KUNII, H. 'Ueber einen Stammbaum der blauen Sklera,' etc., Zeitschr. f. Augenh., 1930, lxxi, p. 328. Abst. in Amer. Jour. Ophth., 1930, xiii, p. 1030.

KUNTZ, A. (i) 'The Development of the Sympathetic Nervous System in Man,' Jour. Comp. Neurol., 1920, xxxii, p. 173.

(ii) 'The Autonomic Nervous System,' 1929, Lea and Febiger, Philadelphia.

KURYROVA, B. 'The Form of the Orbits in Czech Skulls,' Anthropologie, Prague, 1928, vi, p. 268. Abst. in Amer. Jour. Phys. Anthropol., 1930, xv, p. 176.

KÜSEL. 'Contribution à la connaissance des taches pigmentaires dans la conjonctive humaine,' Abstr. in Ann. d'Ocul., 1907, cxxxviii, p. 58.

LAGRANGE, F. Article on the General Anatomy of the Orbit, Encyclopédie Française, 1903, Paris.

LANDOLT, E. Article on the Anomalies of the Motor Apparatus of the Eye, in Norris and Oliver's System of Diseases of the Eye.

LANDSTRÖM, J. 'Über Morbus Basedowii,' Nord. Med. Arkh. Stockholm, 1908. (Abstr. in Nagel's Jahresb. f. Ophth., 1907, p. 472.)

LANG, P. 'Zur Entwicklung des Tränensfuhrapparatus beim Menschen,' Anat. Anzeiger, 1911, xxxviii, p. 561.

LANG, W. and TREACHER COLLINS. Article 'Congenital Malformations and Abnormalities of the Human Eye,' in Norris and Oliver's System of Diseases of the Eye.

LANGER, K. VON. (i) 'Über die Blutgefässe im Augenlide,' Med. Jahrbuch, Wien, 1878, p. 329.

(ii) 'Der Sinus cavernosus der harten Hirnhaut,' Sitzungsb. d. math.-nat. Klassen d. Kaiser-Akad. d. Wiss., Wien, 1885, xci. iii, p. 307. Also Centralbl. f. Vet.-Wiss., 1886, iv, p. 105.

LANGLEY, J. and ANDERSON. (i) 'The Action of Nicotine on the Ciliary Ganglion, &c.,' Jour. of Physiol., 1892, xiii, p. 460.

(ii) 'On the Mechanism of the Movements of the Iris,' Jour. of Physiol., 1892, xiii, p. 554.

LANS. Weekbl. v. Geneesk. Aug. 1901. (Abst. in Ann. d'Ocul., 1902.)

LAW, J. 'Comparative Anatomy of the Eye,' Ophth. Record, 1905, xiv, p. 431.

LAWRENCE, T. 'On the Position of the Optic Commissure,' Proc. Anat. Soc., in Jour. Anat. and Physiol., 1894, xxviii, p. xviii.

LAWSON, A. 'The Bacteriology of the Normal Conjunctival Sac,' Trans. of Jenner Institute, 1899, 2nd series, p. 56. Abst. in Brit. Med. Jour., 1898, ii, p. 486.

LEBENSOHN, J. 'Oculovisceral Reflexes. An Oculogastric Reflex experimentally demonstrated,' Amer. Jour. Ophth., 1929, xii, p. 562.

LEBER, T. 'Die Circulations- und Ernährungsverhältnisse des Auges,' Graefe-Saemisch Handbuch, d. g. Augenh., 1876, ii, pt. 2, p. 302.

LEBOUCQ, G. 'Étude sur les voies lymphatiques de l'œil et de l'orbite,' Arch. de Biologie, 1914–15, xxix, p. 1.

LEDERER, R. 'Über Augenbewegungen und Augendrück,' Klin. Monatsbl. f. Augenh., 1912, l, p. 594.

LEDOUBLE, F. (i) Traité des variétés du système musculaire de l'homme, 1897, Paris.

LEDOUBLE—*continued.*

(ii) Traité des variations des os du crâne de l'homme, 1903, Paris.

(iii) Traité des variations des os de la face de l'homme, 1906, Paris.

LE GROS CLARK. *See* CLARK, LE GROS.

LENZ. 'Untersuchungen über das Kerngebiet des Okulomotorius,' Klin. Monatsbl. f. Augenh., 1924, lxxii, p. 769.

LEPAGE. Thèse doct. (Orbicularis), 1909, Bordeaux. Quoted by Picou.

LESSHAFT, P. 'Über den Musculus orbicularis orbitae und seinen Einfluss auf den Mechanismus der Tränenabsonderung,' Arch. f. Anat. u. Physiol., 1868, p. 265.

LEVINSOHN, G. (i) 'Über das Verhalten der Nervenendigungen in den äusseren Augenmuskeln des Menschen,' Graefe's Arch. f. Ophth., 1901, liii, p. 295.

(ii) 'Über den Einfluss der äusseren Augenmuskeln auf den intraokularen Druck,' Graefe's Arch. f. Ophth., 1910, lxxvi. i, p. 129.

LEWITSKY. 'Die Anatomie und Pathologie der Tenon'schen Kapsel,' Inaugural Dissertation. Odessa, 1910. (Abst. in Arch. of Ophth., 1911, xl, p. 341, and in Nagel's Jahresb. f. Ophth., 1910, p. 30.)

LICHAL, F. 'Anatomy and Histology of the Nasal Tear Duct of the different Domestic Animals,' Anat. Anzeiger, 1915, xlviii. 11–12, p. 296 and p. 341.

LIEBRECHT, K. (i) 'Schädelbruch und Auge,' Arch. f. Augenh., 1906, lv, p. 36.

(ii) 'Schädelbruch und Sehnerv,' Graefe's Arch. f. Ophth., 1912, lxxxiii, p. 525.

LIGHTOLLER, G. S. *See under* BURKITT.

LINDAHL, C. 'Étude sur l'action bactéricide des larmes,' Hygiea, Stockholm, 1907, p. 353. (Abst. in Rev. gén. d'Opht., 1908, xxvii, p. 295.)

LISTER, W. T. *See under* HOLMES, G.

LO CASCIO, G. 'Lo Sviluppo della Guaine del Nervo Ottico nell' Uomo,' Ann. di ottal. e clin. ocul., 1923, vi, p. 879. Abst. in Brit. Jour. Ophth., 1924, viii, p. 286.

LOCKHART, R. (i) 'The Dural Relations of the Gasserian Ganglion with Reference to a New Method of Surgical Approach.' Jour. of Anat., 1927–28, lxii, p. 105.

(ii) 'The Anatomical Relations of the Gasserian (Semilunar) Ganglion.' Brit. Jour. Surg., 1925, xii, p. 558.

LOCKWOOD, C. 'The Anatomy of the Muscles, Ligaments, Fascia of the Orbit, etc.,' Jour. Anat. and Physiol., 1886, xx, p. 1.

LOEB, H. 'A Study of the Anatomical Relations of the Optic Nerve to the Accessory Cavities of the Nose,' Ann. of Otol., Rhin., and Laryngol., 1909, p. 243.

LOEWENTAL, N. (i) 'Beiträge zur Kenntnis der Harder'schen Drüse bei den Säugetieren,' Anat. Anzeiger, 1892, vii, p. 546.

(ii) 'Zur Kenntnis der Glandula infraorbitalis einiger Säugetiere,' Anat. Anzeiger, 1894, x, p. 123.

(iii) 'Nouvelles recherches sur les glandes sous-orbitaires, etc.,' Bibliog. Anatomique, 1909–10, xix, p. 101.

LOR, L. 'Du mécanisme des mouvements palpébraux,' Soc. Belge d'Oph., Séance Avr. 1898. (Abst. in Ann. d'Ocul., 1898, cxx, p. 46.)

LOWMAN and LLOYD MILLS. 'The Effect of Faulty Skeletal Alignment upon the Eyes,' Amer. Jour. Orthopedic Surgery, 1918, xvi, p. 459.

LUCAS KEENE, *see under* KEENE.

LUCIC, H. 'The Bacteriology of the Normal Conjunctival Sac,' Amer. Jour. Ophth., 1927, x, p. 829.

LUTZ, A. 'Explanation of Jaw-winking Phenomenon,' Arch. of Ophth., 1919, xlviii, p. 144.

MACALISTER, A. (i) 'On the Presence of a Lacrymo-jugal Suture in a Human Subject,' Proc. Roy. Irish Academy, 1874, ii. 2, p. 58.

(ii) 'Additional Observations on Muscle Anomalies in Human Anatomy,' 3rd Series (Orbit), Trans. Roy. Irish Academy, 1875, xxv, p. 7.

(iii) 'Notes on the Varieties and Morphology of the Human Lacrimal Bone and its Accessory Ossicles,' Proc. Roy. Society, London, 1884, xxxvii, p. 229.

(iv) Text-book of Human Anatomy, 1889, Griffin, London.

MACKLIN, M. 'Hereditary Abnormalities of the Eye. VII. Inheritable Defects Involving Eye Muscles, Refraction, etc.,' Canadian Med. Assoc. Jour., 1927, xvii, p. 1493.

MADDOX, E. Tests and Studies of the Ocular Muscles, 2nd edition. Keystone Publ. Co., 1907, Philadelphia; also 1898, Bristol.

MAGATH, T. 'A Variation in the Distribution of the Nervus Abducens,' Arch. of Ophth., 1919, xlviii. 1, p. 67.

MAGGIORE, L. 'The Development of the Scleral Canal and of the Lamina Cribrosa in the Human Eye,' Ann. di Ottal. e clin. ocul., 1923, vii, p. 727. Abst. in Brit. Jour. Ophth., 1924, viii, p. 288; and in Amer. Jour. Ophth., 1924, vii, p. 735.

MAGITOT, A. 'L'iris,' 1921, Doin, Paris.

MAGNASCO, M. 'Affezione Oculari d'Origini Dentaria,' Riv. Oto-neuro-oftal., 1928, v, p. 264. Abst. in Amer. Jour. Ophth., 1929, xii, p. 447.

MANN, I. C. (i) 'The Development of the Human Eye,' 1928, Cambridge Univ. Press, London.

(ii) 'The Developing Third Nerve Nucleus,' Jour. of Anat., 1927, lxi, p. 424.

(iii) *See also under* PATON.

MANZ, W. 'Über neue eigentümliche Drüsen am Corneabrande und über den Bau des Limbus conjunctiva,' Zeitschr. f. rationelle Medizin, 1859, 3, v, p. 122.

Marinesco, G. 'Sur la nature du ganglion ciliaire,' Compt. Rend. Soc. Biologie, Paris, 1908, i. lxiv, p. 88.

Marquez, M. 'On the Double Innervation of the Internal Rectus Muscle of the Eye,' Rev. d'oto-neuro-opht., 1930, viii, p. 343. Abst. in Arch. Neurol. and Psychiat., 1931, xxv, 3, p. 612.

Martyn, H. 'Notes on a Case of Anomalous Innervation of the Levator Palpebrae Superioris,' Brit. Jour. of Ophth., 1919, iii, p. 310.

Matys, V. 'Entwickelung der Tränenableitungswege,' Zeitschr. f. Augenh., 1905, xiv. i, p. 222, and 1906, xvi. ii, p. 303.

Maziarski, S. 'Über den Bau und die Einteilung der Drüsen: die Tränendrüse,' Anat. Hefte (Merkel und Bonnet), 1902, xviii, p. 215.

Meltzer, P. E. 'Gradenigo's Syndrome. Anatomic Aspects,' Arch. of Otolaryngol., 1931, xiii, p. 87. Abst. in Arch. Neurol. and Psychiat., 1931, xxvi, p. 412.

Mendel, E. 'Über den Kernursprung des Augen-Facialis,' Neurol. Centralblatt, 1887, xxiii, p. 537.

Merkel, F. (i) 'Der Musculus superciliaris,' Anat. Anzeiger, 1887, ii, p. 17.

(ii) Handbuch der topographischen Anatomie, 2nd ed., 1885.

Merkel and Kallius. Makroskopische Anatomie des Auges, Graefe-Saemisch Handbuch der Augenh., 1901, 2. Aufl., 1. Bd., 1. Kap. (and new ed., 1904, i. 1, p. 1).

Metchnikoff. 'Über die Beschaffenheit der Augenlider bei den Mongolen und Kaukasiern,' Zeitschr. f. Ethnol., 1874, vi, p. 153.

Meyer, A. 'The Connections of the Occipital Lobes and the Present Status of Cerebral Visual Affections,' Trans. Assoc. Amer. Physicians, 1907, xxii, p. 7.

Meyer, F. 'Zur Anatomie der Orbitalarterien,' Morphol. Jahrbuch, 1887, xii, p. 414.

Miceli, I. 'Centres of Ocular Movement,' Giorn. di Ocul., 1923, iv, pp. 69, 107, 121, 161, 173. Abst. in Intern. Survey Ophth., Otol. and Rhinol., 1923, vi, pp. 194, 278; 1924, vii, p. 57.

Mills, Lloyd. 'The Effects of Faulty Cranio-spinal Form and Alignment upon the Eyes,' Amer. Jour. Ophth., 1919, ii. 7, p. 493.

Minkowski, M. (i) 'Sur les conditions anatomiques de la vision binoculaire dans les voies optiques cérébrales, L'encéphale, 1922, xvii, p. 65. Abst. in Arch. Neurol. and Psychiat., 1922, viii, p. 559.

(ii) 'Comparative Anatomy of the Optic Nerve,' Schweiz. Arch. f. Neurol. u. Psychiat., 1920, vi, p. 201; 1920, vii, p. 268.

Mirimanoff, A. 'Two Ocular Symptoms in Oxycephaly,' Rev. gen. d'opht., 1924, xxxviii, p. 165. Abst. in Brit. Jour. Ophth., 1925, ix, p. 414.

Mises, F. von. 'Über die Nerven der menschlichen Augenlider,'

Sitzungsb. d. König. Akad. der Wiss. Wien, 1882, 3, lxxxv, p. 172.

MOLINIE, J. 'Ocular Reflexes of Auditory Origin,' Rev. de laryngol., d'otol. et de rhinol., 1916, i, p. 385. Abst. in Brit. Jour. Ophth., 1920, iv, p. 432.

MOLL, J. 'Bemerkungen über den Bau der Augenlider des Menschen,' Graefe's Arch. f. Ophth., 1857, iii. 2, p. 258.

MONBRUN, A. 'Le centre cortical de la vision et les radiations optiques,' etc., Arch. d'opht., 1918–19, xxxvi, p. 641.

MONESI, L. (i) 'Die Morphologie der fötalen Tränenwege beim Menschen,' Klin. Monatsb. f. Augenh., 1904, xlii. 1, p. 1.

(ii) 'Osservazioni di embriologia e di anatomia comparata sulle vie lacrimali,' Bull. della scienza med., Jan. 1905. (Abst. in Rev. gén. d'Opht., 1907, p. 350.)

(iii) 'Embryology and Comparative Anatomy of the Lacrimal Passages,' Ann. di Ottal., xxxv, p. 868. (Abst. in Rev. gén. d'Opht., 1907, p. 62.)

MONTALTI, M. 'The Anterior Corpora Quadrigemina in Man,' Ann. di Ottal. e Clin. Ocul., 1928, lvi, p. 1. Abst. in Amer. Jour. Ophth., 1928, xi, p. 931.

MORRIS. A Treatise on Human Anatomy. Ed. 8, 1925. Churchill, London.

MOST, A. 'Über die Lymphgefässe und die regionären Lymphdrüsen der Bindehaut und der Lider des Auges,' Arch. f. Anat. u. Physiol., 1905, 2/3, p. 96.

MOTAIS, E. L'Appareil moteur de l'œil. 1887, Paris.

MÜLLER, H. (i) 'Über einen glatten Muskel in der Augenhöhle des Menschen und der Säugetiere,' Zeitschr. f. wiss. Zoologie, 1858, ix, p. 541.

(ii) 'Über glatte Muskeln an den Augenlidern des Menschen und der Säugetiere,' Verhandl. der P. M. G. in Würzburg, 1859, ix, p. 244.

MÜLLER, L. R. 'Die Lebensnerven,' 1924, Springer, Berlin.

MYERS, I. 'Nystagmus: Neuro-otologic Studies concerning its Seat of Origin,' Amer. Jour. Med. Sci., 1925, clxix, p. 742. Abst. in Arch. Neurol. and Psychiat., 1925, xiv, p. 411.

NAKAGAWA, J. 'Über echte Papillen in der normalen Conjunctiva,' Arch. f. Augenh., 1903, xlvii, p. 51.

NATALE, A. 'Congenital Ptosis and Muscular Anomalies of the Eye,' Rev. de Especialidades, 1928, iii, p. 908; also in Arch. de Oftal., Buenos Aires, 1929, iv, p. 14. Abst. in Arch. of Ophth., 1929, ii, p. 485.

NEAL, H. (i) 'The Morphology of the Eye-Muscle Nerves,' Jour. of Morphol., Philadelphia, 1914, xxv, pp. 1–187.

(ii) 'The History of the Eye-Muscles,' Jour. of Morphol., 1918, xxx. 2, p. 433.

NEDDEN. 'Experimentelle Untersuchungen baktericider Substanzen im Auge nicht immunisierter Individuen,' Graefe's Arch. f. Ophth., 1907, lxv, p. 267. (Abst. from Ann. d'Ocul., 1908, p. 369.)

NICHOLLS, G. 'On the Occurrence of an Intra-cranial Ganglion in the Oculomotor Nerve in Scyllium Canicula, etc.' (with observations on the ciliary ganglion), Proc. Roy. Soc., London, Series B, 1915, lxxxviii, p. 553.

NICHOLSON, HELEN. 'On the Presence of Ganglion Cells in the Third and Sixth Nerves of Man,' Jour. Comp. Neurol., 1924, xxxvii, p. 31.

NICOLA, B. Giorn. d. R. Accad. d. Med. d. Torino, 1903. Quoted by Ledouble, 1906, p. 160 (on the Zygomatico-frontal Suture).

NOLL, A. 'Morphologische Veränderungen der Tränendrüsen bei der Sekretion,' Arch. f. mikros. Anat., 1901, lviii, p. 487.

NORDMANN. *See* WEILL.

NORRIS and OLIVER. System of Diseases of the Eye, 1900, Lippincott, Philadelphia.

NUSSBAUM, M. 'Entwicklungsgeschichte des menschlichen Auges,' Graefe-Saemisch-Handbuch d. g. Augenh., 3rd ed., 1912, Pt. I, Kap. viii.

OAST, S. P. 'Blue Sclerotics and Brittle Bones,' Arch. of Ophth., 1928, lvii, p. 254.

OCHI, S. 'Relation of the Ocular Muscles and Sclera in the Etiology of Myopia,' Amer. Jour. Ophth., 1919, ii, p. 675.

OETTEKING, B. 'The Processus Frontosphenoidalis of the Zygoma and its bearing on the Configuration of the Orbit,' Anat. Record, 1919, xvii, p. 25.

OGAWA, K. 'Über Pigmentierung des Sehnerven,' Arch. f. Augenh., 1905, lii, p. 437 and 1906, lv, p. 106.

ONO, R. 'Untersuchungen über die Orbita von Japanern,' Japanese Jour. Med. Sci., Tokyo, 1928, I, Anat., i, no. 4.

ONODI, L. (i) 'Das Verhältnis des Nerven opticus zu der Keilbeinhöhle und ins besondere zu der hintersten Siebbeinzelle,' Arch. f. Laryngol., 1903, xiv, p. 360.

(ii) Der Sehnerv und die Nebenhöhlen der Nase, 1907, Wien.

(iii) The Relations of the Lacrymal Organs to the Nose and Nasal Accessory Sinuses. (English trans. by D. Mackenzie.) 1913, London.

OTTLEY. 'On the Attachment of the Eye Muscles in Mammals,' Proc. Zoological Soc. London, 1879, p. 121.

PARSONS, J. HERBERT. (i) 'The Nerve Supply of the Lacrimal Gland,' Roy. London Ophth. Hosp. Rep. (May 1902), 1899–1903, xv, p. 81.

(ii) 'The Ocular Circulation,' Arris and Gale Lecture, 1903, Royal College of Surgeons, London.

(iii) The Pathology of the Eye, 1904 and 1905, London.

(iv) 'The Innervation of the Pupil,' Roy. London Ophth. Hosp. Rep., 1904, xvi, p. 20.

(v) Diseases of the Eye, 1907, Churchill, London.

(vi) 'The Fourth Cranial Nerve,' Brit. Jour. Ophth., 1921, v, p. 529. Abst. in Amer. Jour. Ophth., 1922, v, p. 53.

(vii) 'Discussion of the Physiology and Pathology of the Pupil Reactions,' Trans. Ophth. Soc. United Kingdom, 1924, xliv, p. 1.

(viii) 'The Foundation of Vision,' Trans. Ophth. Soc. United Kingdom, 1925, xlv, p. 14.

PARSONS SCHAEFFER. *See* SCHAEFFER.

PATON, L. 'The Trigeminal and its Ocular Lesions,' Brit. Jour. Ophth., 1926, x, p. 305.

PATON, L. and MANN, I. C. 'The Development of the Third Nerve Nucleus and its Bearing on the Argyll-Robertson Pupil,' Trans. Ophth. Soc. United Kingdom, 1925, xlv, pt. 2, p. 610.

PATTON, J. M. 'Regional Anatomy of the Tear Sac,' Ann. Otol., Rhinol. and Laryngol., 1923, xxxii, p. 58.

PERGENS, E. 'Les Dépôts pigmentaires dans la conjonctive des nègres,' Ann. d'Ocul., 1898, cxx, p. 42. (Abstract.)

PERLIA. 'Die Anatomie des Oculomotorius Centrums beim Menschen,' Graefe's Arch. f. Ophth., 1889, xxxv. iv, p. 287.

PERNA, G. 'Un musculo trasverso anomalo della cavità orbitaria nell' uomo,' Anat. Anzeiger, Verhand., 1905, xxvii, p. 215.

PESCHEL, M. 'Congenitaler Epidermisüberzug der Tränenkarunkel,' Centralb. f. prakt. Augenh., 1903, xxvii, p. 148.

PETER, L. C. 'The Extra-ocular Muscles,' 1927, Lea and Febiger, Philadelphia.

PETERS, A. 'Epiphora caused by Cornification of Epithelium near Caruncle,' (Bibliography) Klin. Monatsbl. f. Augenh., 1918, lxi, p. 252.

PETZETAKIS, M. 'Le Réflexe Sous-orbitaire,' Paris Médicale, 1929, ii, p. 129.

PFEIFER, R. A. 'Myelogenetisch-anatomische Untersuchungen über den zentralen Abschnitt der Schleitung,' Monograph. a. d. G. d. Neurol. u. Psychiat., 1925, xliii, Springer, Berlin.

PICOU, R. 'Article on the Annexes of the Eye,' in Poirier's Anatomy, 1912, tome V, ii.

PIERSOL. Text-book on Human Anatomy, Ed. 9, 1930. Lippincott, Philadelphia and London.

PINES, J. L. 'Die Morphologie des Ganglion ciliäre beim Menschen,' Zeitschr. f. mikros.-anat. Forschung, 1927, x, p. 313.

PLOMAN, K. G. 'Continued Investigations on the Lacrymal Passages,' Acta Ophth., 1930, viii, p. 155. Abst. in Arch. of Ophth., 1931, v. 2, p. 302.

PLOMAN, K. G., ENGEL, A., and KNUTSSON, F. 'Experimental Studies of the Lacrymal Passageways,' Acta Ophth., 1928, vi, p. 55.

POCKLEY, F. 'Epicanthus and Congenital Ptosis,' Med. Jour. of Australia, 1919, i, sixth year, No. 25, p. 509.

POIRIER. Traité d'anatomie humaine, éd. iii. 1911, tome v, fasc. 2 (Les organes du sens). (Ostéologie, ed. iii. 1912, tome i) Masson, Paris.

POLLOCK, INGLIS. 'The Persistence of the Nerve Plexus of the Iris after Excision of certain Ganglia' (abstract), Brit. Med. Jour., Nov. 1, 1913, p. 1158.

PONCET, F. 'Recherches sur la terminaison des nerfs dans la conjonctive,' Arch. de Physiol. norm. et path., Paris, 1875, ii, p. 545.

POOLE, F. 'The Relations of the Superior Oblique Muscle of the Eye in Mammals,' Jour. Anat. and Physiol., 1905, xxxix, p. 154.

POPOFF, E. 'Contribution à l'étude du repli semilunaire et de la caroncle lacrymale chez l'homme,' Thèse de Paris, 1912. (Abst. in Nagel's Jahresb. f. Ophth., 1912, p. 29.)

POSEY, W. C. 'Concerning some gross Structural Anomalies of the Muscles of the Eye and its Adnexa,' Trans. Amer. Acad. Ophth. and Oto-laryngol., 1923, xxviii, p. 243.

POSEY and SPILLER. 'The Eye and Nervous System,' 1906, Lippincott, Philadelphia and London.

POWER, H. (i) Lectures on Diseases of the Lacrimal Apparatus, Lect. III, Pt. II, Lancet, 1886, ii, p. 907.

(ii) Case of Oxycephaly, Trans. Ophth. Soc. of United Kingdom, 1894, xiv, p. 212.

POYALES, F. 'On the Development of the Eye Muscles, Rectus Internus and Externus, in the Human Embryo' (in Spanish), Anat. Record, 1917, xiii, No. 6, p. 375.

PRANGEN, A. DE H. 'A Study of the Comparative Anatomy of the Extraocular Muscles,' Trans. Amer. Ophth. Soc., 1928, xxvi, p. 353.

PUGLISI-ALLEGRA, S. 'Sui nervi della glandola lagrimale,' Riforma medica Roma, 1903, xix, p. 795. (Abst. in Anat. Anzeiger, 1903, xxiii, p. 392.)

PUTNAM, T. J. 'Studies on the Central Visual Connections,' Parts I to IV, Arch. Neurol. and Psychiat., 1926, xvi, pp. 1, 285, 566, 683.

QUAIN, R. Elements of Anatomy, 2nd edition, 1908–. Longmans, Green, London.

RANGLARET, A. 'Anatomie et pathologie des cellules ethmoïdales,' 1896, Thèse de Paris.

REDSLOB, E. (i) 'On the Dilating Apparatus of the Iris,' Bull. Soc. franç. d'opht., 1928, p. 4. Abst. in Amer. Jour. Ophth., 1929, xii, p. 345.

(ii) 'Retrobulbar Optic Neuritis of Nasal Sinus and Dental Origin: Evolution and Present State of the Question,' Rev. d'oto-neuro-opht., 1929, vii, p. 405. Abst. in Arch. Neurol. and Psychiat., 1930, xxiv, p. 162.

REICH, M. 'Zur Histologie der Conjunctiva des Menschen,' Graefe's Arch. f. Ophth., 1875, xxi. 1, p. 1.

REID, D. 'The Presence of Lachrymo-jugal Sutures in Two Skulls,' Jour. Anat. and Physiol., 1910, xliv, p. 249.

REITSCH, W. 'Anatomische Betrachtungen über Fornix, Karunkel und Semilunarfalte,' Zeitschr. f. Augenh., 1929, lxix, p. 207.

RETZIUS, G. 'Über das Ganglion ciliäre,' Anat. Anzeiger, 1894, ix. 21, p. 633.

RIDDOCH, G. 'Dissociation of Visual Perceptions due to Occipital Injuries,' Brain, 1917, xl, p. 15.

RIDLEY, F. 'Lysozyme: an Antibacterial Body present in Great Concentration in Tears, and its Relation to Infections of the Human Eye,' Trans. Ophth. Sect., Roy. Soc. Med., London. Abst. in Arch. of Ophth., 1928, lvii, p. 647.

RILEY, H. A. 'The Central Nervous System Control of the Ocular Movements and the Disturbances of this Mechanism,' Arch. of Ophth., 1930, iv, pp. 640 and 885.

ROCHAT, G. and BENJAMINS. (i) 'Einige Bemerkungen über die Anatomie der Tränenwege des Kaninchens,' Graefe's Arch. f. Ophth., 1915, xci, p. 66.

(ii) 'Experimente über die Tränenableitung,' Arch. f. Ophth., 1915, xci, p. 92.

(iii) 'Contribution à la connaissance de la physiologie des voies lacrimales,' Nederl. tijdschr. v. geneesk., 1915, p. 2067.

(iv) 'Über eine neue vasomotorische Automatie,' Arch. f. d. g. Physiol., 1916, clxiv, p. 111.

ROCHON-DUVIGNEAUD, A. (i) 'Recherches sur le développement des voies lacrymales,' Arch. d'Opht., 1900, xx, p. 241.

(ii) 'Note sur l'anatomie de l'orbite, etc.,' Arch. d'Opht., 1903, xxiii, p. 769.

ROLLET, E. Section on the Accessory Sinuses of the Nose in the Encyclopédie Française d'Opht., 1903.

ROSENGREN, B. 'The Mechanism of Tear Conduction,' Acta Ophth., 1929, vi. 4, p. 367. Abst. in Amer. Jour. Ophth., 1929, xii, p. 787.

ROUSSEAU, R. 'Les fibres musculaires lisses de l'orbite d'après Karl Hesser,' Ann. d'Ocul., 1916, cliii, p. 299.

ROUVIÈRE, H. 'Le Tendon de Zinn, etc.,' Bibliog. anatomique, 1914, xxiv, p. 92.

ROY, J. (i) 'Anatomie et physiologie comparées de l'œil et ses annexes,' Arch. d' Opht., 1912, xxxii, p. 494.

(ii) 'The Eyesight of the Negroes of Africa,' Arch. of Ophth., 1919, xlviii. 1, p. 72.

RUSH, C. and SCHAEFFER. 'Supernumerary Orbital Muscles,' Arch. of Ophth., 1917, xlvi, p. 524.

RUSKIN, S. L. (i) 'Control of Tearing by Blocking the Nasal Ganglion,' Arch. of Ophth., 1930, iv, p. 208.

RUSKIN—*continued*

(ii) 'Studies of the Sphenopalatine Ganglion,' Laryngoscope, 1925, xxxv, p. 87.

(iii) 'Sensory Field of Facial Nerve,' Arch. of Otolaryngol., 1928, vii, p. 351.

RUSSELL, RISIEN. 'An Experimental Investigation of Eye Movements,' Jour. of Physiol., 1894, xvii, pp. 1, 378.

RUTHERFORD, L. 'The Lacrymal Gland in Surgical Anaesthesia,' Brit. Med. Jour., June 21, 1913, p. 1313.

RYDER, J. Article on the Development of the Eye, Norris and Oliver's System of Diseases of the Eye.

SALZMANN, M. Anatomy and Histology of the Human Eyeball in the Normal State. Brown's translation, 1912, Chicago.

SANTOS-FERNANDEZ, J. (i) Disposition anatomique du canal nasal chez le nègre,' Recueil d'Opht., 1903, p. 501.

(ii) 'The Measurements of the Nasal Canal According to Race,' Amer. Jour. Ophth., 1921, iv, p. 32.

SAPPEY. Anatomie descriptive. 1867, Paris. Ed. iv, 1888.

SÄTTLER, C. (i) 'Beitrag zur Kenntnis der modificirten (Moll'schen) Schweissdrüsen des Lidrandes,' Arch. f. mikros. Anat., 1877, xiii, p. 783.

(ii) 'Über den sogenannten Landströmschen Muskel und seine Bedeutung für den Exophthalmus bei Morbus Basedowii,' Bericht d. ophth. Gesell. Heidelberg, 1911, p. 181. (Abst. in Ann. d'Ocul., 1911, cxlvi, p. 211.)

(iii) 'Über die Markscheidenentwickelung im Tractus opticus, Chiasma und Nervus opticus,' Graefe's Arch. f. Ophth., 1915, xc, p. 271.

SCAMMON, R. E. and ARMSTRONG, E. L. 'On the Growth of the Human Eyeball and Optic Nerve,' Jour. Comp. Neurol., 1925, xxxviii, p. 165.

SCHAEFFER, J. PARSONS. (i) 'The Sinus Maxillaris and its Relations in the Embryo, Child, and Adult Man,' Amer. Jour. Anat., 1910, x, p. 313.

(ii) 'The Lateral Wall of the Cavum Nasi in Man, with Special Reference to the various Developmental Stages,' Jour. of Morphol., 1910, xxi, No. 4, p. 613.

(iii) 'The Genesis and Development of the Naso-lacrimal Passages in Man,' Amer. Jour. Anat., 1912, xiii, p. 1.

(iv) 'The Nose, Paranasal Sinuses, Naso-lacrimal Passages and Olfactory Organ in Man,' 1920, Blakiston, Philadelphia.

(v) 'Anatomic Relations of Optic Nerve and Chiasm to Paranasal sinuses,' Amer. Jour. Ophth., 1921, iv, p. 451.

(vi) 'The Visual Pathway and the Paranasal Sinuses,' Amer. Jour. Ophth., 1922, v, p. 105.

(vii) 'The Modern Conception of the Anatomy of the Naso-lacrimal Passageways in Man,' Amer. Jour. Ophth., 1921, iv, p. 683.

(viii) 'The Sphenoidal Sinus and the Temporal Lobe,' Jour. Amer. Med. Assoc., 1921, lxxvi, p. 1488.

(ix) 'Some Points in the Regional Anatomy of the Optic Pathway, with especial reference to Tumours of the Hypophysis Cerebri and resulting Ocular Changes,' Anat. Record, 1924, xxviii, p. 243.

SCHÄFER, E. Text-book of Microscopic Anatomy in Quain's Elements of Anatomy, vol. ii. 1, 1912, Longmans, London.

SCHAPRINGER, A. 'Ueber Varietäten des Epitarsus.' Centralb. f. prakt. Augenh., 1905, 29th year, p. 129.

SCHIEFFERDECKER, P. (i) 'Eine Eigentümlichkeit im Bau des Augenmuskels,' Sitzungs. d. Nieder. Gesell. zu Bonn, 1904, etc. (Abst. in Zeitschr. f. Augenh., 1905, xiv, p. 186.)

(ii) 'Über die Lidmuskulatur des Menschen,' Sitzungs. d. Niederheimschen Gesell. zu Bonn, 1905, B. xiv, p. 54.

SCHIRMER, O. Section on 'Mikroskopische Anatomie und Physiologie der Tränensorgane,' Graefe-Saemisch Handbuch d. g. Augenh., 1904, 2nd ed., Lief. 25 und 76.

SCHMIDT-RIMPLER, H. (i) 'Kurzsichtigkeit und Augenhöhlenbau,' Graefe's Arch. f. Ophth., 1889, xxxv, p. 1 and p. 200.

(ii) 'Bemerkungen zu Stilling's Aufsatz,' Fortschritte der Medizin, 1889, p. 573.

SCHWALBE, G. 'Lehrbuch der Anatomie des Auges,' 1887. Erlangen.

SCHWEGEL, A. 'Knochenvarietäten' (On the Supra-Orbital Notch), Zeitschr. f. rationelle Medizin, 1859, v. 3, p. 283.

SCHINDLER, E. 'The Anatomy and Physiology of the Glial System of the Intra-cranial Optic Nerves,' Zeitschr. f. Augenh., 1926, lx, p. 15. Abst. in Brit. Jour. Ophth., 1928, xii, p. 543.

SCHWEINITZ, G. DE, 'Concerning certain Ocular Aspects of Pituitary Body Disorders,' Trans. Ophth. Soc. United Kingdom, 1923, xliii, p. 12.

SCHWEINITZ, DE, WILDER, BALL, and WEEKS. 'The Sympathetic and the Eye,' Jour. Amer. Med. Assoc., 1904, xlii, p. 286. Abst. in Ann. d'Ocul., 1904, cxxxii, p. 233.

SCIMEMI, E. 'Sulla conduttura delle lagrime,' Ann. di. Ottal., 1892, xxi. 1, p. 222; and in Arch. f. Physiol., 1892, p. 291, suppl.

SEEFELDER, R. 'Pathologisch-anatomische Beiträge zur Kenntnis der angeborenen Colobome des Auges,' and other papers following. Graefe's Arch. f. Ophth., 1908, lxviii, p. 335; 1909, lxx, pp. 45, 448; 1910, lxxiii, p. 216.

SEGGEL. 'Über die Abhängigkeit der Myopie vom Orbitalbau,' Graefe's Arch. f. Ophth., 1890, xxxvi, 2. Abt., p. i.

SESEMANN, E. 'Die Orbitalvenen des Menschen, etc.,' Arch. f. Anat. u. Physiol. und wiss. Med., 1869, p. 154.

SHARPE, W. 'The Cranial Deformity of Oxycephaly; its Operative

Treatment, with a Report of Cases,' Amer. Jour. Med. Sc., 1916, cli, p. 840.

SHELLSHEAR, J. E. (i) The Arterial Supply of the Cerebral Cortex in the Chimpanzee,' Jour. of Anat., 1930, lxv, p. 45.

(ii) 'Arterial Supply of Cerebral Cortex in Man,' Brain, 1927, l, p. 238.

SHERRINGTON, C. 'Observations on the Sensual Role of the Proprioceptive Nerve-supply of the Extrinsic Ocular Muscles,' Brain, 1918, xli, p. 332.

SHERRINGTON, C. and TOZER. 'Receptors and Afferents of the Third, Fourth and Sixth Cranial Nerves,' Proc. of Roy. Soc. London, 1910, lxxxii, p. 450.

SHOEMAKER, J. 'A Case of Supernumerary Eyelid,' Amer. Jour. Ophth., 1914, xxxi, p. 225.

SIEUR et JACOB. Les Fosses et leurs sinus. 1901, Paris.

SINGER, S. 'Brittle Bone and Blue Sclerotics Question,' Zeitschr. f. klin. Med., 1923, xcvii, p. 43.

SKILLERN, R. The Accessory Sinuses of the Nose, 1913, Philadelphia.

SMITH, ELLIOT. (i) 'Asymmetry of the Brain and Skull,' Jour. Anat. and Physiol., 1907, xli, p. 236.

(ii) 'New Studies on the Folding of the Visual Cortex,' etc., Jour. Anat. and Physiol., 1907, xli, p. 198 and p. 237.

(iii) The Central Nervous System in Cunningham's Text-book of Anatomy.

(iv) 'The New Vision,' Bowman Lecture. Trans. Ophth. Soc. United Kingdom, 1928, xlviii, p. lxiv.

(v) 'New Light on Vision,' Nature (suppl.), 1930, cxxv.

(vi) Proc. Roy. Soc. London, 1904, p. 62.

(vii) 'The Evolution of Man,' 1924, Oxford Univ. Press.

SPECIALE-CIRINCIONE, F. (i) 'Über die Entwicklung der Tränendrüse beim Menschen,' Graefe's Arch. f. Ophth., 1908, lxix. 2, p. 193.

(ii) 'The Development of the Lacrimal Canal in Man,' Ann. di ottal. e clin. ocul., 1929, lvii, p. 435. Abst. in Amer. Jour. Ophth., 1930, xiii, p. 359; also in Arch. of Ophth., 1930, iii, p. 460.

SPERINO. 'La ossificazione e la posizione della trochlea, etc.,' Mem. R. Accad. di Scienze, Ser. III, vi. (Abst. in Schwalbe's Jahresb. f. Anat., 1905, p. 85.)

STANCULEANU, G. (i) 'Recherches sur le développement des voies lacrymales chez l'homme et chez les animaux,' Arch. d'Opht., 1900, xx, p. 141.

(ii) 'Les Rapports anatomiques entre les sinus de la face et l'appareil orbito-oculaire,' Arch. d'Opht., 1902, xxii, p. 108 and p. 248.

STARLING, E. Principles of Human Physiology, ed. iii, 1920, Churchill, London.

STEINER, L. 'Les Taches pigmentaires de la conjunctiva,' Ann. d'Ocul., 1906, cxxxv, p. 466.

STEPHENSON, S. (i) 'Supernumerary carunculae lacrymales,' Ophth. Review, 1896, xv, p. 8.

(ii) 'Congenital Distichiasis,' Ophth. Review, 1901, xx, p. 350.

(iii) 'A Case of Ocular Torticollis,' Proc. Roy. Soc. Med., London, Sect. Ophth., May 7, 1913, vi, p. 90.

(iv) 'Eye Symptoms as the only Indication of Fractured Base of the Skull,' Brit. Jour. Ophth., 1919, iii, No. 2, p. 505.

STEVENS, G. (i) 'Anomalies of the Ocular Muscles,' Series of Papers in Arch. of Ophth. for 1887, 1888, 1889.

(ii) 'Facial Expressions as influenced by the Ocular Muscles,' Jour. Amer. Med. Assoc. of Chicago, 1892, xix, p. 257.

STIBBE, E. P. (i) 'Sensory Components of the Motor Nerves of the Eye,' Jour. of Anat., 1929, lxiv, p. 112 (under Proceedings).

(ii) 'A Comparative Study of the Nictitating Membrane of Birds and Mammals,' Jour. of Anat., 1928, lxii, p. 159.

STIEDA, L. 'Über die Caruncula lacrymalis des Menschen,' Arch. f. mikros. Anat., 1890, xxxvi, p. 291.

STILLING, J. (i) Schädelbau und Kurzsichtigkeit, 1888, Wiesbaden.

(ii) Über das Wachstum der Orbita und dessen Beziehungen zur Refraktion. Arch. f. Augenh., 1890, xxii. 3, 4, p. 47.

(iii) See also Fortschritte der Medizin, 1889, pp. 444 and 647.

STOPFORD, J. B. (i) 'The Arteries of Pons and Medulla Oblongata,' Jour. Anat. and Physiol., 1916, v. l, pp. 131, 255, and 1917, li, p. 250.

(ii) 'The Function of the Spinal Nucleus of the Trigeminal Nerve,' Jour. of Anat., 1925, lix, p. 120.

STRUTHERS, J. 'On the Anatomy and Physiology of the Oblique Muscles of the Eye in Man and Vertebrate Animals,' Monthly Jour. Med. Sci., 1849, ix, p. 1143.

SUNDWALL, J. 'The Lachrymal Gland,' Amer. Jour. Anat., 1916, xx, p. 147.

SUTHERLAND, C. G. 'Oxycephaly; Steeple or Tower Head,' etc. Jour. of Radiol., 1922, iii, p. 465.

SUTTON, J. E. 'The fascia of the Human Orbit,' Anat. Record, 1920, xviii. 2, p. 141.

SWANZY, H. Article in Norris and Oliver's System of Diseases of the Eye, 1900, iv, p. 631.

SWERSCHEWSKY. 'Die anatomischen und pathologischen Verhältnisse der Tränenableitungswege,' Westn. Ophth. S. 549. (Abst. in Nagel's Jahresb. f. Ophth., 1910, p. 37.)

SYME, W. S. 'The Sphenoidal Sinus in Relation to the Optic Nerve,' Jour. Laryngol. and Otol., 1924, xxxix, p. 375.

TARTUFERI, F. 'Anatomie pathologique des dacryocystites catarrhales, etc.,' Arch. d'Opht., 1902, xxii, p. 166.

TAYLOR, J. 'Changes in the sella turcica in family optic atrophy,' Brit. Jour. Ophth., 1919, iii. 5, p. 193.

TENCHINI, L. 'Di un emissario anomalo dell'orbito frontale,' Monit. Zool. ital., 1905, anno xvi. Firenze. (Abst. in Schwalbe's Jahresb. f. Anat., 1905, p. 90.)

TENON. Mémoires et observations sur l'anatomie, 1806, Paris.

TEPLIACHINE, A. 'Recherches sur les nerfs sécretoires de la glande lacrymale,' Arch. d'Opht., 1894, xiv, p. 401.

TERNI, T. 'Researches on the Abducens Nerve and particularly on the Significance of its Accessory Nucleus,' Folia neuro-biolog., 1922, xii, p. 277. Abst. in Arch. Neurol. and Psychiat., 1923, ix, p. 783.

TERSON, A. Section on Anatomy and Physiology of the Eyelids in the Encyclopédie Française d'Opht., i, 1903, Paris.

TESTUT, L. Traité d'anatomie humaine, 5th edition, 1905, Paris.

TESTUT and JACOB. Traité d'anatomie topographique, 3rd ed., 1914, Doin, Paris.

THEOBALD, S. (i) Article on Diseases of the Lacrimal Apparatus in Norris and Oliver's System of Diseases of the Eye.

(ii) 'Chief Functions of the Oblique Muscles of the Eye,' Bull. Johns Hopkins Hosp., 1918, xxix, p. 15.

THOMSON, A. (i) 'The Orbito-Maxillary Frontal Suture in Man and the Apes, with Notes on the Varieties of the Human Lacrymal Bone,' Jour. Anat. and Physiol., 1890, xxiv, p. 349.

(ii) 'The Anatomy of the Human Eye, as illustrated by Enlarged Stereoscopic Photographs,' 1912, Clarendon Press, Oxford.

THOMSON, E. S. (i) 'Ocular Involvement in Sinus Disease,' Laryngoscope, 1928, xxxviii, pp. 439 and 521.

(ii) 'Conditions of the Optic Nerve caused by Disease of the Sinuses,' Arch. of Otolaryngol., 1929, x, p. 248.

THOMSON, ST. CLAIR. (i) 'Cerebral and Ophthalmic Complications in Sphenoidal Sinusitis,' Trans. Med. Soc. London, 1906, xxix, p. 12.

(ii) 'The Frequency of Orbital Manifestations of Nasal Sinusitis,' Ophthalmoscope, 1908, vi, p. 228.

THOMSON, W. E. 'Determination of the Influence of the Eye on the Growth of the Orbit by Experimental Enucleation of one Eye in Young Animals,' Meeting of Ophth. Soc. of United Kingdom, Nov. 1900. (Abst. in Arch. of Ophth., 1901, xxx, p. 93.)

THORSCH, E. 'Beziehungen der Tränensackgrube zur Nase und ihren Nebenhöhlen,' Klin. Monatsbl. f. Augenh., 1909, xlvii, p. 530.

TILLEY, J. H. 'Exophthalmos: Mechanism of its Production in Exophthalmic Goitre,' Ann. of Surgery, 1926, lxxxiv, p. 647.

TOPINARD, P. Éléments d'anthropologie générale, 1885, Paris.

TOZER, F. (i) 'On the Presence of Ganglion Cells in the Roots of the Third, Fourth, and Sixth Cranial Nerves,' Physiol. Soc. Proc., in Jour. of Physiol., 1912, xlv, p. xv.

(ii) See also under Sherrington.

TRAQUAIR, H. 'The Anatomical Relations of the Hypophysis and the Chiasma,' Ophthalmoscope, 1916, xiv, p. 562.

TREACHER COLLINS. *See* COLLINS.

TSUCHIDA, U. 'On the Oculomotor Nucleus,' Arbeiten aus dem hirnanat. Inst. in Zürich, 1906, ii (referred to in Piersol's Anatomy).

TUBBY, A. 'Is there a Connexion between Skeletal Asymmetry and Defects of the Eye?' Trans. Ophth. Soc. of United Kingdom, 1919, p. 335.

TURNER, ALDREN. 'A Note on the Oculofacial Muscular Group,' Roy. London Ophth. Hosp. Rep., 1890–92, xiii, p. 328.

TURNER, LOGAN. (i) 'The Accessory Sinuses of the Nose,' 1901, Green, London.

(ii) 'The Relation of Disease of the Nasal Accessory Sinuses to Disease of the Eye,' Lancet, 1908, ii, p. 396; Brit. Med. Jour., 1908, ii, p. 730.

TURNER, SIR W. (i) 'On a Non-striped Muscle connected with the Orbital Periosteum of Man,' Natural History Review, 1862, p. 107.

(ii) 'The Infra-orbital Suture,' Jour. Anat. and Physiol., 1885, xix, p. 218.

UNDERWOOD, A. 'Anatomy and Pathology of the Maxillary Sinus,' Jour. Anat. and Physiol., 1910, xliv, p. 354.

VAIL, H. H. 'Anatomical Studies of Dorello's Canal,' Laryngoscope, 1922, xxxii, 8, p. 569.

VILLARD, H. 'Recherches sur l'histologie de la conjonctivite normale,' N. Montpellier méd., 1896, v, pp. 651, 672, 693.

VILLIGER, E. Brain and Spinal Cord, edition iii, 1912. Trans. by Piersol, ed. iv, 1931.

VIRCHOW, H. (i) 'Über Tenon'schen Raum und Tenon'sche Kapsel,' Abhandl. d. Kgl. Preuss. Akad. d Wiss., 1902.

(ii) 'Mikroskopische Anatomie der äusseren Augenhaut und des Lidapparatus,' Graefe-Saemisch Handbuch d. g. Augenh., 2nd ed., 1905–10.

VITALI, G. 'Sui Rami Orbitali del Ganglio Sfeno-palatino,' etc., Riv. oto-neuro-oftal., 1929, vi, p. 151. Abst. in Arch. of Ophth., 1930, iii, p. 627.

WALDEYER, W. Article on Microscopical Anatomy in De Wecker and Landolt's Traité complet d'opht., tome i, p 169. 1886, Paris.

WALLIS, G. 'Some Observations upon the Anatomical Relations of the Optic Nerves and Chiasma to the Sphenoidal Bone,' Practitioner, Jan. 1917, xcviii, p. 41.

WARD, F. O. Outlines of Human Osteology, 1858, Renshaw, London.

WEILL, G. and NORDMANN, J. 'Congenital Absence of Abduction,' Arch. d'Opht., 1927, p. 593. Abst. in Brit. Jour. Ophth., 1928, xii, p. 328.

WEISS, L. (i) Beiträge zur Anatomie der Orbita. 1890, Tübingen.
(ii) 'Über das Verhalten von M. rectus externus und rect. internus bei wachsender Divergenz der Orbita, Arch. f. Augenh., 1894, xxix. 3, 4, p. 298.
(iii) 'Über das Wachstum des menschlichen Auges,' Anat. Hefte (Merkel und Bonnet), 1897, viii, p. 191.

WELCKER, H. 'Cribra orbitalia,' Arch. f. Anthropol., 1887, xvii, p. 1.

WERNCKE, T. 'Ein Beitrag zur Anatomie des Tränensackes speziell zur Frage der Tränensackdrüsen,' Klin. Monatsbl. f. Augenh., 1905, xliii, 1, p. 191.

WEST, J. 'Eine Probe zur Feststellung der Funktionsfähigkeit des Tränenröhrchens (eine Canaliculusprobe) und ihre klinische Bedeutung,' Zeitschr. f. Augenh., 1918, xxxix, p. 260.

WESTLAKE, S. B. 'Note on the Relation of the Optic Nerve to the Last Posterior Ethmoid Cell,' Ann. Otol., Rhinol. and Laryngol., 1923, xxxii, p. 729.

WHEELER, J. M. 'Paralysis of the Sixth Cranial Nerve associated with Otitis Media,' Jour. Amer. Med. Assoc., 1918, lxxi, p. 1718.

WHITE, L. E. 'An Anatomic and X-ray Study of the Optic Canal in Cases of Optic Nerve Involvement,' Ann. Otol., Rhinol. and Laryngol., 1924, xxxiii, p. 121. Also in Laryngoscope, 1924, xxxiv, p. 255.

WHITNALL, S. E. (i) 'On a Ligament acting as a Check to the Action of the Palpebrae Superioris Muscle,' Jour. Anat. and Physiol., 1910, xlv, p. 131.
(ii) 'An Instance of the Retractor Bulbi Muscle in Man,' Jour. Anat. and Physiol., 1911, xlvi, p. 36.
(iii) 'On a Tubercle on the Malar Bone,' Jour. Anat. and Physiol., 1911, xlv, p. 426.
(iv) 'The Relations of the Lacrimal Fossa to the Ethmoidal Cells,' Ophth. Review, 1911, xxx, p. 321.
(v) 'The Naso-lacrimal Canal: the Extent to which it is formed by the Maxilla and the Influence of this upon its Calibre,' Ophthalmoscope, 1912, x, p. 557.
(vi) 'The Shape of the Orbit: its Influence upon the Eyeball (Stilling's theory),' Ophth. Review, 1913, xxxii, p. 34.
(vii) 'The Relation of the Naso-lacrimal Canal to the Maxillary Antrum: Formation of the Lacrimal Recess,' Ophth. Review, 1913, xxxii, p. 33.
(viii) 'The Ligamentum Palpebrarum Mediale: Why its Removal in Ablation of the Lacrimal Sac does not necessarily entail Ectropion of the Lower Eyelid,' Ophthalmoscope, 1913, xi, p. 216.
(ix) 'The Levator Palpebrae Superioris: the Attachment and Relations of its Aponeurosis,' Ophthalmoscope, 1914, xii, p. 258.

(x) 'Some Instances of Abnormal Ocular Muscles,' Anat. Record, 1921, xxi, p. 143.

(xi) 'Some Descriptive Errors in the Anatomy of the Orbit,' Trans. Intern. Congress Ophth., Washington, April, 1922.

WIEDERSHEIM, R. 'The Structure of Man,' Bernard's Translation. 1895, London.

WILBRAND, H. (i) 'The Course of the Nerve Fibres in the Chiasma,' Zeitschr. f. Augenh., 1926, lix, 3, p. 135. Abst. in Brit. Jour. Ophth., 1928, xii, p. 542.

(ii) and SAENGER, A. 'Die Neurologie des Auges,' 1921, Bergmann, Munich.

WILKINSON, H. J. (i) 'Experimental Studies on the Innervation of Striated Muscle,' Jour. Comp. Neurol., 1930, li, p. 129; also in Med. Jour. Australia, 1929, 2, p. 768.

(ii) and BURKITT, A. N. 'Nerve Endings in Adipose Tissue,' Med. Jour. Australia, 1926, i, p. 179.

(iii) 'The Argyll-Robertson Pupil: a Contribution towards its Explanation,' Med. Jour. Australia, 1927, i, p. 267.

WILMER, W. H. 'The Relation of the Teeth to Diseases of the Eye,' Arch. of Ophth., 1930, iv, p. 453.

WILSON, J. and PIKE. 'The Mechanism of Labyrinthine Nystagmus, etc.,' Arch. of Internal Medicine, 1915, xv. 1, p. 31.

WINKLER, C. 'Manuel de Neurologie,' 1927, Bohn, Haarlem.

WITT, E. 'Ausbreitung der Stirnhöhlen und Siebenzellen über die Orbita,' Anat. Hefte (Merkel und Bonnet), 1908, xxxvii, p. 145.

WOLFF, E. 'A Bend in the Sixth Cranial Nerve, and its probable Significance,' Brit. Jour. Ophth., 1928, xii, p. 22.

WOLFF, H. 'Über die Sehne des Musculus levator palpebrae superioris,' Zeitschr. f. Augenh., 1905, xiii, p. 440, and 1906, xv, p. 596.

WOLFRING, E. 'Untersuchungen über die Drüsen der Bindhaut des Auges,' Centralblatt f. d. med. Wiss., 1872, x, p. 852.

WOOD, CASEY. American Encyclopedia of Ophthalmology, 1913, Cleveland Press, Chicago.

WOOLLARD, H. H. (i) 'The Innervation of Involuntary Muscle,' Jour. of Anat., 1927, lxi, p. 498 (under Proceedings).

(ii) 'The Innervation of the Ocular Muscles,' Jour. of Anat., 1931, lxv, p. 215.

(iii) and Beattie, J. 'The Comparative Anatomy of the Lateral Geniculate Body,' Jour. of Anat., 1927, lxi, p. 414.

YERGER, C. 'Exophthalmos as a Complication of Nasal Sinusitis,' Arch. of Otolaryngol., 1925, i, p. 415. Abst. in Arch. Neurol. and Psychiat., 1925, xiv, p. 267.

YOUNG, G. 'Relation of the Optic Nerve to the Sphenoidal and Posterior Ethmoidal Sinuses,' Brit. Med. Jour., 1922, ii, p. 1258.

ZABEL, E. 'Varietäten und vollständiges Fehlen des Tränenbeines beim Menschen,' Anat. Hefte (Merkel und Bonnet), 1900, Band xv, p. 153.

ZALUSKOWSKI, K. 'Bemerkungen über den Bau der Bindehaut,' Arch. f. mikros. Anat., 1887, xxx, p. 311.

ZANDER, R. (i) Beiträge zur Kenntnis der mittleren Schädelgrube mit besonderer Berücksichtigung der Lage des Chiasma opticum,' Anat. Anzeiger, 1896, xii, p. 457.

(ii) 'Beiträge zur Kenntnis der Hautnerven des Kopfes,' Anat. Hefte (Merkel and Bonnet), 1897, ix, p. 59.

ZEEMAN, W. P. C. and BROUWER, B. 'Projection of the Retina in the Primary Optic Neuron in Monkeys,' Brain, 1926, xlix, p. 1.

ZEIS. 'Anatomische Untersuchungen der Meibom'schen Drüsen des Menschen und der Tiere,' Von Ammon's Zeitschr. f. d. Ophth., 1835, iv, p. 231.

ZIETZSCHMANN, O. (i) 'Vergleichende Histologie: Untersuchungen über den Bau der Augenlider der Haussäugetiere,' Graefe's Arch. f. Ophth., 1904, lviii, p. 61.

(ii) 'Zur Frage des Vorkommens eines Tarsus im Lide der Haussäugetiere,' Graefe's Arch. f. Ophth., 1904, lix, p. 166.

ZIMMERMANN, K. 'Beiträge zur Kenntnis einiger Drüsen und Epithelien. Die Tränendrüsen des Menschen,' Arch. f. mikros. Anat., 1898, lii, p. 552.

ZINN. Descriptio anatomica oculi humani, 1755, Göttingen.

ZOJA, G. 'Sopra un solco men noto dell' osso frontale,' 1884, Pavia. Researches quoted by Ledouble, 1903, p. 173.

ZUCKERKANDL, E. (i) Normale und pathologische Anatomie der Nasenhöhle, 1882, 1892, Wien.

(ii) Bardelebens Handbuch der Anatomie des Menschen, 1896, Jena. (Quoted by Graf Spee.)

(iii) Atlas der topographischen Anatomie des Menschen, 1904, p. 114, Wien.

ZWEIBACK, S. 'Über die Incisurae supraorbitalis und frontalis des Stirnbeins und ihre Varietäten,' Inaugural Dissertation, 1900, Königsberg. (Quoted by Ledouble, 1903, p. 196.)

The following authors have been mentioned without reference to particular papers:

ALBINUS.
AUERBACH.

BARKER.
BARRET.
BASEDOW.
BEADLES.
BELL, Sir C.
BERT.
BIANCHI.
BLUM.
BLUMBERGER.
BOCH.
BOLTON.
BONNET.
BOURGEOIS.
BRAMWELL.
BRAUNE.
BRAUNSTEIN.
BRISTOWE.
BROWN-SÉQUARD.

Bull.
Burkholder.

Campbell.
Carleton.
Casserius.
Chimanowski.
Chudzinski.
Cohn.
Colombo.
Cooper.

Dahl.
Dalrymple.
Darkschewitsch.
Delbet.
Deniker.
Dohn.
Duchenne.
Dupuy.

Edinger.
Egeberg.
Elschnig.

Ferrall.
Ferrier.
Flechsig.
Foltz.
Frerichs.
Friedenthal.
Friteau.

Gaillard.
Galen.
Geddes.
Geissler.
Generali.
Giannuzzi.
Gradenigo.
Grasset.
Groethuysens.
Gudden.

Halliburton.
Harder.
Harvey.
Hasner.
Henke.
Hewett.
Hirsch.
Hoeve.
Humphry.
Huschka.

Jones, F. Wood.

Kappers.
Keiller.
Koenig.
Kolmus.
Krankow.
Kronlein.
Kuhnt.

Laffay.
Lang.
Lange.
Langer, von.
Latumeten.
Lawford.
Lewin.
Lewis.
Lloyd.
Luschka.

Magaard.
Maier.
Malaigne.
Marshall.
Mathewson.
Meibomius.
Meissner.
Mitchell.
Molinette.
Monakow.
Mott.

Oppenheimer.

Panas.
Pegler.
Peter.
Radinski.
Rauber.
Rawling.
Recklinghausen.
Rex.
Riolan.
Robertson.
Ruge.
Rupter, de.

Salus.
Salzer.
Sandifort.
Sanna.
Schoute.
Shane.
Sommering.
Stenver.
Stromeyer.
Szily.

Taillefer.
Terlinck.
Thane.
Thomsen.
Tiedemann.
Tillaux.
Toldt.
Toti.
Tourneux.
Traube.
Treves.

Valentin.
Verga.
Vesalius.
Vlacovich.
Vomus.
Vossius.

Weber.
Wecker, de.
Weeks.
Wicherkiewicz.
Wilkinson.
Winslow.

INDEX

PRINTED IN GREAT BRITAIN AT THE UNIVERSITY PRESS, OXFORD
BY JOHN JOHNSON, PRINTER TO THE UNIVERSITY

This special edition of

THE ANATOMY OF
THE HUMAN ORBIT
AND ACCESSORY
ORGANS OF VISION

By S. Ernest Whitnall

has been privately printed for the members of The Classics of Ophthalmology Library by The John D. Lucas Printing Company. Film was prepared from the second edition of 1932. New type matter was composed by P&M Typesetting, Inc. in Caslon. The text paper was especially made for this edition by Newton Falls Paper Mill, Inc. The volume has been bound in genuine top-grain cowhide by R. R. Donnelley & Sons Company. Endleaves are a specially commissioned original marbled design of Richard J. Wolfe. Edges are gilded and covers are brass-die stamped in 22-karat gold. Cover stampings and design of the edition by Daniel B. Bianchi and Selma Ordewer.